Contemporary Calculus III

For the
students

A free, color PDF version is available online at
http://scidiv.bellevuecollege.edu/dh/Calculus_all/Calculus_all.html

Dale Hoffman
Bellevue College
Author web page: http://scidiv.bellevuecollege.edu/dh/

This text is licensed under a Creative Commons Attribution-Share Alike 3.0 United States License.

You are **free**:
  to Share – to copy, distribute, display and perform the work
  to Remix – to make derivative works

Under the following conditions:
  Attribution: You must attribute the work in the manner specified by the author
    (but not in any way that suggests that they endorse you or your work)
  Share Alike: If you alter, transform or build upon this work, you must distribute
    the resulting work only under the same, similar or compatible license.

Copyright © 2016 Dale Hoffman   Version: May 2016

CONTEMPORARY CALCULUS III: Contents

Note:
Each section contains Practice Problems throughout the section. The solutions to these Practice Problems are at the end of that section, after the Problem Set for the section.

Each section also contains a Problem Set. The solutions to the odd problems of each Problem Set are at the end of each chapter.

These materials are also available free and in color on the web at:
http://scidiv.bellevuecollege.edu/dh/Calculus_all/Calculus_all.html

Chapter 12: Vector-Valued Functions

12.0 Introduction to Vector-Valued Functions	102
12.1 Vector-Valued Functions and Curves in Space	103
12.2 Derivatives and Antiderivatives of Vector-Valued Functions	119
12.3 Arc Length of Space Curves	129
12.4 Cylindrical & Spherical Coordinate Systems in 3D	141
Odd Numbered solutions for Chapter 12	155

Chapter 13: Functions of Several Variables

13.1 Functions of Two or More Variables	165
13.2 Limits an Continuity	177
13.3 Partial Derivatives	186
13.4 Tangent Planes and Differentials	197
13.5 Directional Derivatives and the Gradient Vector	204
13.6 Maximums and Minimums	213
13.7 Lagrange Multiplier Method	222
13.8 Chain Rule for functions of Several Variables	227
Odd numbered solutions for Chapter 13	233

Chapter 14: Multiple Integrals

14.0 Introduction to Double Integrals and Applications	240
14.1 Double Integrals Over Rectangles	247
14.2 Double Integrals Over General Regions	255
14.3 Double Integrals in Polar Coordinates	264
14.4 Applications of Double Integrals	271
14.5 Surface Areas Using Double Integrals	276
14.6 Triple Integrals and Applications	281
14.7 Triple Integrals in Cylindrical and Spherical Coordinates	288
14.8 Changing Variables in Double ans Triple Integrals	295
Odd numbered solutions for Chapter 14	308

Chapter 15: Vector Calculus

15.0	Introduction to Vector Calculus	316
15.1	Vector Fields	318
15.2	Del Operator, 2D Divergence and Curl	325
15.3	Line Integrals	331
15.4	Fundamental Theorem of Line Integrals and Potential Functions	341
15.5	Theorems of Green, Stokes and Gauss: An Introduction	349
15.6	Green's Theorem	353
15.7	Divergence and Curl in 3D	363
15.8	Parametric Surfaces	369
15.9	Surface Integrals	378
15.10	Stoke's Theorem	387
15.11	Gauss/Divergence Theorem	398
Odd numbered solutions for Chapter 15		406

12.0 INTRODUCTION TO VECTOR–VALUED FUNCTIONS

So far, our excursion into 3–dimensional space has been rather static — we examined points, lines, planes, and vectors, but they did not move (except for points along lines). Those ideas and techniques are important for representing the positions of objects, but objects change position, and calculus is the study of "change." This chapter begins our extension of the ideas of calculus beyond two dimensions, and that extension is the focus for most of the rest of the book.

In earlier chapters, the functions we worked with generally had the form $y = f(x)$ — a single input value x resulted in a single output value y. As we move to higher dimensions, we expand the notion of function keeping the idea that each input produces a single output: f(input) = output. Now, however, we expand the types of objects that can be valid outputs, the range of f, and we expand the types of objects that can be valid inputs, the domain of f.

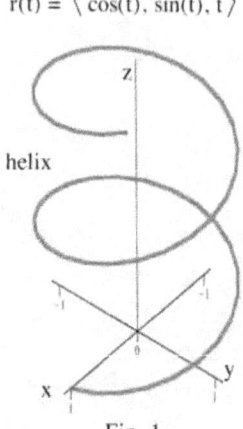

Fig. 1

Chapter 12 focuses on functions whose domains consist of numbers, but whose ranges consist of vectors: functions of the form f(number) = vector such as $\mathbf{r}(t) = \langle \cos(t), \sin(t), t \rangle$. These are called vector–valued functions, and typically their graphs are curves in space (Fig. 1).

Chapters 13 and 14 focus on functions whose domains consist of more than one variable and whose ranges consist of numbers: functions of the type f(number, number) = number or f(x,y) = z. These are called functions of several variables, and typically their graphs are surfaces in space (Fig. 2).

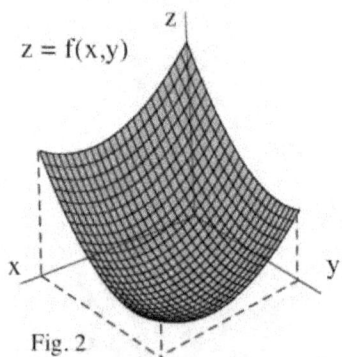

Fig. 2

In this chapter we examine vector–valued functions, and the discussion of vector–valued functions is similar in outline to our discussion of functions $y = f(x)$ in the early chapters of this book. First we discuss the meaning of vector–valued functions and their graphs. Then we look at the calculus ideas of limit, derivative and integral as they apply to vector–valued functions and examine some applications of these calculus ideas. As before, the meaning of these topics is inherently geometric, and you need to be able to work "visually" as well as "analytically."

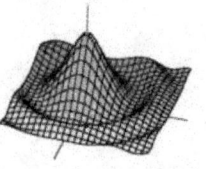
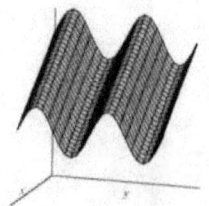

12.1 VECTOR–VALUED FUNCTIONS AND CURVES IN SPACE

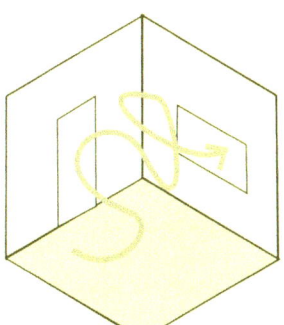

Fig. 1: Bug's path

As a bug buzzes around a room (Fig. 1) or a submarine explores the oceans or a planet orbits a moving star, we could describe the location of the bug or submarine or planet at any particular time as a point (x, y, z) in 3-dimensional space. But to describe the object's movement we need the locations at many times, and that leads very naturally to the idea of representing the path of the object as a set of points $P(t) = (x(t), y(t), z(t))$ given by parametric equations $x = x(t)$, $y = y(t)$, and $z = z(t)$ where the variable t represents the parameter time. Fig. 2 shows the paths of $P(t) = (1 + t, 0 + 3t, 1 + 2t)$ and $Q(t) = (\cos(t), \sin(t), t)$. Such parametric equations and their graphs are fundamental to our study of motion in 3-dimensional space, but for many uses it is more effective to work with vectors rather than points, and that leads to **vector–valued** functions.

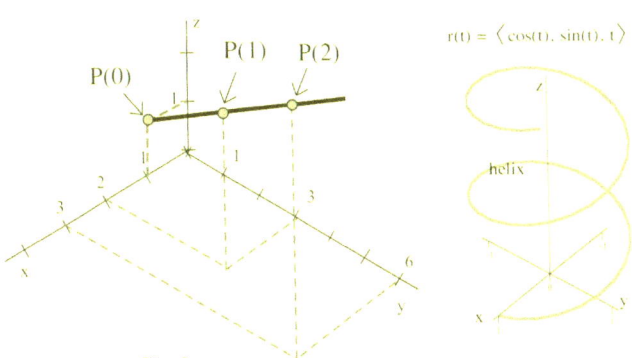

Fig. 2

Definition: A **vector–valued function** is a rule that assigns a vector to each input number.

Typically, a vector–valued function has the form
$$\mathbf{r}(t) = x(t)\mathbf{i} + y(t)\mathbf{j} + z(t)\mathbf{k} = \langle x(t), y(t), z(t) \rangle$$
where x, y, and z are scalar–valued functions.

The domain of a vector–valued function $\mathbf{r}(t)$ is a set of real numbers: the domain of $\mathbf{r}$ consists of those t in the domains of x, y, and z.

The range of a vector–valued function is a collection of vectors.

Vector–valued functions offer two advantages that should become clearer as you work with them. First, vector–valued functions are often notationally simpler than parametric equations — it is easier to write $\mathbf{r}(t)$ than $(x(t), y(t), z(t))$. This is not a big advantage; P(t) is also easy to write, but the $\mathbf{r}(t)$ notation does make some ideas and computations easier. The second advantage of vector–valued functions is that they allow us to use the powerful machinery we have already developed for working with vectors. This vector machinery is particularly useful when we consider **tangent vectors** to curves and **angles** between curves.

Our discussion of vector–valued functions is similar to the discussion of functions of the form $y = f(x)$ in the early chapters. We begin by considering their graphs and the ideas of limits and continuity for vector–valued functions.

Graphs of Vector-Valued Functions — Space Curves

If an object is located at the point $P(t) = (x(t), y(t), z(t))$ at time t, then we say that

$$\mathbf{r}(t) = x(t)\mathbf{i} + y(t)\mathbf{j} + z(t)\mathbf{k} = \langle x(t), y(t), z(t) \rangle ,$$

a vector from the origin to the point P(t), is the object's **position vector**. The functions x(t), y(t), and z(t) are called the **components** of $\mathbf{r}(t)$ or the **component functions** of $\mathbf{r}(t)$. Fig. 3 shows a point P(t) and the position vector $\mathbf{r}(t)$ for that point. It is difficult to draw and difficult to interpret a collection of vectors, so typically when we work with the graphs of vector–valued functions we draw only the path of the endpoints of the vectors: the graph of $\mathbf{r}(t) = \langle x(t), y(t), z(t) \rangle$ is the collection of points $P(t) = (x(t), y(t), z(t))$.

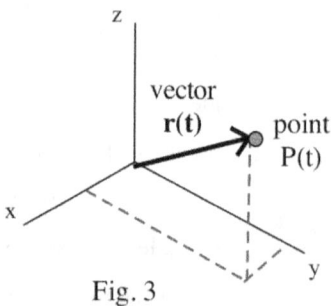
Fig. 3

The graphs of vector–valued functions can be very complex and difficult to sketch, for example Fig. 4, but many are manageable "by hand" and you should practice sketching some of their graphs.

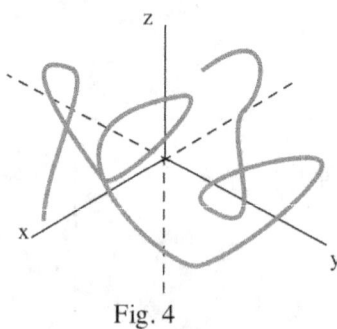
Fig. 4

Example 1: Sketch the graphs of $\mathbf{r}(t) = \langle 3 - t, 2t, 4 - 2t \rangle$ for $0 \le t \le 3$, and $\mathbf{s}(t) = \langle t, 0, \sin(t) \rangle$ for $0 \le t \le 2\pi$.

Solution: The graphs are shown in Fig. 5.

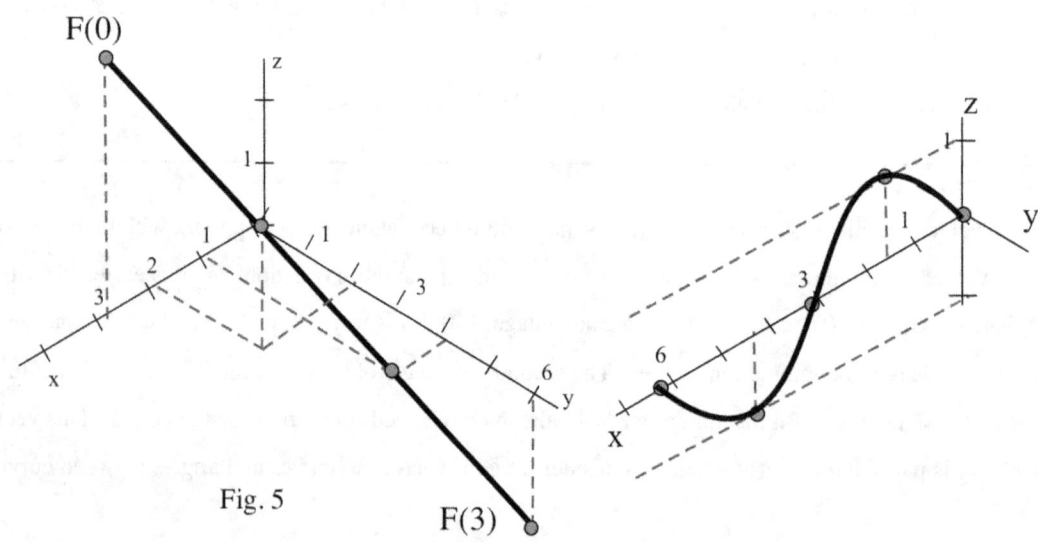
Fig. 5

Practice 1: Sketch the graphs of $\mathbf{r}(t) = \langle t, 4 - 2t, 2 - t \rangle$ for $0 \le t \le 2$, and $\mathbf{s}(t) = \langle t, t^2, 1 \rangle$ for $0 \le t \le 2$.

Sometimes the component functions $x(t), y(t)$, and $z(t)$ may only be given as graphic information, but we can still sketch the graph of the vector–valued function they define.

Example 2: The graphs of $x(t), y(t)$, and $z(t)$ are shown in Fig. 6. Use this information about the component functions of $\mathbf{r}(t) = x(t)\mathbf{i} + y(t)\mathbf{j} + z(t)\mathbf{k}$
$= \langle x(t), y(t), z(t) \rangle$
to sketch a graph of $\mathbf{r}(t)$.

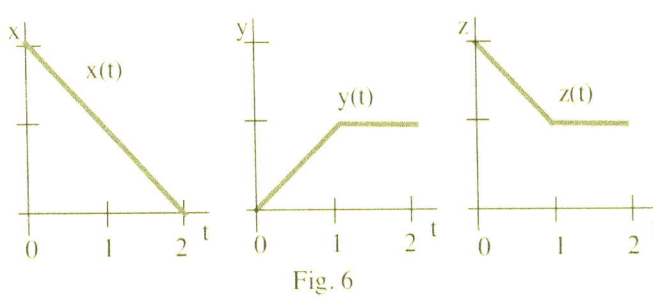

Fig. 6

Solution: The graph of $\mathbf{r}(t)$ is shown in Fig. 7.

Practice 2: The graphs of $x(t), y(t)$, and $z(t)$ are shown in Fig. 8. Use this information about the component functions of
$\mathbf{r}(t) = x(t)\mathbf{i} + y(t)\mathbf{j} + z(t)\mathbf{k} = \langle x(t), y(t), z(t) \rangle$ to sketch a graph of $\mathbf{r}(t)$.

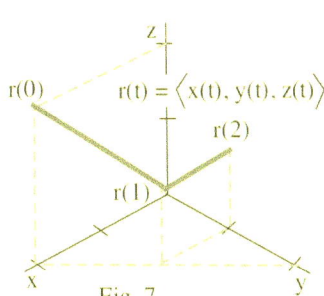

Fig. 7

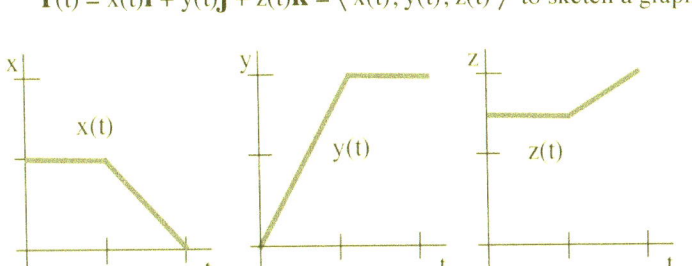

Fig. 8

Limits and Continuity of Vector–Valued Functions

Limits and continuity of vector–valued functions are defined in terms of the components of the function. If you understand the meaning of limits and continuity for functions $y = f(x)$, then these concepts are not difficult for vector–valued functions.

Definition: For $\mathbf{r}(t) = \langle x(t), y(t), z(t) \rangle$,

$$\lim_{t \to a} \mathbf{r}(t) = \langle \lim_{t \to a} x(t), \lim_{t \to a} y(t), \lim_{t \to a} z(t) \rangle$$

provided that each of the limits of the component functions exists.

To calculate the limit of a vector-valued function we simply need the limits of each of the component scalar-valued functions. If any of the limits of the component functions fail to exist, we say that the limit of the vector-valued function does not exist. The various properties of limits of vector-valued functions follow directly from the corresponding properties of scalar-valued functions as they are applied to each component separately.

Example 3: Determine $\lim_{t \to 0} \mathbf{r}(t)$ and $\lim_{t \to \infty} \mathbf{s}(t)$ for $\mathbf{r}(t) = \langle \cos(t), \sqrt{4+t}, 3e^{2t} \rangle$

and $\mathbf{s}(t) = \langle \frac{2t}{t+3}, \frac{5}{t}, 3 + \frac{\sin(t)}{2t} \rangle$

Solution: $\lim_{t \to 0} \mathbf{r}(t) = \langle \lim_{t \to 0} \cos(t), \lim_{t \to 0} \sqrt{4+t}, \lim_{t \to 0} 3e^{2t} \rangle = \langle 1, 2, 3 \rangle$ and

$\lim_{t \to \infty} \mathbf{s}(t) = \langle \lim_{t \to \infty} \frac{2t}{t+3}, \lim_{t \to \infty} \frac{5}{t}, \lim_{t \to \infty} 3 + \frac{\sin(t)}{2t} \rangle = \langle 2, 0, 3 \rangle$

Practice 3: Determine $\lim_{t \to \pi} \mathbf{r}(t)$ and $\lim_{t \to \infty} \mathbf{r}(t)$ for $\mathbf{r}(t) = \langle \cos(t), \frac{\sin(t)}{t}, 3 \rangle$

Continuity of vector-valued functions is stated in terms of limits, so the continuity of a vector-valued function depends on the continuity (and limits) of the component functions.

Definition: A vector-valued function $\mathbf{r}(t) = \langle x(t), y(t), z(t) \rangle$ is
continuous at the point $t = t_0$

if $\lim_{t \to t_0} \mathbf{r}(t) = \mathbf{r}(t_0)$.

The following result about continuity follows directly from the definition and is usually easier to use.

Component Continuity Theorem for Vector-valued Functions

A vector-valued function $\mathbf{r}(t) = \langle x(t), y(t), z(t) \rangle$ is continuous at the point $t = t_0$
if and only if each of the component functions $x(t), y(t)$, and $z(t)$ is continuous at $t = t_0$.

Proof: The proof follows directly from the definitions of continuity and limits of vector-valued functions. If $\mathbf{r}(t)$ is continuous at $t = t_0$, then

$$\lim_{t \to t_0} \mathbf{r}(t) = \mathbf{r}(t_0) = \langle x(t_0), y(t_0), z(t_0) \rangle.$$

But, from the definition of limit of $\mathbf{r}(t)$, $\lim_{t \to a} \mathbf{r}(t) = \langle \lim_{t \to a} x(t), \lim_{t \to a} y(t), \lim_{t \to a} z(t) \rangle$.

Two vectors are equal if and only if their respective components are equal, so

$$\lim_{t \to t_0} x(t) = x(t_0), \lim_{t \to t_0} y(t) = y(t_0), \text{ and } \lim_{t \to t_0} z(t) = z(t_0),$$

and we have shown that $x, y,$ and z are continuous at $t = t_0$.

The proof that "if $x(t), y(t),$ and $z(t)$ are continuous at $t = t_0$, then $\mathbf{r}(t)$ is continuous at $t = t_0$" is similar but starts with the assumption that $x, y,$ and z are continuous at $t = t_0$.

Example 4: Where are $\mathbf{r}(t) = \langle \cos(t), \sin(t), t^2 \rangle$ and $\mathbf{s}(t) = \langle 2 + t, \frac{1}{t-3}, \ln(t) \rangle$ continuous?

Solution: (a) $\mathbf{r}(t) = \langle \cos(t), \sin(t), t^2 \rangle$ is continuous everywhere (for all t) since all of the component functions $x(t) = \cos(t), y(t) = \sin(t),$ and $z(t) = t^2$ are continuous for all values of t.

(b) $\mathbf{s}(t) = \langle 2 + t, \frac{1}{t-3}, \ln(t) \rangle$ is continuous for $0 < t < 3$ and $3 < t$ since $x(t) = 2 + t$ is continuous for all values of t; $y(t) = \frac{1}{t-3}$ is continuous for $t \neq 3$; and $z(t) = \ln(t)$ is continuous for $t > 0$.

Practice 4: Where is $\mathbf{r}(t) = \langle 1/t, e^t, \text{INT}(t) \rangle$ continuous?

Fig. 9 shows the graphs of continuous components $x(t)$ and $y(t)$ and a discontinuous component $z(t)$. It also shows the discontinuous vector–valued function $\mathbf{r}(t) = \langle x(t), y(t), z(t) \rangle$.

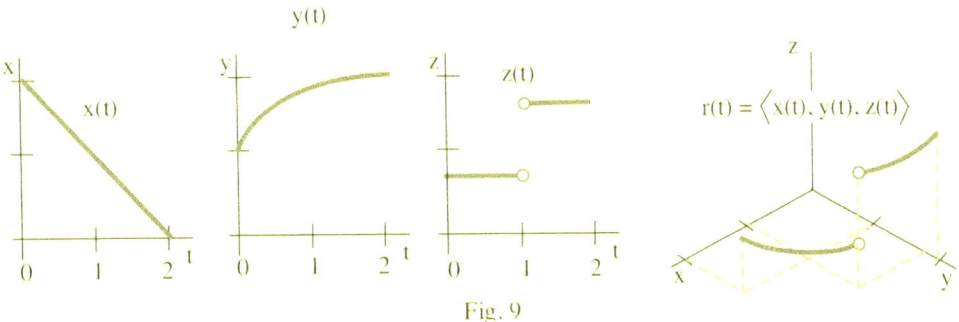

Fig. 9

Some Useful Space Curves: Lines, Helix, and Bezier Curves

By this point in your mathematical development, the graphs of some functions should be almost "automatic." You should be able to visualize the graphs of $y = x^2, y = 3 + 2\sin(x), y = |x-2|$ and a variety of others with little effort. It is useful to **start** to develop similar skills with the graphs of a few vector–valued functions. Lines and helices are useful shapes to begin with, and Bezier curves in three dimensions have a number of useful properties.

Lines

If $\mathbf{r}(t) = \langle x(t), y(t), z(t) \rangle$, and each of the component functions $x(t), y(t)$, and $z(t)$ is a linear function $(at + b)$, then the graph of $\mathbf{r}(t)$ is a straight line in space. In this case we need only evaluate and plot $\mathbf{r}(t)$ for a couple values of t before finishing the graph with a straightedge.

Example 5: Sketch the graphs of $\mathbf{r}(t) = \langle 2+t, 3-t, 1+2t \rangle$ and $\mathbf{s}(t) = \langle 2, t, 3 \rangle$.

Solution: $\mathbf{r}(0) = \langle 2, 3, 1 \rangle$ and $\mathbf{r}(1) = \langle 3, 2, 3 \rangle$. Fig. 10 shows these two points and a graph of $\mathbf{r}(t)$.
$\mathbf{s}(0) = \langle 2, 0, 3 \rangle$ and $\mathbf{s}(1) = \langle 2, 1, 3 \rangle$. Fig. 11 shows these two points and a graph of $\mathbf{s}(t)$.

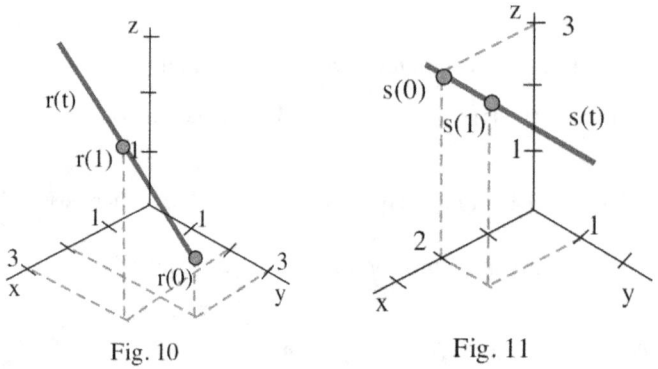

Fig. 10   Fig. 11

Practice 5: Sketch the graphs of $\mathbf{r}(t) = \langle t-1, 0, 1+2t \rangle$ and $\mathbf{s}(t) = \langle 2, 4, 3t \rangle$.

If $x(t), y(t)$, and $z(t)$ are linear functions, then they are continuous for all values of t so $\mathbf{r}(t) = \langle x(t), y(t), z(t) \rangle$ is also continuous for all values of t (by the Component Continuity Theorem).

The Helix and Some Variations

The graph of $\mathbf{r}(t) = \langle \cos(t), \sin(t), t \rangle$ is shown in Fig. 12, and it is called a **helix**, or a **circular helix around the z–axis**. The parametric graph of $(\cos(t), \sin(t))$ is a circle in the xy–plane (Fig. 13), and $z(t) = t$ then "stretches" this circle into the z direction to create the spiral shape (Fig. 12). The helix is sometimes useful, and it provides a focus to investigate the effects on the shape of the graph of certain changes in the component functions.

$r(t) = \langle \cos(t), \sin(t), t \rangle$

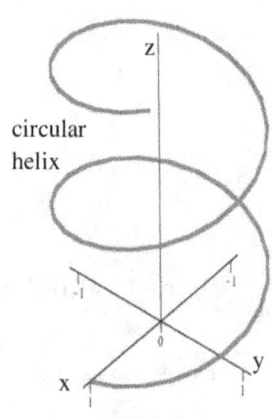

circular helix

Fig. 12

$P(t) = (\cos(t), \sin(t))$
a circle in the xy-plane

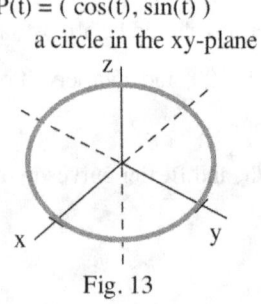

Fig. 13

Example 6: (a) Without plotting any points, describe the
shapes of the graphs of $r(t) = \langle \cos(t), t, \sin(t) \rangle$ and $s(t) = \langle \cos(t), 2\cdot\sin(t), t \rangle$.

(b) Then sketch the graphs of $r(t)$ and $s(t)$.

Solution: (a) The graph of $r(t) = \langle \sin(t), t, \cos(t) \rangle$ is a circular helix around the y–axis (Fig. 14)

(b) The parametric graph of $(\cos(t), 2\cdot\sin(t))$ in the xy–plane is an ellipse (Fig. 15a), so the graph of $s(t) = \langle \cos(t), 2\cdot\sin(t), t \rangle$ is an elliptical helix around the z–axis (Fig. 15b)

$r(t) = \langle \sin(t), t, \cos(t) \rangle$

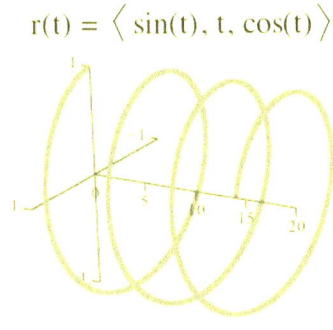

Fig. 14

$P(t) = (\cos(t), 2\sin(t))$
ellipse in the xy-plane

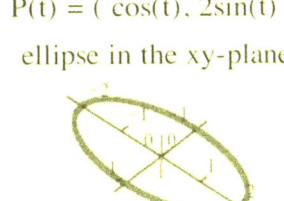

(a)

Fig. 15

$r(t) = \langle \cos(t), 2\sin(t), t \rangle$

elliptical helix

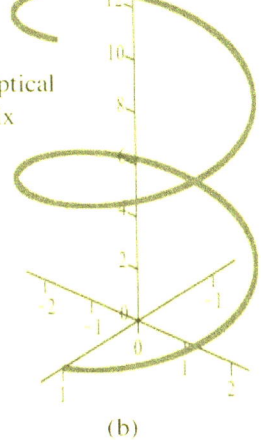

(b)

Practice 6: (a) Without plotting any points, describe the shapes of the
graphs of $r(t) = \langle t, \cos(t), \sin(t) \rangle$ and $s(t) = \langle 2\cdot\cos(t), t, 3\cdot\sin(t) \rangle$.
(b) Then sketch the graphs of $r(t)$ and $s(t)$.

Figs. 16 and 17 show the graphs of two more variations on the component functions of the original helix.

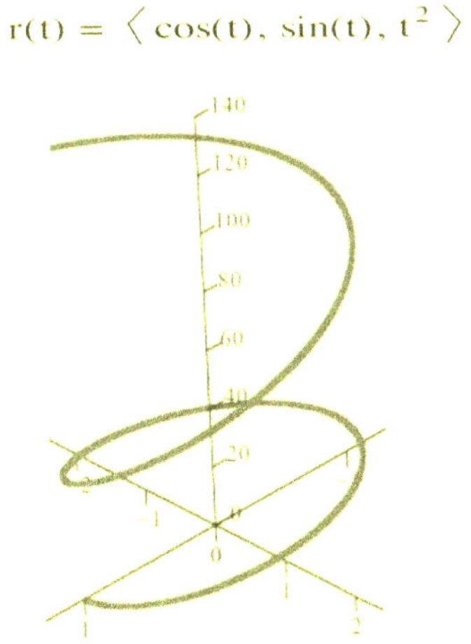

$r(t) = \langle \cos(t), \sin(t), t^2 \rangle$

$r(t) = \langle t\cos(t), t\sin(t), t \rangle$

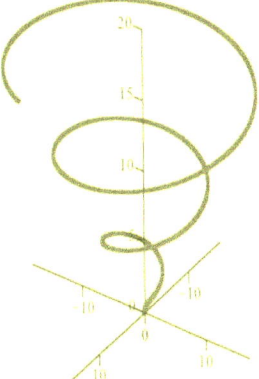

Fig. 16

Fig. 17

Bezier Curves in Three Dimensions

One powerful aspect of the Bezier curves we saw in Section 9.4.5 is that the ideas and even the formulas for these curves extend very easily to Bezier curves in three dimensions. In Section 9.4.5 we saw that if we start with four points $P_0 = (x_0, y_0), P_1 = (x_1, y_1), P_2 = (x_2, y_2)$, and $P_3 = (x_3, y_3)$ then, for $0 \leq t \leq 1$, the curve given by

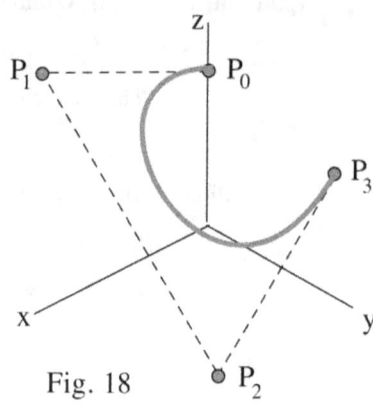

Fig. 18

$$x(t) = (1-t)^3 \cdot x_0 + 3(1-t)^2 t \cdot x_1 + 3(1-t)t^2 \cdot x_2 + t^3 \cdot x_3 \text{ and}$$
$$y(t) = (1-t)^3 \cdot y_0 + 3(1-t)^2 t \cdot y_1 + 3(1-t)t^2 \cdot y_2 + t^3 \cdot y_3$$

(or more simply as $B(t) = (1-t)^3 \cdot P_0 + 3(1-t)^2 t \cdot P_1 + 3(1-t)t^2 \cdot P_2 + t^3 \cdot P_3$)

has several useful properties:

(1) $B(0) = P_0$ and $B(1) = P_3$ so P_0 and P_3 are the "endpoints" of $B(t)$ for $0 \leq t \leq 1$.

(2) $B(t)$ is a cubic polynomial so it is continuous and differentiable.

(3) $B'(0) = $ slope of the line segment from P_0 to P_1:
$B'(1) = $ slope of the line segment from P_2 to P_3.

(4) For $0 \leq t \leq 1$, the graph of $B(t)$ is in the region whose corners are the control points.

A Bezier curve $B(t)$ for four given points P_0, P_1, P_2 and P_3 is shown in Fig. 18.

This extends very simply to three dimensions.

Bezier Curve in Three Dimensions

If P_0, P_1, P_2 and P_3 are four points in 3–dimensional space, then the Bezier curve in three dimensions for those points is

$$B(t) = (1-t)^3 \cdot P_0 + 3(1-t)^2 t \cdot P_1 + 3(1-t)t^2 \cdot P_2 + t^3 \cdot P_3 \text{ for } 0 \leq t \leq 1:$$

$$x(t) = (1-t)^3 \cdot x_0 + 3(1-t)^2 t \cdot x_1 + 3(1-t)t^2 \cdot x_2 + t^3 \cdot x_3,$$
$$y(t) = (1-t)^3 \cdot y_0 + 3(1-t)^2 t \cdot y_1 + 3(1-t)t^2 \cdot y_2 + t^3 \cdot y_3, \text{ and}$$
$$z(t) = (1-t)^3 \cdot z_0 + 3(1-t)^2 t \cdot z_1 + 3(1-t)t^2 \cdot z_2 + t^3 \cdot z_3.$$

This 3-dimensional Bezier curve has the same properties as those listed for the 2-dimensional Bezier curve with two modifications.

(3') When $t = 0$ the direction of $B(t)$ is the same as the direction of the line segment from P_0 to P_1

(Fig. 19): when $t = 1$ the direction of $B(t)$ is the same as the direction of the line segment from P_2 to P_3. (We will define the "direction" of a space curve in Section 12.2, but it is similar to "slope" in two dimensions.)

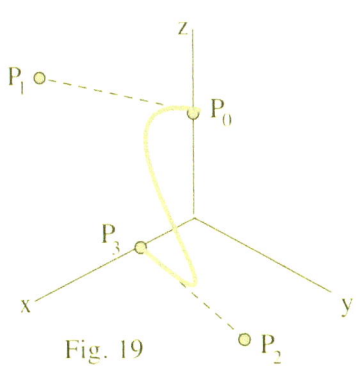

Fig. 19

(4') For $0 \leq t \leq 1$, the graph of $B(t)$ is "on the rubber sheet" whose corners are the control points (Fig. 20). (This is not very precise, but it should convey the idea without the need for more technical vocabulary.)

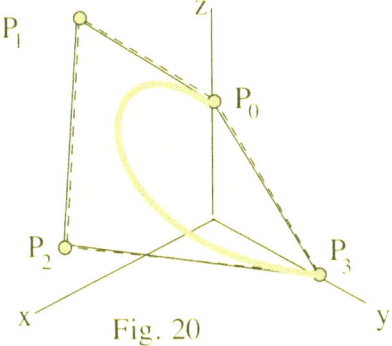

Fig. 20

PROBLEMS

In practice, most 3-dimensional graphs are created using computers. The point of many of the following problems is to help develop your 3-dimensional visualization skills.

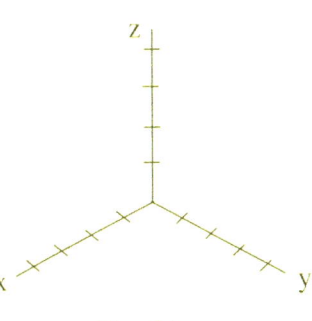

Fig. 21

In problems 1 – 10, sketch the graph of each vector-valued function. Many of these graphs lie on a coordinate plane so it is not difficult to sketch relatively good graphs by hand.

1. Sketch $\mathbf{r}(t) = \langle t, t^2, 0 \rangle$ on the axes system in Fig. 21 for $0 \leq t \leq 2$.

2. Sketch $\mathbf{s}(t) = \langle t, 0, t^2 \rangle$ on the axes system in Fig. 21 for $0 \leq t \leq 2$.

3. Sketch $\mathbf{r}(t) = \langle 0, t^2, t \rangle$ on the axes system in Fig. 22 for $0 \leq t \leq 2$.

4. Sketch $\mathbf{s}(t) = \langle t, 2t, 0 \rangle$ on the axes system in Fig. 22 for $0 \leq t \leq 3$.

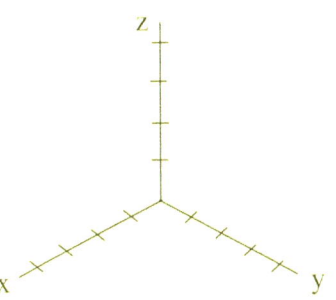

Fig. 22

5. Sketch $\mathbf{r}(t) = \langle 3t, 0, t \rangle$ on the axes system in Fig. 23 for $0 \le t \le 3$.

6. Sketch $\mathbf{s}(t) = \langle 0, t, \sin(t) \rangle$ on the axes system in Fig. 23 for $0 \le t \le 2\pi$.

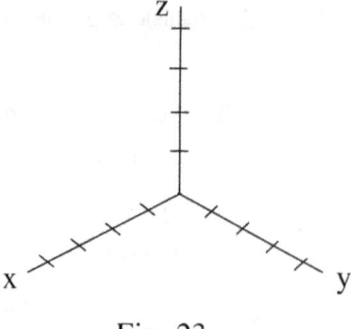

Fig. 23

7. Sketch $\mathbf{r}(t) = \langle 0, \sin(t), t \rangle$ on the axes system in Fig. 24 for $0 \le t \le 2\pi$.

8. Sketch $\mathbf{s}(t) = \langle \cos(t), \sin(t), 1 \rangle$ on the axes system in Fig. 24 for $0 \le t \le 2\pi$.

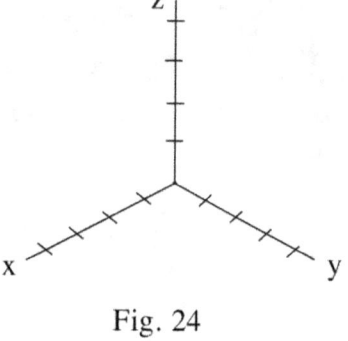

Fig. 24

9. Sketch $\mathbf{r}(t) = \langle 2, \sin(t), \cos(t) \rangle$ on the axes system in Fig. 25 for $0 \le t \le 2\pi$.

10. Sketch $\mathbf{s}(t) = \langle 0, \sin(t), \cos(t) \rangle$ on the axes system in Fig. 25 for $0 \le t \le 2\pi$.

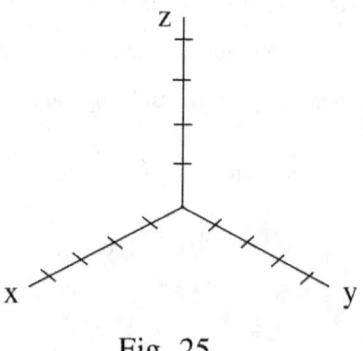

Fig. 25

In problems 11 – 16, calculate and carefully plot three points on each vector–valued linear function and then complete the graph of the line.

11. $\mathbf{r}(t) = \langle 2 - t, t, 1 + 2t \rangle$ in Fig. 26.

12. $\mathbf{s}(t) = \langle 2, t, 1 + 2t \rangle$ in Fig. 26.

13. $\mathbf{r}(t) = \langle t, 2, 3 \rangle$ in Fig. 27.

14. $\mathbf{s}(t) = \langle 1, t, 3 \rangle$ in Fig. 27.

15. $\mathbf{r}(t) = \langle t + 1, t - 2, 2t \rangle$ in Fig. 28.

16. $\mathbf{s}(t) = \langle 1 + 2t, t, 3 \rangle$ in Fig. 28.

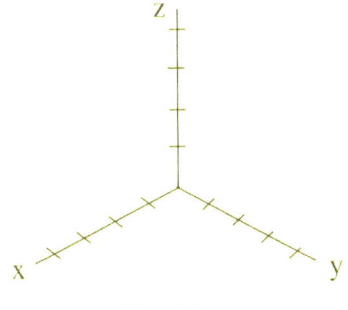

Fig. 27

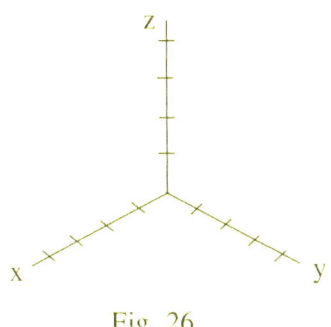

Fig. 26

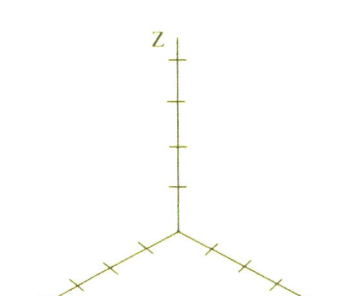

Fig. 28

In problems 17 – 20, graphs are given for x(t), y(t), and z(t). Use the information in these graphs to graph the vector–valued function $\mathbf{r}(t) = \langle x(t), y(t), z(t) \rangle$ on the given coordinate system.

17. The graphs of x(t), y(t), and z(t) are in Fig. 29.

18. The graphs of x(t), y(t), and z(t) are in Fig. 30.

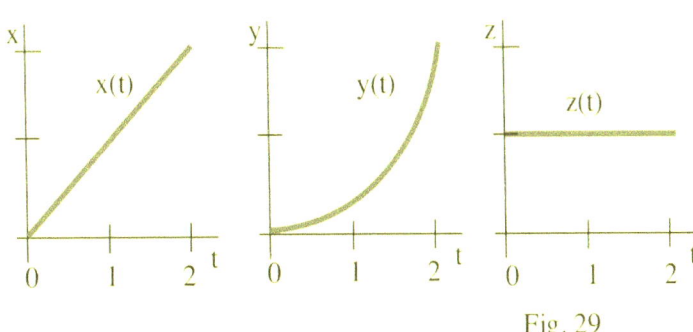

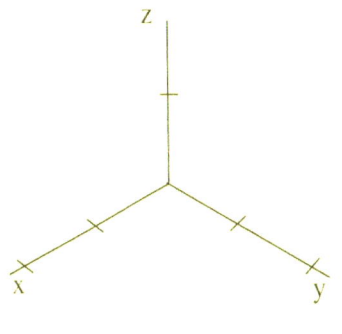

Fig. 29

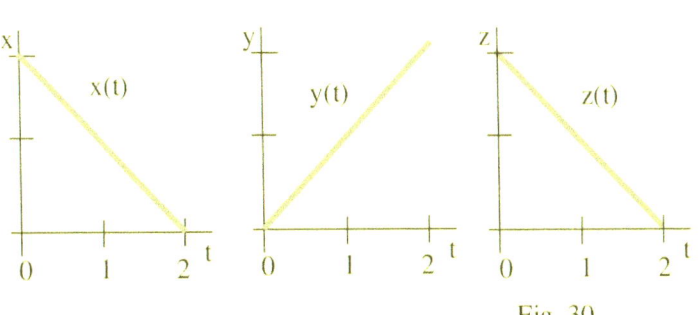

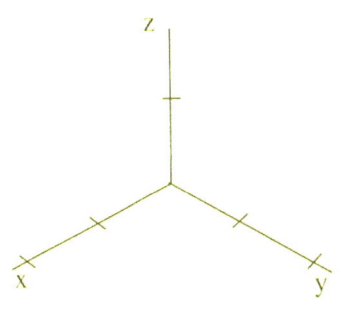

Fig. 30

19. The graphs of x(t), y(t), and z(t) are in Fig. 31.   20. The graphs of x(t), y(t), and z(t) are in Fig. 32.

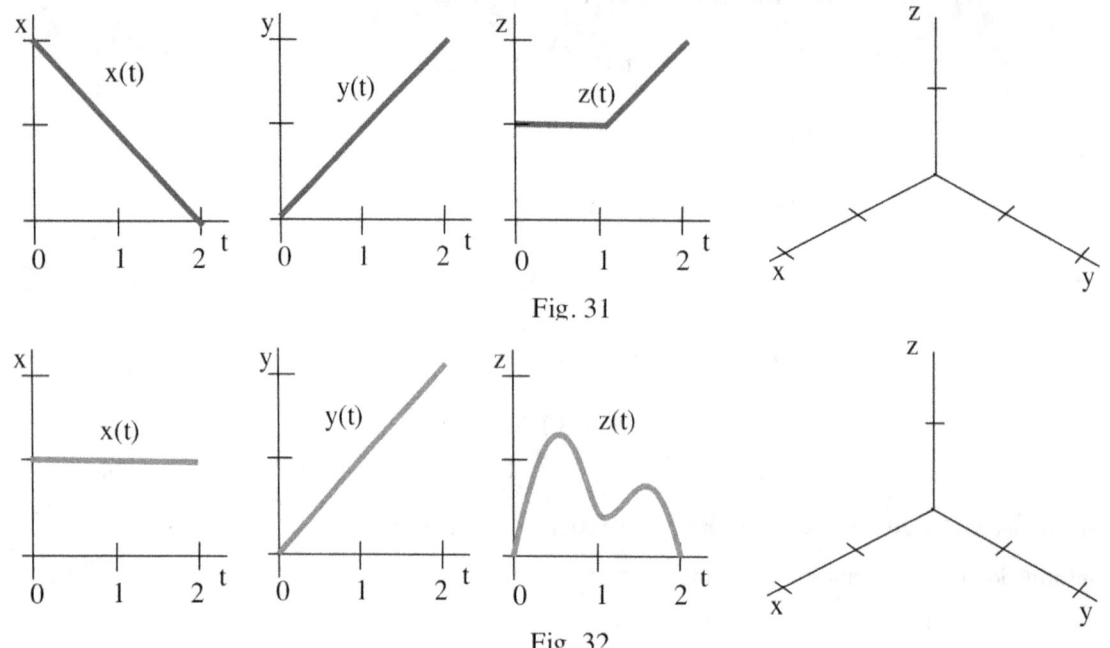

Fig. 31

Fig. 32

In problems 21 – 28, calculate the limits.

21. $\lim\limits_{t \to 3} \langle t, \frac{t^2}{2}, \frac{1}{t-1} \rangle$.

22. $\lim\limits_{t \to 0} \langle \cos(t), \frac{t}{3-t}, t^2 \rangle$.

23. $\lim\limits_{t \to 2} \langle 4, \frac{t}{t-2}, \sqrt{7+t} \rangle$.

24. $\lim\limits_{t \to 2} \langle \sqrt{t-1}, t^3, 1 \rangle$.

25. $\lim\limits_{t \to \infty} \langle \frac{4}{t+1}, \frac{3t+1}{t-2}, \frac{\sin(t)}{t} \rangle$.

26. $\lim\limits_{t \to \infty} \langle \frac{5t^2+2t+1}{t^2+t-6}, \sin(t), \frac{1}{1+t^2} \rangle$.

27. $\lim\limits_{t \to \infty} \langle \arctan(t), \frac{\ln(t)}{t}, \frac{3}{t^2} \rangle$.

28. $\lim\limits_{t \to \infty} \langle \sin(\arctan(t)), 1, 5^{1/t} \rangle$.

In problems 29 – 36, determine where the given vector–valued functions are continuous.

29. $\mathbf{r}(t) = \langle t, \frac{t^2}{2}, \frac{1}{t-1} \rangle$.

30. $\mathbf{s}(t) = \langle \cos(t), \frac{t}{3-t}, t^2 \rangle$.

31. $\mathbf{u}(t) = \langle 4, \frac{t}{t-2}, \sqrt{7+t} \rangle$.

32. $\mathbf{v}(t) = \langle \sqrt{t-1}, t^3, 1 \rangle$.

33. $\mathbf{w}(t) = \langle \frac{4}{t+1}, \frac{3t+1}{t-2}, \frac{\sin(t)}{t} \rangle$.

34. $\mathbf{r}(t) = \langle \frac{5t^2+2t+1}{t^2+t-6}, \sin(t), \frac{1}{1+t^2} \rangle$.

35. $\mathbf{r}(t) = \langle \arctan(t), \frac{\ln(t)}{t}, \frac{3}{t^2} \rangle$.

36. $\mathbf{s}(t) = \langle \sin(\arctan(t)), 1, 5^{1/t} \rangle$.

In problems 37 – 40, determine the Bezier curve B(t) for the given control points P_0, P_1, P_2 and P_3. If you have access to a computer with the appropriate software, graph B(t).

37. $P_0 = (0, 0, 1), P_1 = (1, 0, 1), P_2 = (1, 2, 0),$ and $P_3 = (0, 2, 0)$.

38. $P_0 = (0, 0, 2), P_1 = (1, 0, 1), P_2 = (0, 2, 0),$ and $P_3 = (1, 3, 0)$.

39. $P_0 = (0, 1, 2), P_1 = (0, 0, 2), P_2 = (2, 1, 0),$ and $P_3 = (2, 0, 0)$.

40. $P_0 = (0, 1, 2), P_1 = (0, 0, 2), P_2 = (2, 0, 0),$ and $P_3 = (1, 1, 0)$.

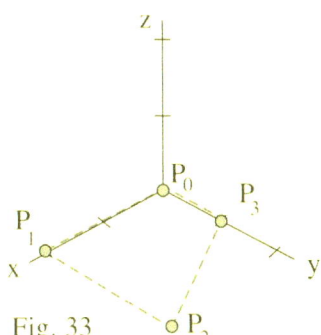
Fig. 33

In problems 41 – 44, control points P_0, P_1, P_2 and P_3 are shown on a graph. For the given control points sketch a curve with the properties of the Bezier curve. (In some graphs additional lines are included with the control points to help with your sketch.)

41. P_0, P_1, P_2 and P_3 are given in Fig. 33.

42. P_0, P_1, P_2 and P_3 are given in Fig. 34.

43. P_0, P_1, P_2 and P_3 are given in Fig. 35.

44. P_0, P_1, P_2 and P_3 are given in Fig. 36.

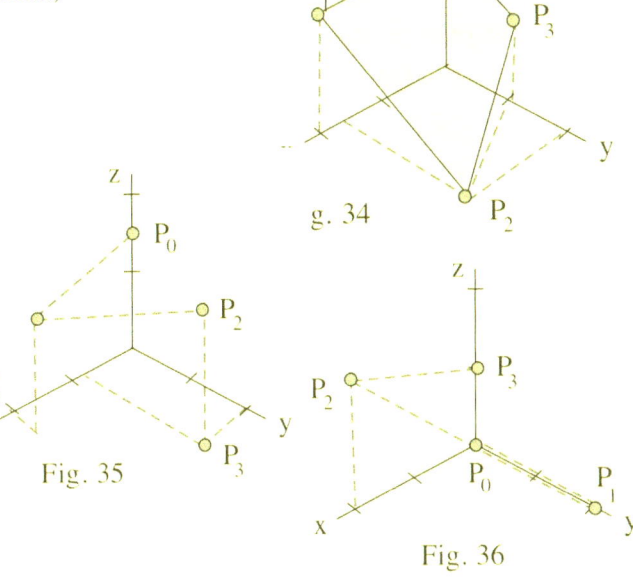

In problems 45 – 48, graphs of vector-valued functions are given. At the points labeled A, B, and C on each curve determine whether each variable function x(t), y(t), and z(t) is increasing (I) or decreasing (D) and fill in the table for each function.

45. $\mathbf{r}(t) = \langle x(t), y(t), z(t) \rangle$ is given in Fig. 37.

46. $\mathbf{s}(t) = \langle x(t), y(t), z(t) \rangle$ is given in Fig. 38.

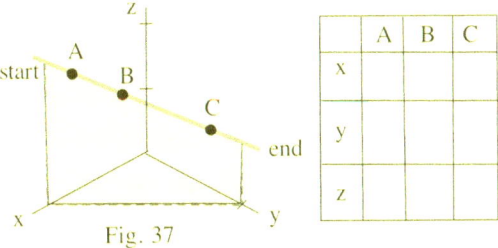

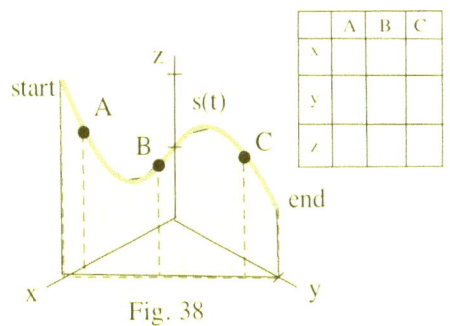

47. $\mathbf{u}(t) = \langle x(t), y(t), z(t) \rangle$ is given in Fig. 39.

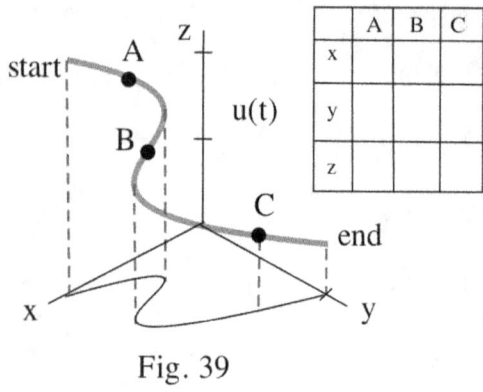

Fig. 39

48. $\mathbf{v}(t) = \langle x(t), y(t), z(t) \rangle$ is given in Fig. 40.

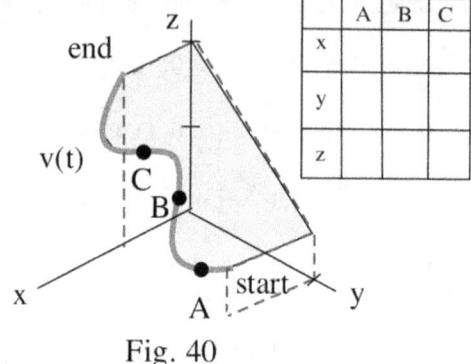

Fig. 40

Practice Answers

Practice 1: The graphs of $\mathbf{r}(t) = \langle t, 4 - 2t, 2 - t \rangle$ and $\mathbf{s}(t) = \langle t, t^2, 1 \rangle$ for $0 \le t \le 2$ are shown in Fig. 41.

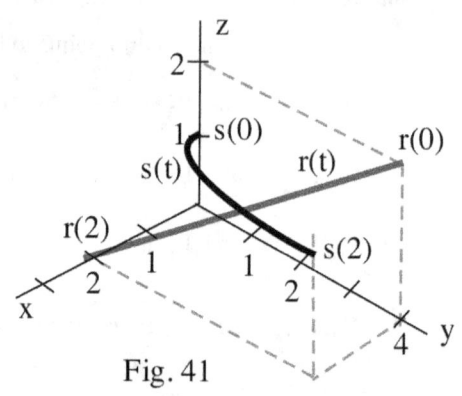

Fig. 41

Practice 2: The graphs of $x(t), y(t),$ and $z(t)$ are shown in Fig. 42a, 42b, and 42c. The graph of $\mathbf{r}(t) = x(t)\mathbf{i} + y(t)\mathbf{j} + z(t)\mathbf{k} = \langle x(t), y(t), z(t) \rangle$ is shown in Fig. 42d.

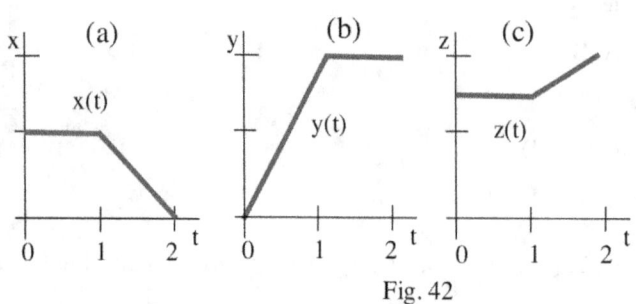
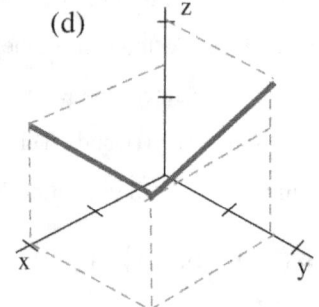

Fig. 42

Practice 3: For $\mathbf{r}(t) = \langle \cos(t), \frac{\sin(t)}{t}, 3 \rangle$, $\lim_{t \to \pi} \mathbf{r}(t) = \langle -1, 0, 3 \rangle$, and $\lim_{t \to \infty} \mathbf{r}(t)$ does not exist because $\lim_{t \to \infty} \cos(t)$ does not exist.

Practice 4: $\mathbf{r}(t) = \langle 1/t, e^t, \text{INT}(t) \rangle$ is continuous everywhere except where t is an integer since INT(t) is not continuous where t is an integer. $1/t$ is not continuous when $t = 0$, but we have already excluded $t = 0$ because 0 is an integer.

12.1 Vector–valued Functions and Curves in Space   Contemporary Calculus   117

Practice 5: The graphs of $\mathbf{r}(t) = \langle t-1, 0, 1+2t \rangle$ and $\mathbf{s}(t) = \langle 2, 4, 3t \rangle$ are shown in Fig. 43.

Practice 6: The graph of $\mathbf{r}(t) = \langle t, \cos(t), \sin(t) \rangle$ is a circular helix around the x–axis. This graph is shown in Fig. 44.
The graph of $\mathbf{s}(t) = \langle 2 \cdot \cos(t), t, 3 \cdot \sin(t) \rangle$ is an elliptical helix around the y–axis. This graph is shown in Fig. 45.

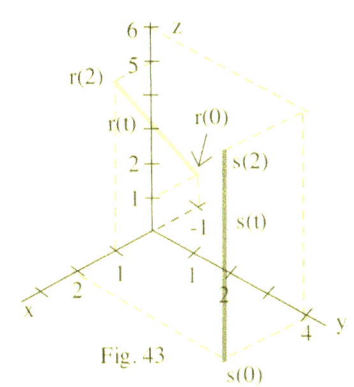
Fig. 43

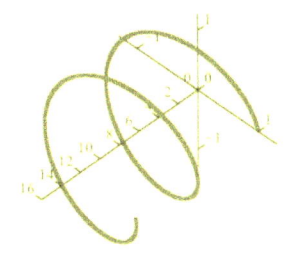
circular helix around the x-axis
Fig. 44

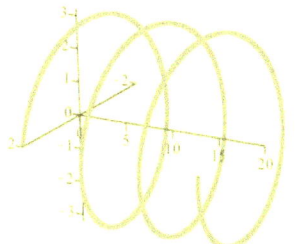
elliptical helix around the y-axis
Fig. 45

Appendix: MAPLE and graphs of vector–valued functions — spacecurve( )

The computer language MAPLE has a number of commands for creating graphs of 3–dimensional objects including vector–valued functions. In order to access these commands, we first need to load the "plots" package:

with(plots); *(then press the <enter> key)*   This loads the "plots" package and lists the new commands available for us to use.

The command to create graphs of vector–valued functions is

   spacecurve([x(t), y(t), z(t)] , t = a..b, *options*);

where   x(t), y(t), and z(t) are formulas for the x, y, and z coordinates,
   a and b are the starting and stopping values for the variable t
   and *options* includes commands for the type of axes, the color, the thickness, and the number of points to be plotted.

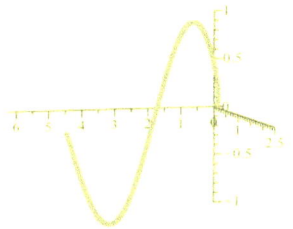
Fig. M1

For example, the command

   spacecurve([t, sqrt(t), sin(t)], t=0..2*Pi, axes=NORMAL, numpoints=200, color=red, thickness=3);

creates the graph in Fig. M1.

By positioning the cursor on the graph, a "hand" appears and, while holding down the mouse button, we can rotate the graph by slowing moving the mouse and then releasing the button. Fig. M2 shows another view of the graph.

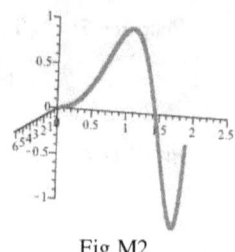

Fig M2

Fig. M3 shows three views created using the command

spacecurve([sin(t)*cos(25*t), t, sin(t)*sin(25*t)], t=0..Pi, axes=NORMAL, numpoints=300);

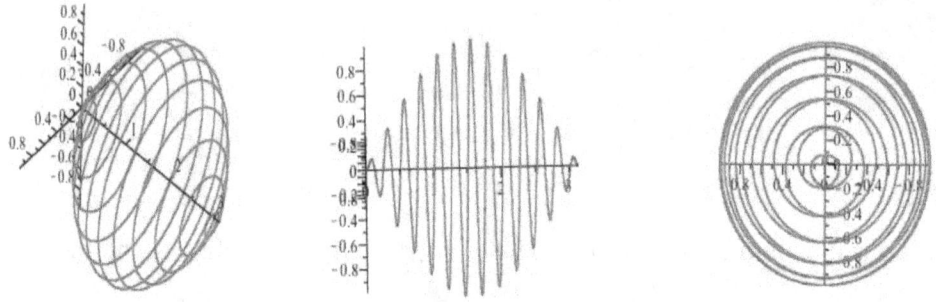

Fig. M3: three views of [sin(t)*cos(25t), t , sin(t)*sin(25t)]

12.2 DERIVATIVES AND ANTIDERIVATIVES OF VECTOR–VALUED FUNCTIONS

Derivatives of Vector–valued Functions

The derivative of a vector–valued function is another vector–valued function, and this derivative is defined much like the derivative of a scalar function. Derivatives of vector–valued functions are generally easy to compute, component–by–component, and they have a useful geometric interpretation as the vectors tangent to the graph of the vector–valued function.

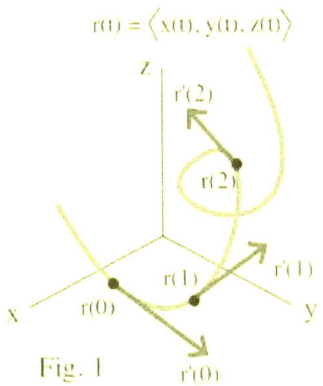

Definition: The **derivative** of $\mathbf{r}(t) = \langle x(t), y(t), z(t) \rangle$, denoted $\frac{d}{dt}\mathbf{r}(t)$ or $\mathbf{r}'(t)$, is

$$\mathbf{r}'(t) = \lim_{\Delta t \to 0} \frac{\mathbf{r}(t+\Delta t) - \mathbf{r}(t)}{\Delta t} = \langle x'(t), y'(t), z'(t) \rangle$$

provided the limit exists and is finite. (Fig. 1)

A vector–valued function $\mathbf{r}(t)$ is differentiable at a point $t = t_0$ if and only if each of its component functions is differentiable at $t = t_0$, and we can calculate the derivative $\mathbf{r}'(t)$ by calculating the three derivatives $x'(t)$, $y'(t)$, and $z'(t)$.

Visualizing $\mathbf{r}'(t)$: If $\mathbf{r}(t)$ is the position of an object at time t, then the difference vector $\mathbf{r}(t + \Delta t) - \mathbf{r}(t)$ represents the **change** in position from time t to time $t + \Delta t$ (Fig. 2), and the ratio $\frac{\mathbf{r}(t + \Delta t) - \mathbf{r}(t)}{\Delta t}$ is a vector measuring the average rate of change of position during the time interval from t to $t + \Delta t$. The limit $\mathbf{r}'(t)$ of this "average rate of change" vector has two useful geometric properties (Fig. 3):

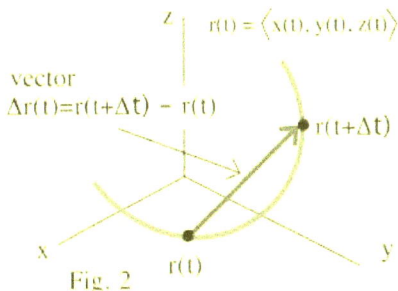

- $\mathbf{r}'(t)$ is **tangent** to the graph of $\mathbf{r}(t)$, and
- the magnitude of $\mathbf{r}'(t)$ is the **speed** of the object along the path at the time t.

The vector $\mathbf{r}'(t)$ is called the velocity of $\mathbf{r}(t)$.
The vector $|\mathbf{r}'(t)|$ is called the speed of $\mathbf{r}(t)$.

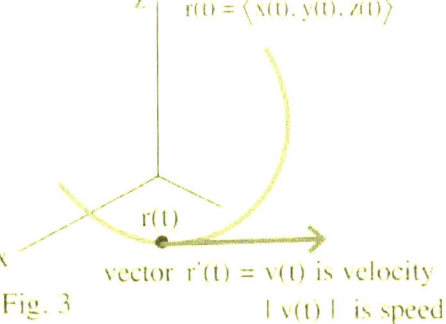

Definitions: Velocity, Speed, Direction, and Acceleration

If $\mathbf{r}(t)$ is the **position** of an object at time t, then

the **velocity** of the object is $\mathbf{v}(t) = \frac{d}{dt} \mathbf{r}(t) = \mathbf{r}'(t)$ (a vector tangent to $\mathbf{r}(t)$),

the **speed** of the object is $|\mathbf{v}(t)|$ (a scalar),

the **direction** of travel is $\mathbf{T}(t) = \frac{\mathbf{v}(t)}{|\mathbf{v}(t)|}$ (the **unit tangent vector**), and

the **acceleration** is $\mathbf{a}(t) = \mathbf{v}'(t) = \mathbf{r}''(t)$.

Example 1: A ladybug is crawling up a helix so its position vector is $\mathbf{r}(t) = \langle \cos(t), \sin(t), t \rangle$ as shown in Fig. 4.

(a) At the points labeled A, B, and C on the graph of $\mathbf{r}(t)$, estimate the sign (positive or negative) of each component of $\mathbf{r}'(t)$.

(b) The values of t for the point A, B, and C are $t = \pi/6, 3\pi/4$, and $7\pi/4$ respectively. Calculate $\mathbf{r}'(t)$ at the given values of t and compare the results with your estimates in part (a).

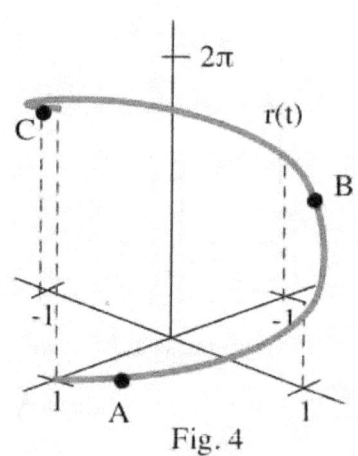

Fig. 4

Solution: (a) At A, $\mathbf{r}'(t)$ is $\langle x'(t), y'(t), z'(t) \rangle = \langle -, +, + \rangle$. At B, $\mathbf{r}'(t)$ is $\langle -, -, + \rangle$. At C, $\mathbf{r}'(t)$ is $\langle +, +, + \rangle$.

(b) $\mathbf{r}'(\pi/6)$ is $\langle x'(\pi/6), y'(\pi/6), z'(\pi/6) \rangle = \langle -\sin(\pi/6), \cos(\pi/6), 1 \rangle \approx \langle -0.5, 0.867, 1 \rangle$.

$\mathbf{r}'(3\pi/4)$ is $\langle -\sin(3\pi/4), \cos(3\pi/4), 1 \rangle \approx \langle -0.707, -0.707, 1 \rangle$.

$\mathbf{r}'(7\pi/4)$ is $\langle -\sin(7\pi/4), \cos(7\pi/4), 1 \rangle \approx \langle 0.707, 0.707, 1 \rangle$.

Practice 1: The position vector of an object at time t is $\mathbf{r}(t) = \langle t, t^2, t^3 \rangle$ as shown in Fig. 5. Calculate the position, velocity, speed, direction, and acceleration of the object when $t = 0, 1,$ and 2.

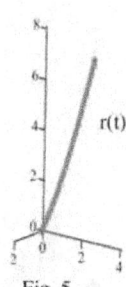

Fig. 5

Angles of Intersection Between Space Curves

The angle of intersection between two curves at a point in space is the angle between their tangent vectors (velocities) at that point of intersection, and the dot product of the tangent vectors can be used to find this angle.

12.2 Derivatives and Antiderivatives of Vector–valued Functions   Contemporary Calculus   121

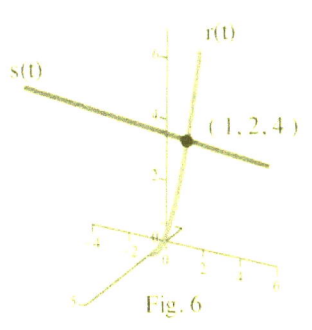

Fig. 6

Example 2: The parabolic path $r(t) = \langle 1, t, t^2 \rangle$ intersects the line
$s(t) = \langle -2 + 3t, 6 - 4t, 2 + 2t \rangle$ (Fig. 6) at the point $(1, 2, 4)$.
Find the angle of intersection of the curves at that point.

Solution: The parabola goes through $(1, 2, 4)$ when $t = 2$, and the line goes through $(1, 2, 4)$ when $t = 1$. Then $r'(2) = \langle 0, 1, 4 \rangle$ and $s'(1) = \langle 3, -4, 2 \rangle$ so

$$\cos(\theta) = \frac{r'(2) \cdot s'(1)}{|r'(2)||s'(1)|}$$

$$\approx \frac{4}{\sqrt{17}\sqrt{29}} \approx 0.180 \text{ and}$$

and $\theta \approx 1.390$ (or about $79.6°$).

Practice 2: The parabolic paths $r(t) = \langle 0, t, t^2 \rangle$ and
$s(t) = \langle 2 - t, 1, 5 - t^2 \rangle$ (Fig. 7) intersect at the point $(0, 1, 1)$.
Find the angle of intersection of the curves at that point.

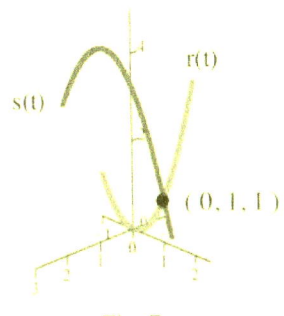

Fig. 7

Differentiation of Combinations of Vector–valued Functions

For scalar functions we have patterns for differentiating sums, differences, products, and compositions, and there are similar rules for differentiating combinations of vector–valued functions. In fact, the rules for vector–valued functions are almost identical to the corresponding rules for scalar function derivatives.

Differentiation Patterns for Vector–valued Functions

Constant: If C is a constant vector, then $\frac{d}{dt} C = 0$ vector.

If $u(t)$ and $v(t)$ are differentiable vector–valued functions, k is a scalar number, and $f(t)$ is a scalar function, then

Sum: $\frac{d}{dt}(u(t) + v(t)) = \frac{d}{dt} u(t) + \frac{d}{dt} v(t) = u'(t) + v'(t)$

Difference: $\frac{d}{dt}(u(t) - v(t)) = \frac{d}{dt} u(t) - \frac{d}{dt} v(t) = u'(t) - v'(t)$

Products: $\frac{d}{dt}(ku(t)) = k\frac{d}{dt}(u(t)) = k\, u'(t)$

---

scalar	$\frac{d}{dt}(f(t)\mathbf{u}(t)) = f(t)\frac{d}{dt}(\mathbf{u}(t)) + \frac{df(t)}{dt}\mathbf{u}(t)$	$= f(t)\mathbf{u}'(t) + f'(t)\mathbf{u}(t)$
dot	$\frac{d}{dt}(\mathbf{u}(t)\cdot\mathbf{v}(t)) = \mathbf{u}(t)\cdot\frac{d\mathbf{v}(t)}{dt} + \frac{d\mathbf{u}(t)}{dt}\cdot\mathbf{v}(t)$	$= \mathbf{u}(t)\cdot\mathbf{v}'(t) + \mathbf{u}'(t)\cdot\mathbf{v}(t)$
cross	$\frac{d}{dt}(\mathbf{u}(t) \times \mathbf{v}(t)) = \mathbf{u}(t) \times \mathbf{v}'(t) + \mathbf{u}'(t) \times \mathbf{v}(t)$	
Chain Rule:	$\frac{d}{dt}\mathbf{u}(f(t)) = f'(t)\mathbf{u}'(f(t))$	

You should notice that all of the **product** differentiation patterns have the form

(first function) "times" (derivative of the second) plus (derivative of the first) "times" (second)

where the "times" is the appropriate type of multiplication, either scalar, dot, or cross.

Proofs: The proofs are very straightforward (componentwise) for the results for the derivatives of constant vectors, sums, differences and a scalar times a vector–valued function, and they are left for you.

The proofs given below for the product rules all follow the pattern of rewriting the original function as components, using our usual product rule or chain rule to differentiate the component functions, and then rewriting the results as the appropriate product of vectors.

Scalar: $\frac{d}{dt}(f(t) \mathbf{u}(t)) = \langle \frac{d}{dt} f(t)u_1(t), \frac{d}{dt} f(t)u_2(t), \frac{d}{dt} f(t)u_3(t) \rangle$

$= \langle f(t)u_1'(t) + f'(t)u_1(t), f(t)u_2'(t) + f'(t)u_2(t), f(t)u_3'(t) + f'(t)u_3(t) \rangle$

$= \langle f(t)u_1'(t), f(t)u_2'(t), f(t)u_3'(t) \rangle + \langle f'(t)u_1(t), f'(t)u_3(t), f'(t)u_3(t) \rangle$

$= f(t) \langle u_1'(t), u_2'(t), u_3'(t) \rangle + f'(t) \langle u_1(t), u_2(t), u_3(t) \rangle$

$= f(t) \mathbf{u}'(t) + f'(t) \mathbf{u}(t)$.

Dot: $\frac{d}{dt}(\mathbf{u}(t)\cdot\mathbf{v}(t)) = \langle \frac{d}{dt} u_1(t)v_1(t), \frac{d}{dt} u_2(t)v_2(t), \frac{d}{dt} u_3(t)v_3(t) \rangle$

$= \langle u_1(t)v_1'(t) + u_1'(t)v_1(t), u_2(t)v_2'(t) + u_2'(t)v_2(t), u_3(t)v_3'(t) + u_3'(t)v_3(t) \rangle$

$= \langle u_1(t)v_1'(t), u_2(t)v_2'(t), u_3(t)v_3'(t) \rangle + \langle u_1'(t)v_1(t), u_2'(t)v_2(t), u_3'(t)v_3(t) \rangle$

$= \mathbf{u}(t)\cdot\mathbf{v}'(t) + \mathbf{u}'(t)\cdot\mathbf{v}(t)$.

The pattern for the derivative of a cross product can also be proved by resorting to the definition of the cross product and showing that the components of $\frac{d}{dt}(\mathbf{u}(t) \times \mathbf{v}(t))$ match the components of $\mathbf{u}(t) \times \mathbf{v}'(t) + \mathbf{u}'(t) \times \mathbf{v}(t)$, but the process is algebraically long and is omitted.

Chain rule: $\frac{d}{dt}\mathbf{u}(f(t)) = \left\langle \frac{d}{dt} u_1(f(t)), \frac{d}{dt} u_2(f(t)), \frac{d}{dt} u_3(f(t)) \right\rangle$

$= \left\langle f'(t) u_1'(f(t)), f'(t) u_2'(f(t)), f'(t) u_3'(f(t)) \right\rangle$

$= f'(t) \left\langle u_1'(f(t)), u_2'(f(t)), u_3'(f(t)) \right\rangle = f'(t) \mathbf{u}'(f(t))$.

These differentiation patterns simply provide alternate, and sometimes easier, ways to compute derivatives. Occasionally they are useful for deriving results about the behavior of vector-valued functions such as the one given in the next example.

Example 3: Suppose a differentiable position vector $\mathbf{r}(t)$ of an object has constant length so $|\mathbf{r}(t)| = k$ for all t. Show that the direction of travel of the object is always perpendicular to its position.

Solution: $\mathbf{r}(t) \bullet \mathbf{r}(t) = |\mathbf{r}(t)|^2 = k^2$ for all t, so $\mathbf{r}(t) \bullet \mathbf{r}(t)$ is a constant

so $\frac{d}{dt} \mathbf{r}(t) \bullet \mathbf{r}(t) = 0$.

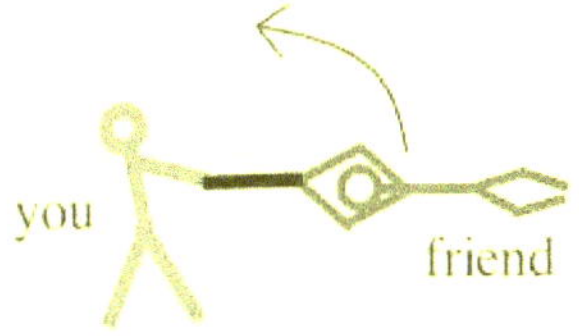

But we also know that

$\frac{d}{dt}\{\mathbf{r}(t) \bullet \mathbf{r}(t)\} = \mathbf{r}(t) \bullet \mathbf{r}'(t) + \mathbf{r}'(t) \bullet \mathbf{r}(t) = 2\mathbf{r}(t) \bullet \mathbf{r}'(t)$,

so we can conclude that $\mathbf{r}(t) \bullet \mathbf{r}'(t) = 0$ for all t.

Fig. 8: Twirling a friend

But $\mathbf{r}(t) \bullet \mathbf{r}'(t) = 0$ means that $\mathbf{r}(t)$ is perpendicular to $\mathbf{r}'(t)$ for all t, and that means the position, $\mathbf{r}(t)$, is always perpendicular to the velocity, $\mathbf{r}'(t)$. The velocity vector $\mathbf{r}'(t)$ points in the direction of travel of the object so we have shown that the direction of travel of the object is always perpendicular to its position.

The result in Example 3 also has straightforward physical interpretations. If we are twirling someone around a central point (Fig. 8), we can take the central point to be the origin. Then the twirled person is

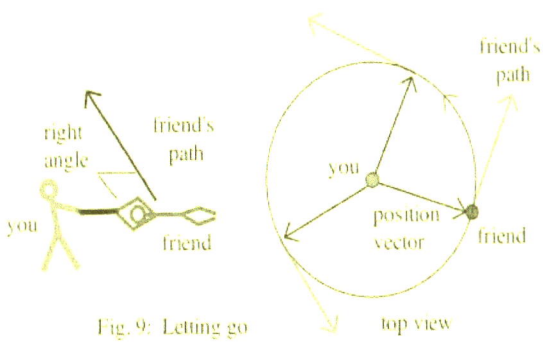

Fig. 9: Letting go

always a constant distance from the central point and the magnitude of their position vector is a constant. The result of this example says that the twirled person's velocity vector is always perpendicular to their position vector. If we let go of the person, their motion will be a straight line that is perpendicular to the circular path they were following (Fig. 9). An equivalent situation in three dimensions is an object moving on the surface of a sphere (Fig. 10) such as the earth (almost). If gravity is "turned off," this object will travel along a path perpendicular to the vector from the center of the earth to its position.

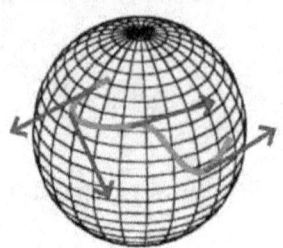

Fig. 10: Path on a sphere

Antiderivatives of Vector-valued Functions

Since the derivative of a vector-valued function is defined to be the vector formed by the derivative of each of the component functions, the antiderivative of a vector-valued function is also defined component by component.

> **Definition:** If $\mathbf{r}(t) = \langle x(t), y(t), z(t) \rangle$,
>
> then the antiderivative of $\mathbf{r}(t)$ is $\int \mathbf{r}(t)\,dt = \langle \int x(t)\,dt, \int y(t)\,dt, \int z(t)\,dt \rangle$
>
> provided that the antiderivatives of $x(t)$, $y(t)$, and $z(t)$ exist.

Example 4: The velocity of an object is $\mathbf{v}(t) = \langle 4t, -\sin(t), e^t \rangle$ and its position at time $t = 0$ is $\mathbf{r}(0) = \langle 2, 3, 4 \rangle$. Find a formula for $\mathbf{r}(t)$, its position at time t.

Solution: $\mathbf{v}(t) = \mathbf{r}'(t)$ so

$$\mathbf{r}(t) = \int \mathbf{r}'(t)\,dt = \langle \int 4t\,dt, \int -\sin(t)\,dt, \int e^t\,dt \rangle = \langle 2t^2 + A, \cos(t) + B, e^t + C \rangle.$$

Then we can use the initial condition that $\mathbf{r}(0) = \langle 2, 3, 4 \rangle$, to determine that $A = 2$, $B = 2$, and $C = 3$ so $\mathbf{r}(t) = \langle 2t^2 + 2, \cos(t) + 2, e^t + 3 \rangle$.

Practice 3: The velocity of an object is $\mathbf{v}(t) = \langle 6t^2, \cos(t), 12e^{3t} \rangle$ and its position at time $t = 0$ is $\mathbf{r}(0) = \langle 1, -5, 2 \rangle$. Find a formula for $\mathbf{r}(t)$, its position at time t.

The inertial guidance system on an airplane uses antiderivatives of vector-valued functions to determine the location of the airplane. The inertial guidance system starts with the initial location and velocity of the airplane and then uses lasers to measure the acceleration of the airplane in each of the x, y, and z directions several times per second. From this acceleration (change in velocity) data, the computer in the system calculates the new velocity in each direction several times per second and then uses the velocities (changes in positions) to calculate the new position of the airplane relative to the starting position.

12.2 Derivatives and Antiderivatives of Vector-valued Functions

Example 5: Fig. 11 shows the initial acceleration, velocity and position of an object along the x–axis as well as its acceleration at 1 second time intervals. Fill in the empty spaces in the table and determine the position of the object on the x–axis after 9 seconds.

time (sec)	acceleration (ft/sec^2)	velocity (ft/sec)	position (ft)
0	0	2	5
1	4		
2	6		
3	4		
4	2		
5	8		
6	6		
7	0		
8	0		
9	0		

Fig. 11

Solution: Acceleration is the $\frac{\text{change in velocity}}{\text{change in time}} = \frac{\text{change in velocity}}{1 \text{ second}}$ so each entry in the velocity column is the previous velocity plus the change in velocity (acceleration):

at t = 1, velocity = (previous velocity) + (change in velocity) = 2 + 4 = 6

at t = 2, velocity = (previous velocity) + (change in velocity) = 6 + 6 = 12

at t = 3, velocity = (previous velocity) + (change in velocity) = 12 + 4 = 16.

The rest of the entries in the velocity column are calculated in the same way and the velocity values are shown in Fig. 12.

Velocity is

$\frac{\text{change in position}}{\text{change in time}} = \frac{\text{change in position}}{1 \text{ second}}$

so each entry in the position column is the previous position value plus the change in position (velocity):

at t = 1, position = (previous position) + (change in position) = 5 + 6 = 11

at t = 2, position = (previous position) + (change in position) = 11 + 12 = 23

at t = 3, position = (previous position) + (change in position) = 23 + 16 = 39.

The rest of the entries in the position column are calculated in the same way.

time (sec)	acceleration (ft/sec^2)	velocity (ft/sec)	position (ft)
0	0	2	5
1	4	6	11
2	6	12	23
3	4	16	39
4	2	18	57
5	8	26	83
6	6	32	115
7	0	32	147
8	0	32	179
9	0	32	211

Fig. 12

Practice 4: Fig. 13 shows the initial acceleration, velocity and position of an object along the y–axis as well as its acceleration at 1 second time intervals. Fill in the empty

time (sec)	acceleration (ft/sec^2)	velocity (ft/sec)	position (ft)
0	0	2	5
1	4		
2	6		
3	8		
4	−3		
5	0		
6	5		
7	−2		
8	1		
9	0		

Fig. 13

spaces in the table and determine the position of the object on the y–axis after 9 seconds?

PROBLEMS

In problems 1 – 4, fill in each component of $\mathbf{r}'$ with " + ", " – ", or " 0 ."

1. For $\mathbf{r}(t)$ in Fig. 14, $\mathbf{r}'(1) = \langle\ ,\ ,\ \rangle$, $\mathbf{r}'(2) = \langle\ ,\ ,\ \rangle$, and $\mathbf{r}'(3) = \langle\ ,\ ,\ \rangle$.

2. For $\mathbf{r}(t)$ in Fig. 14, $\mathbf{r}'(4) = \langle\ ,\ ,\ \rangle$, $\mathbf{r}'(5) = \langle\ ,\ ,\ \rangle$, and $\mathbf{r}'(6) = \langle\ ,\ ,\ \rangle$.

3. For $\mathbf{r}(t)$ in Fig. 15, $\mathbf{r}'(1) = \langle\ ,\ ,\ \rangle$, $\mathbf{r}'(2) = \langle\ ,\ ,\ \rangle$, and $\mathbf{r}'(3) = \langle\ ,\ ,\ \rangle$.

4. For $\mathbf{r}(t)$ in Fig. 15, $\mathbf{r}'(4) = \langle\ ,\ ,\ \rangle$, $\mathbf{r}'(5) = \langle\ ,\ ,\ \rangle$, and $\mathbf{r}'(6) = \langle\ ,\ ,\ \rangle$.

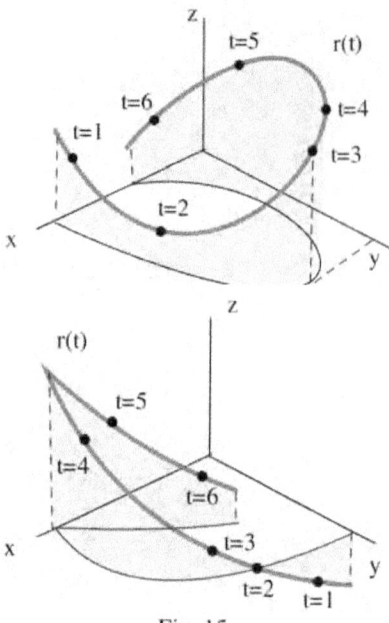

Fig. 15

In problems 5 – 8, the position vector $\mathbf{r}(t)$ is given for an object at time t. Calculate the velocity, speed, direction, and acceleration of the object at the given times.

5. $\mathbf{r}(t) = \langle t^3, 3 + 2t, t^2 \rangle$ and t = 1 and 2.

6. $\mathbf{r}(t) = \langle 5 + 3t^2, \sqrt{t}, t - t^3 \rangle$ and t = 1 and 2.

7. $\mathbf{r}(t) = (2 - t)\mathbf{i} + (4/t)\mathbf{j} + (3)\mathbf{k}$ and t = 1 and 2.

8. $\mathbf{r}(t) = (2 - t^3)\mathbf{i} + (5t)\mathbf{j} + (3 + t)\mathbf{k}$ and t = 1 and 2.

9. $\mathbf{r}(t) = \langle t^3, 7, 1 + 5t \rangle$. Calculate $\frac{d}{dt}\mathbf{r}(2t)$.

10. $\mathbf{r}(t) = \langle 1/t, 6 + 5t, t^3 \rangle$. Calculate $\frac{d}{dt}\mathbf{r}(t^2)$.

11. $\mathbf{r}(t) = \langle t, 2t^2, 3t^3 \rangle$. Calculate $\frac{d}{dt}\{\sin(t)\,\mathbf{r}(t)\}$.

12. $\mathbf{r}(t) = \langle 7 - t^2, 4, t^3 - t \rangle$. Calculate $\frac{d}{dt}\{t^3\,\mathbf{r}(t)\}$.

13. $\mathbf{r}(t) = (2 - 5t^3)\mathbf{i} + (7t)\mathbf{j} + (1 + t)\mathbf{k}$. Calculate $\frac{d}{dt}\mathbf{r}(3t)$.

14. $\mathbf{r}(t) = (1 - t^2)\mathbf{i} + (5t^3)\mathbf{j} + (3 + 2t)\mathbf{k}$. Calculate $\frac{d}{dt}\mathbf{r}(t^3)$.

In problems 15 – 18, determine $\frac{d}{dt}\{\mathbf{u} + 2\mathbf{v}\}$, $\frac{d}{dt}\{\mathbf{u}\cdot\mathbf{v}\}$, and $\frac{d}{dt}\{\mathbf{u}\times\mathbf{v}\}$ for the given vectors $\mathbf{u}(t)$ and $\mathbf{v}(t)$.

15. $\mathbf{u}(t) = \langle 0, t, t^3 \rangle$ and $\mathbf{v}(t) = \langle 1 + 5t, 4 - t, 3 \rangle$.

16. $\mathbf{u}(t) = \langle 4t, 1, 5 - t \rangle$ and $\mathbf{v}(t) = \langle t^2, 2 + 3t, t \rangle$

12.2 Derivatives and Antiderivatives of Vector–valued Functions

17. $\mathbf{u}(t) = (5t^3)\mathbf{i} + (2-7t)\mathbf{j} + (t+2)\mathbf{k}$ and $\mathbf{v}(t) = (1-2t)\mathbf{i} + (3t)\mathbf{j} + (4)\mathbf{k}$

18. $\mathbf{u}(t) = (2t)\mathbf{i} + (4)\mathbf{k}$ and $\mathbf{v}(t) = (1)\mathbf{i} + (2t)\mathbf{j} + (3t^2)\mathbf{k}$

In problems 19 – 22, find the point and angle of intersection for the given curves.

19. $\mathbf{u}(t) = \langle 3-t, t, t^2 \rangle$ and $\mathbf{v}(t) = \langle 0, t, 9 \rangle$

20. $\mathbf{u}(t) = \langle 4-t^2, t, t^2 \rangle$ and $\mathbf{v}(t) = \langle 3, t^2, \sqrt{t} \rangle$

21. $\mathbf{u}(t) = (5t^2)\mathbf{i} + (9)\mathbf{j} + (2-t)\mathbf{k}$ and $\mathbf{v}(s) = (2+s)\mathbf{i} + (3s)\mathbf{j} + (6-s)\mathbf{k}$

22. $\mathbf{u}(t) = (2+t)\mathbf{i} + (7-t)\mathbf{j} + (t+4)\mathbf{k}$ and $\mathbf{v}(s) = (3s)\mathbf{i} + (s+1)\mathbf{j} + (s^3)\mathbf{k}$

23. The vectors $\mathbf{u}(t) = \langle 0, t, t^2 \rangle$ and $\mathbf{v}(t) = \langle t, 2t, 0 \rangle$ form two sides of a parallelogram. How fast is the area of the parallelogram changing when $t = 1$. When $t = 2$?

24. The vectors $\mathbf{r}(t) = \langle 2t, 1, 0 \rangle$ and $\mathbf{s}(t) = \langle 1, 0, 3 \rangle$ form two sides of a parallelogram. How fast is the area of the parallelogram changing when $t = 1$. When $t = 2$?

25. The vectors $\mathbf{u}(t) = \langle 1, t, 3 \rangle$ and $\mathbf{v}(t) = \langle 2t, 0, 0 \rangle$ form two sides of a triangle. How fast is the area of the triangle changing when $t = 1$. When $t = 2$?

26. The vectors $\mathbf{r}(t) = \langle t^2, t, 1 \rangle$ and $\mathbf{s}(t) = \langle t, t^2, 0 \rangle$ form two sides of a triangle. How fast is the area of the triangle changing when $t = 1$. When $t = 2$?

27. The vectors $\mathbf{u}(t) = \langle 1, 0, 0 \rangle$, $\mathbf{v}(t) = \langle 1, t, 0 \rangle$ and $\mathbf{s}(t) = \langle 0, t, 3t \rangle$ form three sides of a tetrahedron. (a) How fast is the volume of the tetrahedron changing when $t = 1$. When $t = 2$?
(b) How fast is the surface area of the tetrahedron changing when $t = 1$. When $t = 2$?

28. The vectors $\mathbf{u}(t) = \langle 2t, 0, 0 \rangle$, $\mathbf{v}(t) = \langle 0, 3t, 0 \rangle$ and $\mathbf{s}(t) = \langle 0, 0, 4t \rangle$ form three sides of a tetrahedron. (a) How fast is the volume of the tetrahedron changing when $t = 1$. When $t = 2$?
(b) How fast is the surface area of the tetrahedron changing when $t = 1$. When $t = 2$?

In problems 29 – 32, use the given information to find a formula for $\mathbf{r}(t)$.

29. $\mathbf{r}'(t) = \langle 12t, 12t^2, 6e^t \rangle$ and $\mathbf{r}(0) = \langle 1, 2, 3 \rangle$.

30. $\mathbf{r}'(t) = \langle 3+4t, \cos(t), 1-6t \rangle$ and $\mathbf{r}(0) = \langle 7, 2, 5 \rangle$.

31. $\mathbf{r}'(t) = (6t^2)\mathbf{i} + (4)\mathbf{j} + (8t-5)\mathbf{k}$ and $\mathbf{r}(1) = 6\mathbf{i} + 2\mathbf{j} - 3\mathbf{k}$.

32. $\mathbf{r}'(t) = (5t^2)\mathbf{i} + (8t)\mathbf{j} + (2-t)\mathbf{k}$ and $\mathbf{r}(2) = (3)\mathbf{i} + (7)\mathbf{j} + (0)\mathbf{k}$.

12.2 Derivatives and Antiderivatives of Vector–valued Functions

33. Fill in the rest of the **i** coordinate entries for **r** and **r** ' in the table in Fig. 16.

34. Fill in the rest of the **j** coordinate entries for **r** and **r** ' in the table in Fig. 16.

35. Fill in the rest of the **i** coordinate entries for **r** and **r** ' in the table in Fig. 17.

36. Fill in the rest of the **j** coordinate entries for **r** and **r** ' in the table in Fig. 17.

37. State and prove a differentiation rule for $\frac{d}{dt}\left(\frac{\mathbf{u}(t)}{f(t)}\right)$.

38. Prove that
$$\frac{d}{dt}(\mathbf{u}(t) \times \mathbf{v}(t)) = -\frac{d}{dt}(\mathbf{v}(t) \times \mathbf{u}(t)).$$

t	**r** ''(t)	**r** '(t)	**r**(t)
0	⟨0, 2, 5⟩	⟨1, 2, 3⟩	⟨0, 3, 1⟩
1	⟨4, 1, 3⟩	⟨ , , 6⟩	⟨ , , 7⟩
2	⟨6, 0, 1⟩	⟨ , , 7⟩	⟨ , , 14⟩
3	⟨4, –2, 0⟩	⟨ , , 7⟩	⟨ , , 21⟩
4	⟨2, 0, 2⟩	⟨ , , 9⟩	⟨ , , 30⟩
5	⟨8, 3, 4⟩	⟨ , , 13⟩	⟨ , , 43⟩

Fig. 16

t	**r** ''(t)	**r** '(t)	**r**(t)
0	⟨1, 2, 3⟩	⟨ , , 4⟩	⟨ , , 2⟩
1	⟨4, 2, 2⟩	⟨ , , 6⟩	⟨ , , 8⟩
2	⟨3, 1, 0⟩	⟨8, 9, 6⟩	⟨30, 20, 14⟩
3	⟨2, 3, 1⟩	⟨ , , 7⟩	⟨ , , 21⟩
4	⟨1, 4, 0⟩	⟨ , , 7⟩	⟨ , , 28⟩
5	⟨0, 1, 3⟩	⟨ , , 10⟩	⟨ , , 38⟩

Fig. 17

Practice Answers

Practice 1: $\mathbf{r}(t) = \langle t, t^2, t^3 \rangle$. $\mathbf{v}(t) = $ velocity $= \mathbf{r}'(t) = \langle 1, 2t, 3t^2 \rangle$ so
$\mathbf{v}(0) = \langle 1, 0, 0 \rangle$, $\mathbf{v}(1) = \langle 1, 2, 3 \rangle$, $\mathbf{v}(2) = \langle 1, 4, 12 \rangle$.

$\mathbf{sp}(t) = $ speed $= |\mathbf{v}(t)|$ so $\mathbf{sp}(0) = 1$, $\mathbf{sp}(1) = \sqrt{14}$, $\mathbf{sp}(2) = \sqrt{161}$.

$\mathbf{dir}(t) = $ direction $= \frac{\mathbf{v}(t)}{|\mathbf{v}(t)|}$. $\mathbf{dir}(0) = \langle 1, 0, 0 \rangle$, $\mathbf{dir}(1) = \frac{1}{\sqrt{14}}\langle 1, 2, 3 \rangle$, $\mathbf{dir}(2) = \frac{1}{\sqrt{161}}\langle 1, 4, 12 \rangle$.

$\mathbf{a}(t) = $ acceleration $= \mathbf{r}''(t) = \langle 0, 2, 6t \rangle$ so $\mathbf{a}(0) = \langle 0, 2, 0 \rangle$, $\mathbf{a}(1) = \langle 0, 2, 6 \rangle$, $\mathbf{a}(2) = \langle 0, 2, 12 \rangle$.

Practice 2: The paths $\mathbf{r}(t) = \langle 0, t, t^2 \rangle$ and $\mathbf{s}(t) = \langle 2 - t, 1, 5 - t^2 \rangle$ intersect at $\mathbf{r}(1) = \mathbf{s}(2) = (0, 1, 1)$.
$\mathbf{r}'(t) = \langle 0, 1, 2t \rangle$ and $\mathbf{s}'(t) = \langle -1, 0, -2t \rangle$ so $\mathbf{r}'(1) = \langle 0, 1, 2 \rangle$ and $\mathbf{s}'(2) = \langle -1, 0, -4 \rangle$.

$\cos(\theta) = \frac{\mathbf{r}'(1) \cdot \mathbf{s}'(2)}{|\mathbf{r}'(1)||\mathbf{s}'(2)|} = \frac{-8}{\sqrt{5}\sqrt{17}} = -0.868$ so the angle between $\mathbf{r}'(1)$ and $\mathbf{s}'(1)$ is

$\theta \approx 2.922$ (or $150.2°$)

Practice 3: $\mathbf{r}'(t) = \mathbf{v}(t) = \langle 6t^2, \cos(t), 12e^{3t} \rangle$ so $\mathbf{r}(t) = \langle 2t^3 + A, \sin(t) + B, 4e^{3t} + C \rangle$.
Since $\mathbf{r}(0) = \langle 1, -5, 2 \rangle = \langle 2(0)^3 + A, \sin(0) + B, 4e^{3(0)} + C \rangle = \langle A, B, 4 + C \rangle$, we have
$A = 1$, $B = -5$, and $C = -2$. Then $\mathbf{r}(t) = \langle 2t^3 + 1, \sin(t) - 5, 4e^{3t} - 2 \rangle$.

Practice 4: Velocity entries: 2, 6, 12, 20, 17, 17, 22, 20, 21, 21
Acceleration entries: 5, 11, 23, 43, 60, 77, 99, 119, 140, 161

12.3 ARC LENGTH AND CURVATURE OF SPACE CURVES

In earlier sections we have emphasized the dynamic nature of vector–valued functions by considering them as the path of a moving object. This is a very fruitful approach, but sometimes it is useful to consider a space curve as a static object and to investigate some of its geometric properties. This section considers two geometric aspects of space curves: arc length (how long is it along the curve from one point to another point?) and curvature (how quickly does the curve bend?).

Arc Length

In Section 9.4 we went through a careful derivation of an integral formula for finding the length of a parametric curve $(x(t), y(t))$ from $t = a$ to $t = b$ (Fig. 1): we

(1) partitioned the interval $[a, b]$ for the variable t and found the points $(x(t_i), y(t_i))$,

(2) found the lengths of the line segments between consecutive points along the curve

(3) added the lengths of the line segments (an approximation of the length of the curve) and got a Riemann sum

(4) took the limit of the Riemann sum to an integral formula for the length of the curve

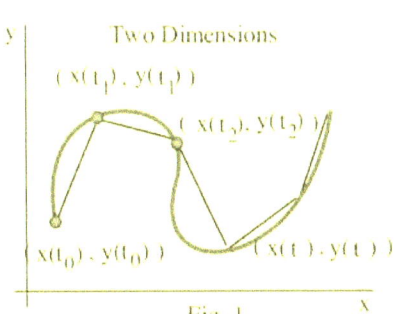

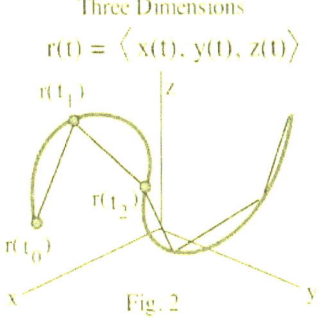

A very similar process also works for finding the length of a curve given by a vector–valued function in three dimensions, a space curve (Fig. 2), and we define the result of that process to be the length of a space curve.

Definition: **Distance Along the Graph of a Vector–Valued Function**
 (Arc Length of a Space Curve)

If $\mathbf{r}(t) = \langle x(t), y(t), z(t) \rangle$ and $x'(t), y'(t), z'(t)$ are continuous

then the **distance traveled, L, along the graph** of $\mathbf{r}(t)$ from $t = a$ to $t = b$ is

$$\text{distance traveled } L = \int_{t=a}^{t=b} \sqrt{\left(\frac{dx}{dt}\right)^2 + \left(\frac{dy}{dt}\right)^2 + \left(\frac{dz}{dt}\right)^2} \, dt$$

If we travel along each part of the curve $\mathbf{r}(t)$ exactly once, then the arc length of the curve is the distance traveled:

$$\text{arc length} = \text{distance traveled } L = \int_{t=a}^{t=b} |\mathbf{r}'(t)| \, dt = \int_{t=a}^{t=b} |\mathbf{v}(t)| \, dt .$$

Example 1: Represent the length of the helix $\mathbf{r}(t) = \langle \cos(t), \sin(t), t \rangle$ from $t = 0$ to $t = 2\pi$. (Fig. 3)

Solution:
$$L = \int_{t=0}^{t=2\pi} \sqrt{(-\sin(t))^2 + (\cos(t))^2 + (1)^2} \, dt$$

$$= \int_{t=0}^{t=2\pi} \sqrt{\sin^2(t) + \cos^2(t) + 1} \, dt = \int_{t=0}^{t=2\pi} \sqrt{2} \, dt = 2\pi\sqrt{2} \approx 8.89.$$

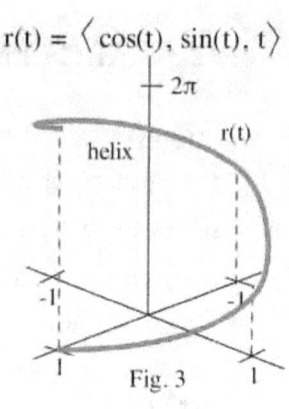
Fig. 3

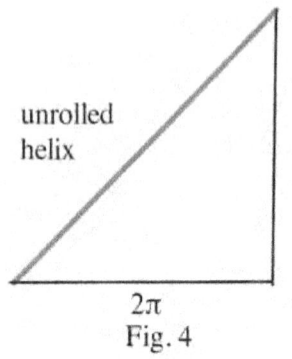
unrolled helix
Fig. 4

(Actually, we can do this particular problem without calculus. If we "unroll" the helix (Fig. 4) we get a right triangle with base 2π (the circumference of the circle with radius 1) and height 2π (the value of $z(t)$ when $t = 2\pi$). The length of the helix is the length of the hypotenuse of this triangle: hypotenuse = $\sqrt{(2\pi)^2 + (2\pi)^2} = 2\pi\sqrt{2}$.)

Practice 1: Represent the length of the graph of $\mathbf{r}(t) = \langle t, t^2, t^3 \rangle$ from $t = 0$ to $t = 2$ as an integral and use Simpson's Rule with $n = 20$ to approximate the value of the integral.

Example 2: Represent the length of the graph of $\mathbf{r}(t) = \langle \cos(t), \cos^2(t), 0 \rangle$ from $t = 0$ to $t = 2\pi$ as an integral and use numerical integration on your calculator to approximate the value of the integral.

Solution: The graph of $\mathbf{r}(t)$ is part of a parabola (Fig. 5), and the distance traveled along the parabola is

$$\text{distance} = \int_{t=0}^{t=2\pi} \sqrt{(-\sin(t))^2 + (-2\cos(t)\sin(t))^2 + (0)^2} \, dt$$

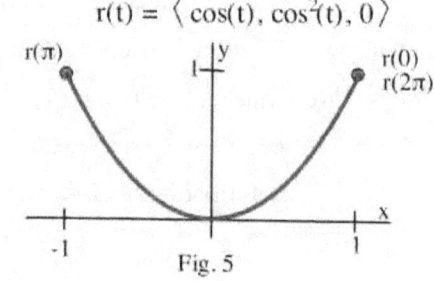
Fig. 5

$$= \int_{t=0}^{t=2\pi} \sqrt{\sin^2(t) + 4\cos^2(t)\sin^2(t)} \, dt \approx 5.916 \text{ (using numerical integration on a calculator).}$$

But in this example, the distance is NOT the length of the graph. As t goes from 0 to π we travel along the parabola from the point $(1, 1, 0)$ to the origin and on to $(-1, 1, 0)$. As t goes from π to 2π we travel back along the parabola to the starting point $(1, 1, 0)$. As t goes from 0 to 2π we cover the parabola twice so the length of the parabola is half of the distance traveled:

$$\text{length} = \frac{1}{2}(\text{distance travelled}) \approx \frac{1}{2}(5.916) = 2.958.$$

We could have calculated the length of the curve as the value of integral from $t = 0$ to $t = \pi$.

Parameterizing a Curve with respect to Arc Length

So far in our dealings with parametric space curves and vector–valued functions we have treated the curves of functions of the variable t and often thought about t as representing time. We have referred to the position vector $\mathbf{r}(t) = \langle x(t), y(t), z(t) \rangle$ as representing the position $(x(t), y(t), z(t))$ of an object at time t. With a space curve, however, it is sometimes more useful to represent a location on the curve as a function of "how far along the curve" we are. For example, if we are giving someone directions to a good picnic spot in the mountains, we might describe the location (Fig. 6) as "drive 5.3 miles along the road from the turnoff" indicating that the driver should travel 5.3 miles from the beginning of the mountain road. This "how far along the road or curve" method avoids the obvious drawbacks of directions such as "drive 9 minutes at 35 miles per hour." Similarly, interstate highways are often marked with signs indicating how far we are from the the beginning of the road or the point where the road entered our state. It is usually more useful to describe the location of a knot in a wire as "17 inches from the end of the wire." That description does not depend on how fast we move along the wire or the orientation of the wire in space or even on the shape of the curve. The benefits of giving directions in terms of "how far along a road or wire" are the same for describing a location on a curve as "how far along the curve." The description of locations along a curve in terms of distance along the curve is called a **parameterization of the curve in terms of arc length**.

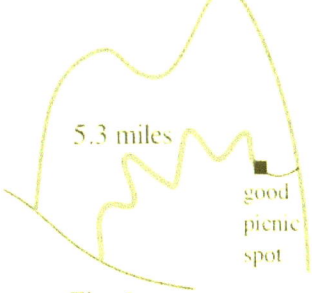

Fig. 6

Definition: **Arc Length Function s(t)**

For a curve that begins at $\mathbf{r}(a) = \langle x(a), y(a), z(a) \rangle$ with continuous x', y' and z', the distance along the curve $\mathbf{r}(t) = \langle x(t), y(t), z(t) \rangle$ at time t is the arc length function s(t) with

$$s(t) = \int_{u=a}^{u=t} \sqrt{\left(\frac{dx(u)}{du}\right)^2 + \left(\frac{dy(u)}{du}\right)^2 + \left(\frac{dz(u)}{du}\right)^2}\, du = \int_{u=a}^{u=t} |\mathbf{r}'(u)|\, du.$$

Example 3:   Fig. 7 illustrates the path $\mathbf{r}$ of a salmon swimming up a river marked with dots at 1 mile intervals.
    (a) Label the location of $\mathbf{r}(4)$ with an "X" for $\mathbf{r}$ parameterized in terms of arc length.
    (b) Label the location of $\mathbf{r}(4)$ with an "O" for $\mathbf{r}$ parameterized in terms of time.

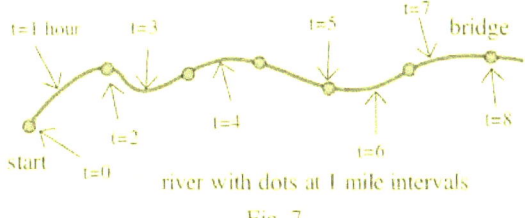

Fig. 7

(c) For an arc length parameterization, find A so r(A) = bridge.

(d) During which time interval was the fish swimming the fastest?

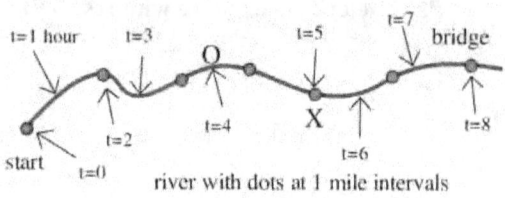

Fig. 8

Solution: (a) and (b) Fig. 8 shows the correct locations of the "X" and the "O."

(c) The bridge is 6 miles from the beginning of the river so A = 6.

(d) The fish swims the greatest distance between t = 4 and t = 5, so it was swimming fastest during that 1 hour time interval.

Practice 2: For the salmon in Fig. 7

(a) label the location of **r**(3) with an "S" for **r** parameterized in terms of arc length.

(b) Label the location of **r**(3) with an "T" for **r** parameterized in terms of time.

(c) For a time parameterization, find B so **r**(B) = bridge.

(d) During which time interval was the fish swimming the slowest?

For most curves, it is difficult to find a simple formula for the arc length function s(t). But sometimes we do get such a nice result for s(t) that we can solve for t(s), t in terms of s, and then we can rewrite the original parameterization $\mathbf{r}(t) = \langle x(t), y(t), z(t) \rangle$ as $\mathbf{r}(t(s)) = \langle x(t(s)), y(t(s)), z(t(s)) \rangle$.

Example 4: Write an arc length parameterization of the helix $\mathbf{r}(t) = \langle 3t, 4\cos(t), 4\sin(t) \rangle$ using $\mathbf{r}(0) = \langle 0, 0, 0 \rangle$ as the starting point.

Solution: $\mathbf{r}'(t) = \langle 3, -4\sin(t), 4\cos(t) \rangle$ for all t so

$$|\mathbf{r}'(t)| = \sqrt{(3)^2 + (-4\sin(t))^2 + (4\cos(t))^2} = \sqrt{9 + 16\sin^2(t) + 16\cos^2(t)} = 5 \text{ for all t.}$$

Then $s = s(t) = \int_{u=0}^{u=t} |\mathbf{r}'(u)| \, du = \int_{u=0}^{u=t} 5 \, du = 5t$ so $t = \frac{s}{5}$. By substituting $t(s) = \frac{s}{5}$ for t,

the original parameterization $\mathbf{r}(t) = \langle 3t, 4\cos(t), 4\sin(t) \rangle$ becomes

$\mathbf{r}(t(s)) = \langle 3t(s), 4\cos(t(s)), 4\sin(t(s)) \rangle = \langle 3\frac{s}{5}, 4\cos(\frac{s}{5}), 4\sin(\frac{s}{5}) \rangle$, a function of s alone.

Practice 3: Write an arc length parameterization of the line $\mathbf{r}(t) = \langle 8t, t, 4t \rangle$ using $\mathbf{r}(0) = \langle 0, 0, 0 \rangle$ as the starting point.

The conversions from "time parameterization" to "arc length parameterization" in the example and practice problems were relatively easy because the object was moving along each curve at a constant speed ($|\mathbf{r}'(t)|$ was constant). Usually this conversion is not that easy, but most of the time the arc length parameterization for a curve will be given so we will not need to translate to get from a "time parameterization" to an "arc length parameterization."

Curvature

Fig. 9 shows a space curve $\mathbf{r}(t)$ and unit tangent vectors $\mathbf{T}(t) = \dfrac{\mathbf{r}'(t)}{|\mathbf{r}'(t)|}$ at several equally spaced (in terms of arc length) points along $\mathbf{r}(t)$. When the curve twists and bends sharply in the left part of the graph, the unit tangent vectors change direction rapidly from point to point. When the curve is almost straight and bends slowly, the unit tangent vectors also change direction slowly. This geometric pattern between the "bendedness" of a curve and the rate of change (with respect to arc length) of the direction of the unit tangent vectors leads to our definition of the **curvature** of a space curve at a point.

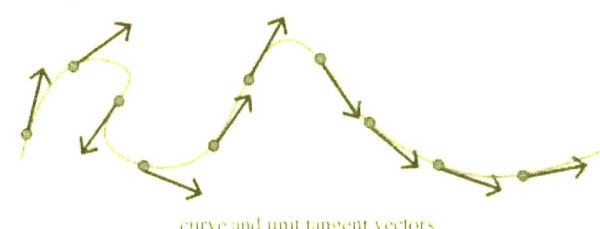

curve and unit tangent vectors
Fig. 9

Definition: Curvature

If $\mathbf{r}$ is a space curve with unit tangent vector $\mathbf{T}$ and arc length parameterization s,

then the **curvature** of $\mathbf{r}$ is

$$\kappa = \left| \frac{d\mathbf{T}}{ds} \right|.$$ (κ is the Greek letter "kappa")

The curvature of a space curve is defined to be the magnitude of the rate of change of direction of the unit tangent vectors with respect to arc length. This definition of curvature is nicely motivated geometrically, but it is difficult to use for computations if we do not have an arc length parameterization of $\mathbf{r}$. However, the Chain Rule and the Fundamental Theorem of Calculus provide us with an easier way to actually calculate the curvature of a space curve $\mathbf{r}(t)$.

By the Chain Rule, $\dfrac{d\mathbf{T}}{dt} = \dfrac{d\mathbf{T}}{ds} \cdot \dfrac{ds}{dt}$ so $\left|\dfrac{d\mathbf{T}}{ds}\right| = \left|\dfrac{d\mathbf{T}/dt}{ds/dt}\right| = \left|\dfrac{\mathbf{T}'(t)}{ds/dt}\right|$. From the Fundamental Theorem of Calculus and the definition of $s(t) = \int_{u=a}^{u=t} |\mathbf{r}'(u)|\, du$, we know that $\dfrac{ds(t)}{dt} = |\mathbf{r}'(t)|$.

By putting these two results together, we get a much easier to use formula for the curvature of a space curve.

> **A Formula for Curvature:** $\kappa = \dfrac{|\mathbf{T}'(t)|}{|\mathbf{r}'(t)|}$

Example 5: For a positive number A, the graph of $\mathbf{r}(t) = \langle A\cdot\cos(t), A\cdot\sin(t), 0 \rangle$ is a circle with radius A in the xy-plane. Find the curvature of this circle.

Solution: $\mathbf{r}'(t) = \langle -A\cdot\sin(t), A\cdot\cos(t), 0 \rangle$ so $|\mathbf{r}'(t)| = \sqrt{A^2\sin^2(t) + A^2\cos^2(t) + 0} = A$.

$\mathbf{T}(t) = \dfrac{\mathbf{r}'(t)}{|\mathbf{r}'(t)|} = \dfrac{\mathbf{r}'(t)}{A} = \langle -\sin(t), \cos(t), 0 \rangle$ so $\mathbf{T}'(t) = \langle -\cos(t), -\sin(t), 0 \rangle$ and $|\mathbf{T}'(t)| = 1$.

Then, for all t, $\kappa = \left| \dfrac{\mathbf{T}'(t)}{\mathbf{r}'(t)} \right| = \dfrac{1}{A}$. The curvature of a circle of radius A is $\kappa = \dfrac{1}{A}$.

This agrees with our geometric idea of curvature:

- when the radius of the circle is large (Fig. 10a), the circle bends slowly and the curvature κ is small

- when the radius of the circle is small (Fig. 10b), the circle bends quickly and the curvature κ is large.

This pattern for curvature of circles leads to the definition of the radius of curvature of a curve.

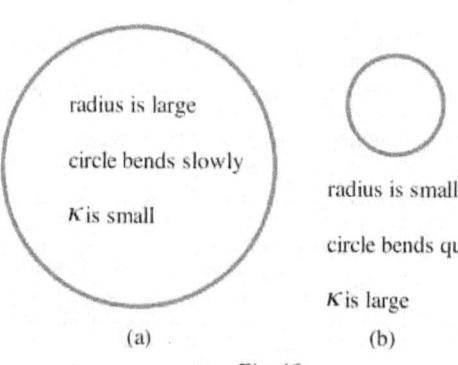

Fig. 10

> **Definition:** The **radius of curvature** of $\mathbf{r}(t)$ is $\dfrac{1}{\text{curvature of } \mathbf{r}(t)} = \dfrac{1}{\kappa}$.

Practice 4: For A, B, and C not equal to 0, show that the **line** $\mathbf{r}(t) = \langle At, Bt, Ct \rangle$ has curvature $\kappa = 0$.

It was relatively straightforward to calculate the curvature in the example and practice problem because $|\mathbf{r}'(t)|$ was a constant. When $|\mathbf{r}'(t)|$ is not constant, it can be difficult to calculate $\mathbf{T}'(t)$, and some other formulas for curvature are often easier. The following formula for curvature looks complicated, but in practice it is often the easiest one to use.

> **"Easiest" Formula for Curvature in 3D:** $\kappa = \dfrac{|\mathbf{r}' \times \mathbf{r}''|}{|\mathbf{r}'|^3}$

A proof that this formula follows from the formula $\kappa = \dfrac{|\mathbf{T}'(t)|}{|\mathbf{r}'(t)|}$ is given in an appendix after the problem set.

Example 6: Use the formula $\kappa = \dfrac{|\mathbf{r}' \times \mathbf{r}''|}{|\mathbf{r}'|^3}$ to determine the curvature and the radius of curvature of $\mathbf{r}(t) = \langle t, t^2, t^3 \rangle$ when $t = 0, 1,$ and 2.

Solution: $\mathbf{r}'(t) = \langle 1, 2t, 3t^2 \rangle$ so $|\mathbf{r}'(t)| = \sqrt{1 + 4t^2 + 9t^4}$ and $\mathbf{r}''(t) = \langle 0, 2, 6t \rangle$.

Then $\mathbf{r}' \times \mathbf{r}'' = \begin{vmatrix} \mathbf{i} & \mathbf{j} & \mathbf{k} \\ 1 & 2t & 3t^2 \\ 0 & 2 & 6t \end{vmatrix} = \mathbf{i}\begin{vmatrix} 2t & 3t^2 \\ 2 & 6t \end{vmatrix} - \mathbf{j}\begin{vmatrix} 1 & 3t^2 \\ 0 & 6t \end{vmatrix} + \mathbf{k}\begin{vmatrix} 1 & 2t \\ 0 & 2 \end{vmatrix}$

$= (6t^2)\mathbf{i} - (6t)\mathbf{j} + (2)\mathbf{k}$ and

$|\mathbf{r}' \times \mathbf{r}''| = \sqrt{(6t^2)^2 + (-6t)^2 + (2)^2} = \sqrt{36t^4 + 36t^2 + 4}$.

Putting this all together, we have $\kappa = \dfrac{\sqrt{36t^4 + 36t^2 + 4}}{(1 + 4t^2 + 9t^4)^{3/2}}$.

When $t = 0$, $\kappa = \dfrac{\sqrt{4}}{1} = 2$ so the radius of curvature is $\dfrac{1}{2}$.

When $t = 1$, $\kappa = \dfrac{\sqrt{76}}{(14)^{3/2}} \approx 0.166$ so the radius of curvature is approximately $\dfrac{1}{0.166} \approx 6.02$.

When $t = 2$, $\kappa = \dfrac{\sqrt{724}}{(161)^{3/2}} \approx 0.013$ so the radius of curvature is approximately 76.9.

For $\mathbf{r}(t) = \langle t, t^2, t^3 \rangle$, as t grows larger (and is positive), the curvature κ becomes smaller.

Practice 5: Use the formula $\kappa = \dfrac{|\mathbf{r}' \times \mathbf{r}''|}{|\mathbf{r}'|^3}$ to determine the curvature of $\mathbf{r}(t) = \langle t, \sin(t), 0 \rangle$ when $t = 0, \pi/4,$ and $\pi/2$.

Curvature in Two Dimensions: $\mathbf{r}(t) = \langle x(t), y(t), 0 \rangle$ and $y = f(x)$

Every curve confined to the xy–plane can be thought of as a curve in space whose z–coordinate is always 0, and that approach leads to alternate formulas for the curvature of the graph.

If the curve we are dealing with is given parametrically in two dimensions as $(x(t), y(t))$ then we can consider it as the vector–valued function $\mathbf{r}(t) = \langle x(t), y(t), 0 \rangle$ in three dimensions, and the curvature formula $\kappa = \dfrac{|\mathbf{r}' \times \mathbf{r}''|}{|\mathbf{r}'|^3}$ "simplifies" to the following.

> The graph of $(x(t), y(t))$ has curvature $\kappa = \dfrac{|x'y'' - x''y'|}{((x')^2 + (y')^2)^{3/2}}$
>
> where the derivatives of x and y are with respect to t.

If $y = f(x)$, then the curve can be parameterized in two dimensions using $x(t) = t$ and $y(t) = f(x) = f(t)$. With this parameterization $\mathbf{r}(t) = \langle x(t), y(t), 0 \rangle = \langle t, y(t), 0 \rangle$ and we have $x'(t) = 1$ and $x''(t) = 0$ so the previous pattern reduces to

> If $y = f(x)$, then $\kappa = \dfrac{|y''|}{(1 + (y')^2)^{3/2}}$ where the derivatives are with respect to x.

These last two formulas for curvature are typically easier to use than the previous ones, but **they are only valid for two–dimensional graphs**.

Example 7: Use the appropriate formula to determine the curvature of $y = x^2$ when $x = 0, 1$ and 2.

Solution: We can use the formula $\kappa = \dfrac{|y''|}{(1 + (y')^2)^{3/2}}$ with $y' = 2x$ and $y'' = 2$. Then

$$\kappa = \frac{|2|}{(1 + (2x)^2)^{3/2}} = \frac{2}{(1 + 4x^2)^{3/2}}.$$

When $x = 0$, $\kappa = \dfrac{2}{(1 + 4x^2)^{3/2}} = \dfrac{2}{(1)^{3/2}} = 2$. When $x = 1$, $\kappa = \dfrac{2}{(1 + 4x^2)^{3/2}} = \dfrac{2}{(5)^{3/2}} \approx 0.179$.

When $x = 2$, $\kappa = \dfrac{2}{(1 + 4x^2)^{3/2}} = \dfrac{2}{(17)^{3/2}} \approx 0.029$.

Practice 6: Use the appropriate 2–dimensional formula to determine the curvature of $y = \sin(x)$ when $x = 0, 1$ and 2. (Your answers should agree with your answers to Practice 5.)

PROBLEMS

In problems 1 – 4, determine the length of the helices for $0 \leq t \leq 2\pi$.

1. $\mathbf{r}(t) = \langle 2 \cdot \cos(t), 2 \cdot \sin(t), t \rangle$

2. $\mathbf{r}(t) = \langle 3 \cdot \cos(t), 3 \cdot \sin(t), t \rangle$

3. $\mathbf{r}(t) = \langle 4 \cdot \cos(t), 4 \cdot \sin(t), t \rangle$

4. $\mathbf{r}(t) = \langle R \cdot \cos(t), R \cdot \sin(t), t \rangle$

In problems 5 – 12, determine the length of the "modified helices" for $0 \leq t \leq 2\pi$. If necessary, use a calculator to approximate the arc length integrals.

5. $\mathbf{r}(t) = \langle 2 \cdot \cos(t), 3 \cdot \sin(t), t \rangle$

6. $\mathbf{r}(t) = \langle 2 \cdot \cos(t), 5 \cdot \sin(t), t \rangle$

7. $\mathbf{r}(t) = \langle A \cdot \cos(t), B \cdot \sin(t), t \rangle$

8. $\mathbf{r}(t) = \langle \cos(2t), \sin(2t), t \rangle$

9. $\mathbf{r}(t) = \langle t \cdot \cos(t), t \cdot \sin(t), t \rangle$

10. $\mathbf{r}(t) = \langle 2t \cdot \cos(t), 2t \cdot \sin(t), t \rangle$

11. $\mathbf{r}(t) = \langle 2t \cdot \cos(t), t \cdot \sin(t), t \rangle$

12. $\mathbf{r}(t) = \langle t^2 \cdot \cos(t), t^2 \cdot \sin(t), t \rangle$

In problems 13 – 16, determine the length of the Bezier curves for $0 \leq t \leq 1$. If necessary, use a calculator to approximate the arc length integrals.

13. $\mathbf{r}(t) = \langle 3(1-t)^2 t + 3(1-t)t^2, 9(1-t)t^2 + 2t^3, (1-t)^3 + 6(1-t)^2 t \rangle$

14. $\mathbf{r}(t) = \langle 2(1-t)^3 + 6(1-t)t^2, 3(1-t)^2 t + 3(1-t)t^2 + 3t^3, 3(1-t)^2 t + 3(1-t)t^2 \rangle$

15. The Bezier curve determined by the control points $P_0 = (2, 0, 0)$, $P_1 = (0, 1, 1)$, $P_2 = (2, 1, 1)$, and $P_3 = (0, 3, 0)$.

16. The Bezier curve determined by the control points $P_0 = (2, 0, 0)$, $P_1 = (0, 3, 2)$, $P_2 = (2, 0, 3)$, and $P_3 = (0, 3, 0)$.

In problems 17 – 22, determine the curvature of the given curves at the specified points.

17. $\mathbf{r}(t) = \langle \cos(t), \sin(t), t \rangle$ when $t = 0, \pi/4$, and $\pi/2$.

18. $\mathbf{r}(t) = \langle 3 \cdot \cos(t), 3 \cdot \sin(t), t \rangle$ when $t = 0, \pi/4$, and $\pi/2$.

19. $\mathbf{r}(t) = \langle R \cdot \cos(t), R \cdot \sin(t), t \rangle$ when $t = 0, \pi/4$, and $\pi/2$.

20. $\mathbf{r}(t) = \langle 5 + 3t, 2 - t, 3 - 2t \rangle$ when $t = 0, 2$, and 7.

21. $\mathbf{r}(t) = \langle 3(1-t)^2 t + 3(1-t)t^2, 9(1-t)t^2 + 2t^3, (1-t)^3 + 6(1-t)^2 t \rangle$ when $t = 0.2$ and 0.5.

22. $\mathbf{r}(t) = \langle 2(1-t)^3 + 6(1-t)t^2, 3(1-t)^2 t + 3(1-t)t^2 + 3t^3, 3(1-t)^2 t + 3(1-t)t^2 \rangle$ when $t = 0.2$ and 0.5.

In problems 23 – 26, determine the curvature and the radius of curvature of the given curves at the specified points.

23. $\mathbf{r}(t) = \langle 3\cdot\cos(t), 5\cdot\sin(t) \rangle$ when $t = 0, \pi/4$, and $\pi/2$.

24. $\mathbf{r}(t) = \langle 2\cdot\cos(t), 7\cdot\sin(t) \rangle$ when $t = 0, \pi/4$, and $\pi/2$.

25. $\mathbf{r}(t) = \langle A\cdot\cos(t), B\cdot\sin(t) \rangle$ when $t = 0, \pi/4$, and $\pi/2$.

26. $\mathbf{r}(t) = \langle t\cdot\cos(t), t\cdot\sin(t) \rangle$ when $t = 1, 2$, and 3.

27. Determine the curvature of $y = 3x + 5$ when $x = 1, 2$, and 3. For what value of x is the curvature of $y = 3x + 5$ maximum?

28. Determine the curvature of $y = Ax + B$ when $x = 1, 2$, and 3. For what value of x is the curvature of $y = Ax + B$ maximum?

29. Determine the curvature of $y = x^2$ when $x = 1, 2$, and 3. For what value of x is the curvature of $y = x^2$ maximum? For what value of x is the radius of curvature of $y = x^2$ minimum?

30. Determine the curvature of $y = x^3 - x$ when $x = 0, 1$, and 2. For what value of x is the curvature of $y = x^3 - x$ maximum? For what value of x is the radius of curvature of $y = x^3 - x$ minimum?

Practice Answers

Practice 1: $|\mathbf{r}'(t)| = \sqrt{1^2 + (2t)^2 + (3t^2)^2} = \sqrt{1 + 4t^2 + 9t^4}$. Then

$$L = \int_{t=0}^{t=2} |\mathbf{r}'(t)| \, dt = \int_{t=0}^{t=2} \sqrt{1 + 4t^2 + 9t^4} \, dt \approx 9.57 \text{ (using Simpson's Rule with } n = 20).$$

Practice 2:   (a) and (b) Fig. 11 shows the correct locations of the "S" and the "T."

(c) The bridge is 8 hours from the beginning of the river so $B = 8$.

(d) The fish swims the smallest distance between $t = 2$ and $t = 3$, so it was swimming slowest during that 1 hour time interval.

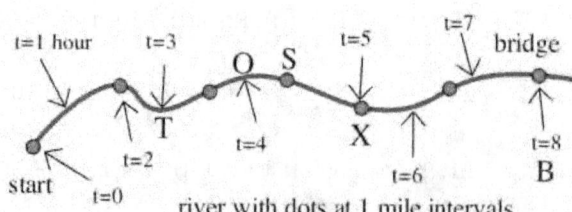

river with dots at 1 mile intervals
Fig. 11

Practice 3: $|\mathbf{r}'(t)| = \sqrt{(8)^2 + (1)^2 + (4)^2} = 9$ for all t.

Then $s = s(t) = \int_{u=0}^{u=t} |\mathbf{r}'(u)|\,du = \int_{u=0}^{u=t} 9\,du = 9t$ so $t = \frac{s}{9}$. By substituting $t(s) = \frac{s}{9}$ for t,

the original parameterization $\mathbf{r}(t) = \langle 8t, t, 4t \rangle$ becomes

$\mathbf{r}(t(s)) = \langle 8t(s), t(s), 4t(s) \rangle = \langle 8\frac{s}{9}, \frac{s}{9}, 4\frac{s}{9} \rangle$, a function of s alone.

Practice 4: $\mathbf{r}'(t) = \langle A, B, C \rangle$ so $|\mathbf{r}'(t)| = \sqrt{A^2 + B^2 + C^2}$.

$\mathbf{T}(t) = \dfrac{\mathbf{r}'(t)}{|\mathbf{r}'(t)|} = \dfrac{1}{\sqrt{A^2 + B^2 + C^2}} \langle A, B, C \rangle$ which is a constant vector so $\mathbf{T}'(t) = \mathbf{0}$.

Then, for all t, $\kappa = \left|\dfrac{\mathbf{T}'(t)}{\mathbf{r}'(t)}\right| = \dfrac{0}{\sqrt{A^2 + B^2 + C^2}} = 0$.

The curvature of a the line $\mathbf{r}(t) = \langle At, Bt, Ct \rangle$ is $\kappa = 0$.

Practice 5: $\mathbf{r}'(t) = \langle 1, \cos(t), 0 \rangle$ so $|\mathbf{r}'(t)| = \sqrt{1 + \cos^2(t)}$ and $\mathbf{r}''(t) = \langle 0, -\sin(t), 0 \rangle$.

Then $\mathbf{r}' \times \mathbf{r}'' = \begin{vmatrix} \mathbf{i} & \mathbf{j} & \mathbf{k} \\ 1 & \cos(t) & 0 \\ 0 & -\sin(t) & 0 \end{vmatrix} = 0\mathbf{i} - 0\mathbf{j} + (-\sin(t))\mathbf{k}$ and $|\mathbf{r}' \times \mathbf{r}''| = |-\sin(t)|$.

Putting this together, we have $\kappa = \dfrac{|-\sin(t)|}{(1 + \cos^2(t))^{3/2}}$.

When $t = 0$, $\kappa = \dfrac{0}{(1+1)^{3/2}} = 0$. When $t = \pi/4$, $\kappa = \dfrac{\sqrt{2}/2}{(1 + 1/2)^{3/2}} \approx 0.385$.

When $t = \pi/2$, $\kappa = \dfrac{1}{(1+0)^{3/2}} = 1$

Practice 6: $y = \sin(x)$, $y'(x) = \cos(x)$, and $y''(x) = -\sin(x)$. Then

$\kappa = \dfrac{|y''|}{(1 + (y')^2)^{3/2}} = \dfrac{|-\sin(x)|}{(1 + \cos^2(x))^{3/2}}$ which is the same result we got in Practice 5.

Appendix: A proof that $\kappa = \dfrac{|\mathbf{T}'(t)|}{|\mathbf{r}'(t)|} = \dfrac{|\mathbf{r}' \times \mathbf{r}''|}{|\mathbf{r}'|^3}$

Since $\mathbf{T}(t) = \dfrac{\mathbf{r}'(t)}{|\mathbf{r}'(t)|}$ and $|\mathbf{r}'(t)| = \dfrac{ds}{dt}$ we know that $\mathbf{r}'(t) = |\mathbf{r}'(t)|\mathbf{T}(t) = \dfrac{ds}{dt}\mathbf{T}(t)$.

Then, by the Product rule for Derivatives, $\mathbf{r}''(t) = \dfrac{d}{dt}\left\{ \dfrac{ds}{dt}\mathbf{T}(t) \right\} = \dfrac{d^2 s}{dt^2}\mathbf{T}(t) + \dfrac{ds}{dt}\mathbf{T}'(t)$.

Replacing $\mathbf{r}'$ and $\mathbf{r}''$ with these results and using the distributive pattern $\mathbf{A} \times (\mathbf{B} + \mathbf{C}) = \mathbf{A} \times \mathbf{B} + \mathbf{A} \times \mathbf{C}$ for the cross product, we have

$\mathbf{r}' \times \mathbf{r}'' = \dfrac{ds}{dt}\mathbf{T}(t) \times \left\{ \dfrac{d^2 s}{dt^2}\mathbf{T} + \dfrac{ds}{dt}\mathbf{T}' \right\} = \dfrac{ds}{dt}\dfrac{d^2 s}{dt^2}\{\mathbf{T} \times \mathbf{T}\} + \left\{ \dfrac{ds}{dt} \right\}^2 \{\mathbf{T} \times \mathbf{T}'\}$.

We know for every vector $\mathbf{V}$ that $\mathbf{V} \times \mathbf{V} = \mathbf{0}$ (the zero vector) so $\mathbf{T} \times \mathbf{T} = \mathbf{0}$.

We also know that $|\mathbf{T}(t)| = 1$ for all t so from Example 3 of Section 12.2 we can conclude that $\mathbf{T}$ is perpendicular to $\mathbf{T}'$. Then $|\mathbf{T} \times \mathbf{T}'| = |\mathbf{T}||\mathbf{T}'||\sin(\theta)| = |\mathbf{T}||\mathbf{T}'| = |\mathbf{T}'|$. Using these results together with the previous result for $\mathbf{r}' \times \mathbf{r}''$ we have

$|\mathbf{r}' \times \mathbf{r}''| = \left\{ \dfrac{ds}{dt} \right\}^2 |\mathbf{T}'| = |\mathbf{r}'|^2 |\mathbf{T}'|$ so $|\mathbf{T}'| = \dfrac{|\mathbf{r}' \times \mathbf{r}''|}{|\mathbf{r}'|^2}$.

Finally, $\dfrac{|\mathbf{T}'|}{|\mathbf{r}'|} = \dfrac{|\mathbf{r}' \times \mathbf{r}''|}{|\mathbf{r}'|^3}$, the result we wanted.

12.4 CYLINDRICAL & SPHERICAL COORDINATE SYSTEMS IN 3D

Most of our work in two dimensions used the rectangular coordinate system, but we also examined the polar coordinate system (Fig. 1), and for some uses the polar coordinate system was more effective and efficient. A similar situation occurs in three dimensions. Mostly we use the 3–dimensional xyz–coordinate system, but there are two alternate systems, called cylindrical coordinates and spherical coordinates, that are sometimes better. In two dimensions, the rectangular and the polar coordinate systems each located a point by means of two numbers, but each system used those two numbers in different ways. In three dimensions, each of the coordinate systems locates a point using three numbers, and each system uses those three numbers in different ways. Fig. 2 illustrates how three numbers are used to locate the point P in each of the different systems, and the rest of this section examines the cylindrical and spherical coordinate systems.

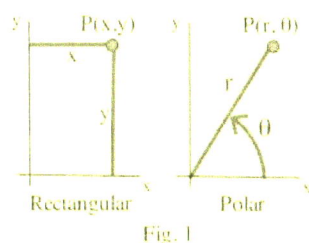

Fig. 1

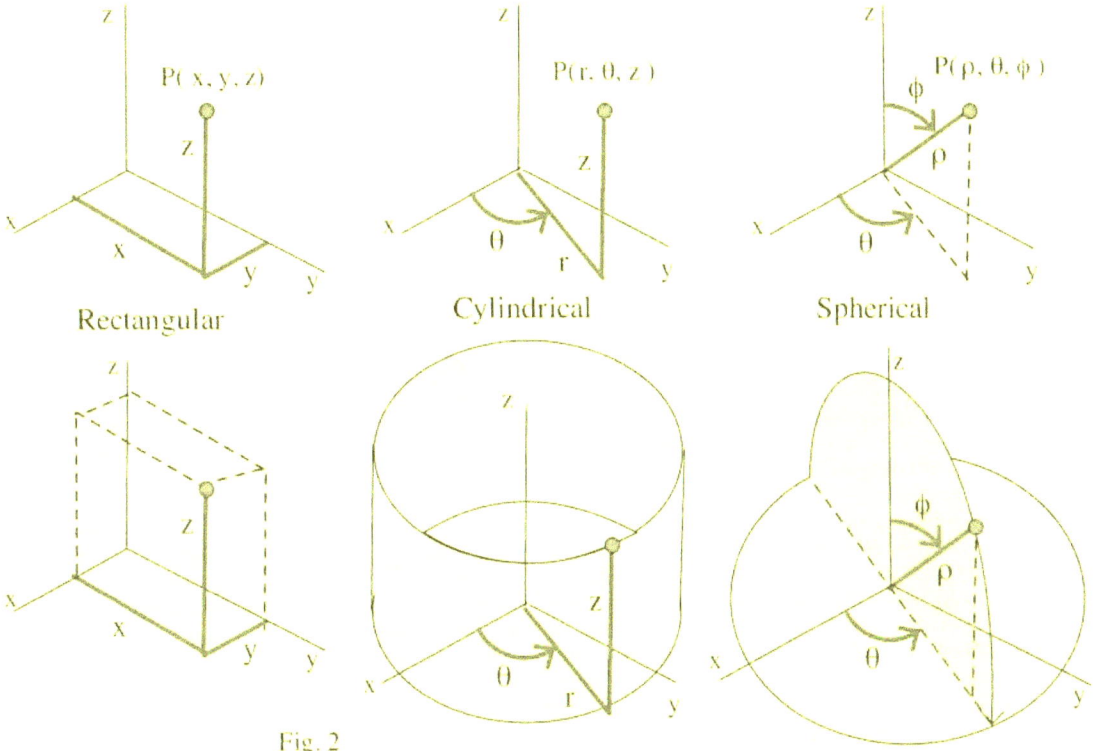

Fig. 2

CYLINDRICAL COORDINATES

Cylindrical coordinates are basically "polar coordinates with altitude." The cylindrical coordinates (r, θ, z) specify the point P that is z units above the point on the xy–plane whose polar coordinates are r and θ (Fig. 3).

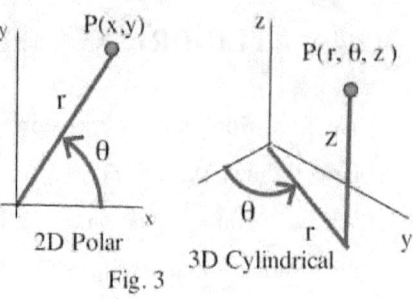

Fig. 3

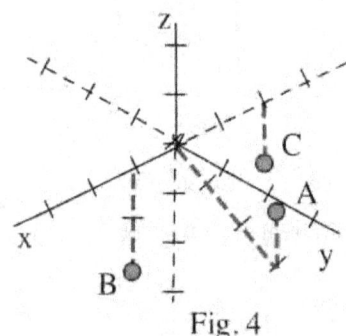

Fig. 4

Example 1: Plot the points given by the cylindrical coordinates $A(3, \pi/3, 1)$, $B(1, 0, -2)$, and $C(2, 180°, -1)$.

Solution: The points are plotted in Fig. 4.

Practice 1: On Fig. 4, plot the points given by the cylindrical coordinates $P(3, \pi/6, -1)$, $Q(3, \pi/2, 2))$, and $R(0, \pi, 3)$.

We can start to develop an understanding of the effect of each variable in the ordered triple by holding two of the variables fixed and letting the other one vary. Fig. 5 shows the results in the rectangular coordinate system of fixing x and y $(x = 1, y = 2)$ and letting z vary: we get a vertical line parallel to the z–axis. Similarly, in the rectangular coordinate system, when we fix x and z $(x = 1, z = 3)$ and let y vary we get a line parallel to the y–axis. We can try the same process in the cylindrical coordinate system.

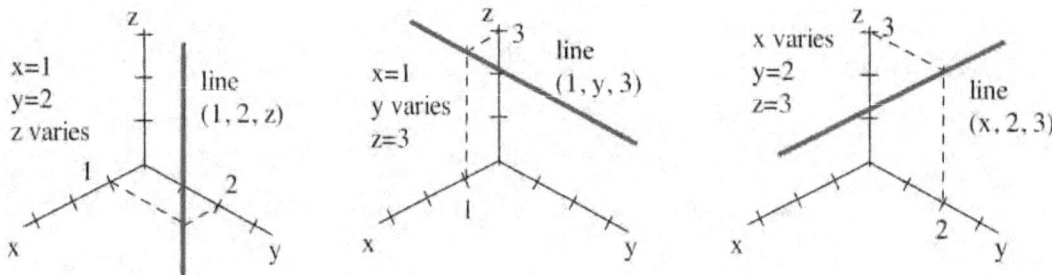

Fig. 5: Rectangular coordinates, two variables fixed

Fixing two of the variables and letting the other one vary:

In the cylindrical coordinate system, if we fix r and θ ($r = 2, \theta = \pi/3 = 60°$) and let z vary (Fig. 6a) we get a line parallel to the z–axis.

If we fix r and z ($r = 2, z = 3$) and let θ vary (Fig. 6b) we get a circle of radius 1 centered around the z–axis at a height of 3 units above the xy plane.

It we fix θ and z ($= \pi/3 = 60°, z = 3$) and let r vary (Fig. 6c) we get a line that is always 3 units above the xy plane.

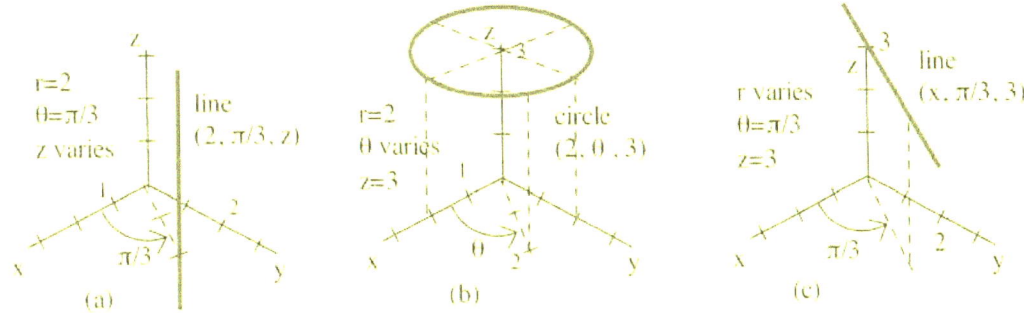

Fig. 6: Cylindrical coordinates, two variables fixed

Fixing the value of only one of the variables and letting the other two vary can also be informative.

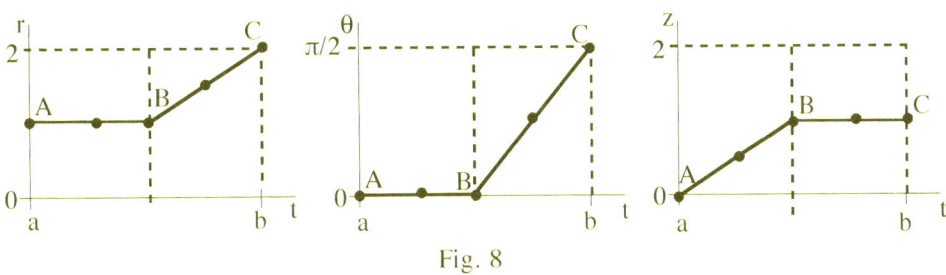

Fig. 8

If we fix r (r = 2) and let θ and z vary, the result (Fig. 7a) is a cylinder, the reason this is called the cylindrical coordinate system.

If we fix θ (θ = π/3) and let r and z vary, the result (Fig. 7b) is a plane.

If we fix z (z = 3) and let r and θ vary, the result (Fig. 7c) is a plane parallel to the xy plane.

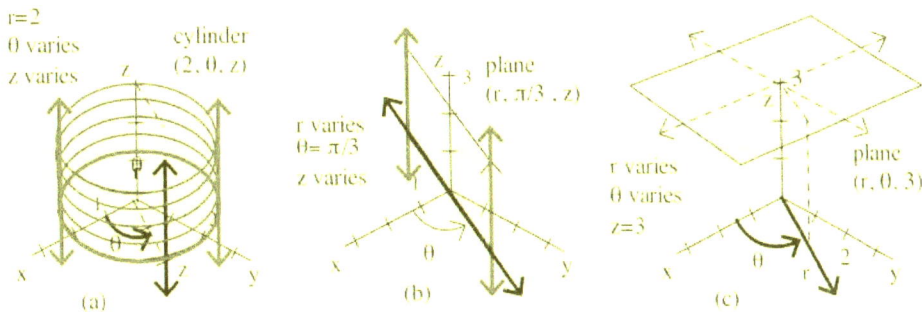

Fig. 7: Cylindrical coordinates, one variable fixed

Practice 2: An Alternate View: Cylindrical coordinates are "rectangular coordinates on a door." We began the discussion of cylindrical coordinates by describing them as "polar coordinates with altitude." A student said she found it easier to think of cylindrical coordinates as "rectangular coordinates on a door." By means of a labeled sketch show how this new description makes good sense.

Parametric Curves Using Cylindrical Coordinates

If we consider each of the rectangular coordinate variables x, y, and z to be a function of a parameter t, then (x, y, z) = (x(t), y(t), z(t)) describes the location of a point at time t. When the values of t are an interval of numbers, the graph of (x(t), y(t), z(t)) is a path or curve in three dimensional space. Similarly, if we consider each of the cylindrical coordinate variables r, θ, and z to be a function of a parameter t, then (r, θ, z) = (r(t), θ(t), z(t)) describes the location of a point, in cylindrical coordinates, at time t. When the values of t are an interval of numbers, the graph of (r(t), θ(t), z(t)) is again a path or curve in three dimensional space.

Example 2: The graphs in Fig. 8 show values of r, θ, and z as functions of t. Use those values to graph the path (r(t), θ(t), z(t)) for $a \le t \le b$.

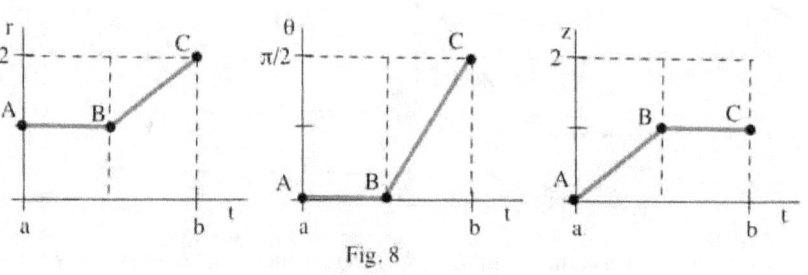

Fig. 8

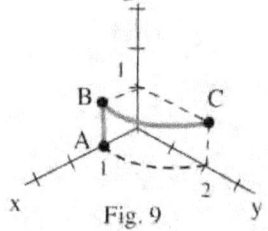

Fig. 9

Solution: Fig. 9 shows the result of plotting several points (r(t), θ(t), z(t)) and connecting them.

Practice 3: The graphs in Fig. 10 show values of r, θ, and z as functions of t. Use those values to graph the path (r(t), θ(t), z(t)) for $a \le t \le b$.

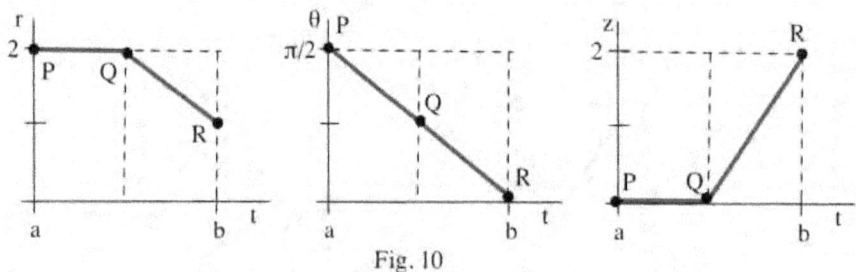

Fig. 10

Converting Cylindrical To And From Rectangular Coordinates

Typically we stay in one coordinate system for any particular use, but occasionally it may be necessary to convert information from one system to another, and it is rather straightforward to convert between rectangular and cylindrical coordinates. The important conversion information is contained in Fig. 11 and is summarized below.

Converting Between Cylindrical and Rectangular Systems

If $P(x, y, z)$ and $P(r, \theta, z)$ represent the same point in rectangular and cylindrical coordinates (Fig. 11),

then (cylindrical to rectangular) $x = r \cdot \cos(\theta)$ $y = r \cdot \sin(\theta)$

(rectangular to cylindrical) $r^2 = x^2 + y^2$ $\tan(\theta) = y/x$

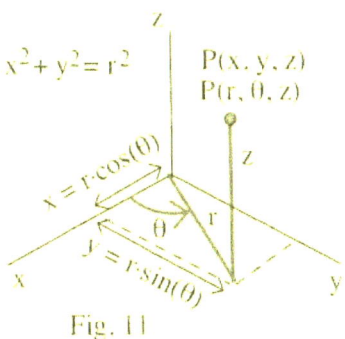

Fig. 11

Example 3: (a) Write the rectangular coordinate location $A(3, 4, 2)$ in the cylindrical coordinate system.

(b) Write the cylindrical coordinate location $B(2, \pi/6, 3)$ in the rectangular coordinate system.

Solution: (a) $r^2 = x^2 + y^2 = 3^2 + 4^2 = 5^2$. $\tan(\theta) = y/x = 4/3$ so $\theta = \arctan(4/3) \approx 0.927$ $(53.1°)$. The cylindrical coordinates of A are approximately $(5, 0.927, 2)$.

(b) $x = r \cdot \cos(\pi/6) = 2(\sqrt{3}/2) \approx 1.73$, $y = r \cdot \sin(\pi/6) = 2(0.5) = 1$. The rectangular coordinates of B are approximately $(1.73, 1, 3)$.

Practice 4: (a) Write the rectangular coordinate location $P(-5, 12, 1)$ in the cylindrical coordinate system.

(b) Write the cylindrical coordinate location $Q(4, 3\pi/4, 3)$ in the rectangular coordinate system.

Example 4: (a) Rewrite the rectangular coordinate equation $x^2 + y^2 = 6z$ as an equation in cylindrical coordinates.

(b) Rewrite the cylindrical coordinate equation $r = 6 \cdot \cos(\theta)$ as an equation in rectangular coordinates.

Solution: (a) $x^2 + y^2 = r^2$ so $x^2 + y^2 = 6z$ becomes $r^2 = 6z$.

(b) Multiplying each side of $r = 6 \cdot \cos(\theta)$ by r we have $r^2 = 6r \cdot \cos(\theta)$ so $x^2 + y^2 = 6x$ or, moving the $6x$ to the left side and completing the square, $(x-3)^2 + y^2 = 9$. The graph of $(x-3)^2 + y^2 = 9$ is a circular cylinder generated by moving the circle with radius 3 and center $(3, 0, 0)$ parallel to the z–axis.

SPHERICAL COORDINATES

A point P with spherical coordinates (ρ, θ, φ) (the Greek letters ρ = rho and φ = phi are pronounced as "row" and "fee" or "fie") is located in three dimensions as shown in Fig. 12: ρ is the distance of P from the origin, and φ is the angle the segment from the origin to P makes with the positive z–axis.

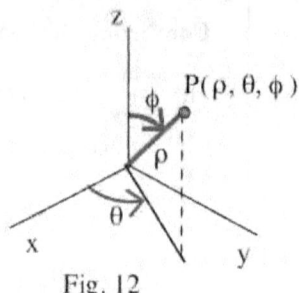

Fig. 12

Example 5: Plot the points given by the spherical coordinates $A(3, \pi/3, \pi/2)$, $B(2, 0, \pi/3)$, and $C(2, 180°, 0)$.

Solution: The points are plotted in Fig. 13.

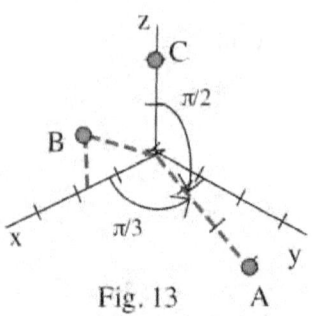

Fig. 13

Practice 5: On Fig. 13, plot the points given by the spherical coordinates $P(3, 3\pi/4, \pi/6)$, $Q(3, 0, \pi/4)$, and $R(2, \pi/6, 0)$.

Fixing two of the variables and letting the other one vary:

In the spherical coordinate system, if we fix ρ and θ ($\rho = 2, \theta = \pi/3$) and let φ vary (Fig. 14a) we get a circle with radius 2 and center at the origin.

If we fix ρ and φ ($\rho = 2, \varphi = \pi/4$) and let θ vary (Fig. 14b) we get a circle with radius $\sqrt{2}$ and center at $(0, 0, \sqrt{2})$.

If we fix θ and φ ($\theta = \pi/3, \varphi = \pi/4$) and let ρ vary (Fig. 14c) we get a straight line through the origin.

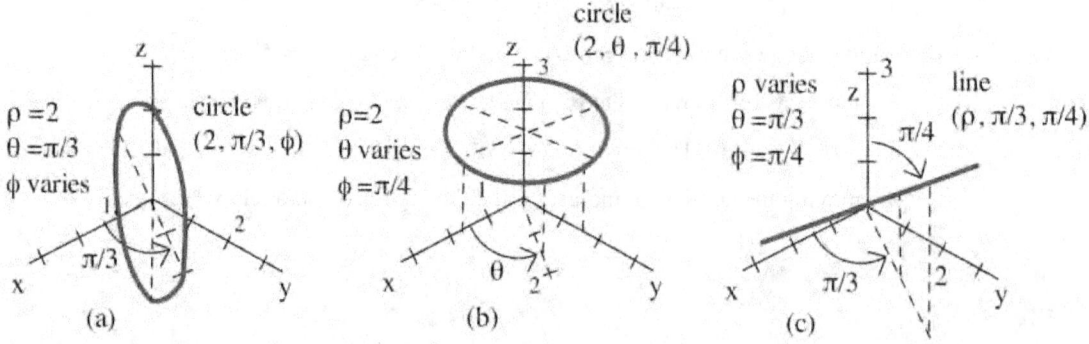

Fig. 14: Spherical coordinates, two variables fixed

Fixing one of the variables and letting the other two vary:

If we fix ρ (ρ = 2) and let θ and φ vary (Fig. 15a), we get a sphere, the reason this is called
the spherical coordinate system, with radius 2 and center at the origin.

If we fix θ (θ = π/3) and let ρ and φ vary (Fig. 15b), we get a plane.

If we fix φ (φ = π/4) and let ρ and θ vary (Fig. 15c), we get a cone around the z–axis.

It is possible to sketch the path of a point when the variables ρ, θ and φ are functions of a parameter t, but it is difficult to do "by hand" and we omit it. The Appendix at the end of this section illustrates some commands in the language Maple to create these graphs.

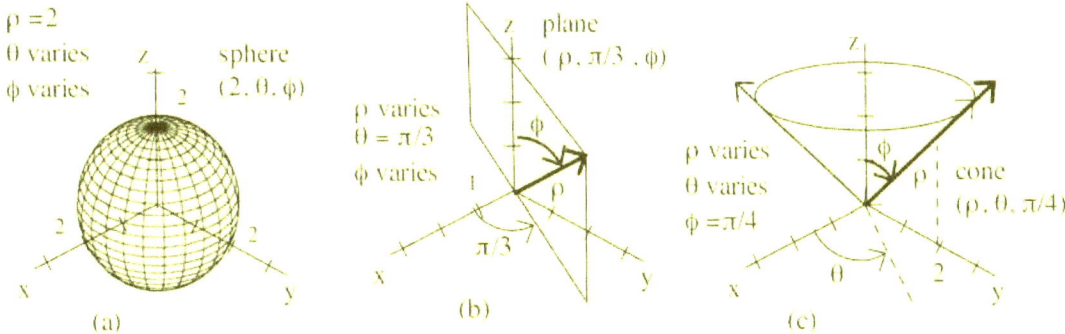

Fig. 15: Spherical coordinates, one variable fixed

Converting Spherical To And From Rectangular Coordinates

The conversions between rectangular and spherical coordinates are mostly a matter of applied trigonometry, and the essential conversion information is contained in Fig. 16.

The conversion from rectangular coordinates (x, y, z) to spherical coordinates (ρ, θ, φ) is straightforward:

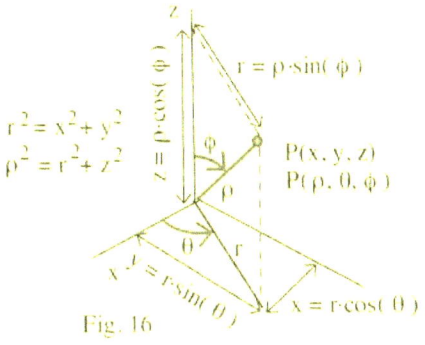

Fig. 16

r is the distance of the point (x, y, z) from the origin so

$$\rho^2 = x^2 + y^2 + z^2,$$

$$\tan(\theta) = \frac{y}{x}, \text{ and}$$

$$\cos(\varphi) = \frac{z}{\sqrt{x^2+y^2+z^2}} = \frac{z}{\rho}.$$

For the spherical to rectangular conversion it helps to calculate the value of $r = \rho \cdot \sin(\varphi)$ in Fig. 16. Then

$$x = r \cdot \cos(\theta) = \rho \cdot \sin(\varphi) \cdot \cos(\theta), \ y = r \cdot \sin(\theta) = \rho \cdot \sin(\varphi) \cdot \sin(\theta), \text{ and } z = \rho \cdot \cos(\varphi).$$

Converting Between Spherical and Rectangular Systems

If $P(x, y, z)$ and $P(\rho, \theta, \varphi)$ represent the same point in rectangular and spherical coordinates (Fig. 16),
then (spherical to rectangular) $x = \rho \cdot \sin(\varphi) \cdot \cos(\theta)$ $y = \rho \cdot \sin(\varphi) \cdot \sin(\theta)$ $z = \rho \cdot \cos(\varphi)$

(rectangular to spherical) $\rho^2 = x^2 + y^2 + z^2$ $\tan(\theta) = \frac{y}{x}$ $\cos(\varphi) = \frac{z}{\sqrt{x^2+y^2+z^2}} = \frac{z}{\rho}$

Example 6: (a) Write the rectangular coordinate location $A(3, 6, 2)$ in the spherical coordinate system.

(b) Write the spherical coordinate location $B(2, \pi/6, \pi/4)$ in the rectangular coordinate system.

Solution: (a) $\rho^2 = x^2 + y^2 + z^2 = 3^2 + 6^2 + 2^2 = 7^2$. $\tan(\theta) = 6/3 = 2$ so $\theta = \arctan(2) \approx 1.107$
($63.4°$). $\cos(\varphi) = z/\rho = 2/7$ so $\varphi = \arccos(2/7) \approx 1.281$ ($73.4°$). The spherical coordinates of A are approximately $(7, 1.107, 1.281)$.

(b) $x = 2 \cdot \sin(\pi/4) \cdot \cos(\pi/6) = 2\left(\frac{\sqrt{2}}{2}\right)\left(\frac{\sqrt{3}}{2}\right) = \frac{\sqrt{6}}{2}$

$y = 2 \cdot \sin(\pi/4) \cdot \sin(\pi/6) = 2\left(\frac{\sqrt{2}}{2}\right)\left(\frac{1}{2}\right) = \frac{\sqrt{2}}{2}$. $z = 2\cos(\pi/4) = 2\left(\frac{\sqrt{2}}{2}\right) = \sqrt{2}$

The rectangular coordinates of B are approximately $(\sqrt{6}/2, \sqrt{2}/2, \sqrt{2}) \approx (1.225, 0.707, 1.414)$.

Practice 6: (a) Write the rectangular coordinate location $P(2, 9, 6)$ in the spherical coordinate system.

(b) Write the spherical coordinate location $Q(4, 3\pi/4, \pi/2)$ in the rectangular coordinate system.

Example 7: (a) Rewrite the rectangular coordinate equation $x^2 + y^2 = 6z$ as an equation in spherical coordinates.

(b) Rewrite the spherical coordinate equation $\rho = 6 \cdot \cos(\varphi)$ as an equation in rectangular coordinates.

Solution: (a) $x^2 + y^2 = 6z$ becomes $\rho^2 \cdot \sin^2(\varphi) \cdot \cos^2(\theta) + \rho^2 \cdot \sin^2(\varphi) \cdot \sin^2(\theta) = 6\rho \cdot \cos(\varphi)$ or
$\rho^2 \cdot \sin^2(\varphi) = 6\rho \cdot \cos(\varphi)$.

(b) Multiplying each side of $\rho = 6 \cdot \cos(\varphi)$ by ρ we have $\rho^2 = 6\rho \cdot \cos(\theta)$ so
$x^2 + y^2 + z^2 = 6z$ or, moving the 6z to the left side and completing the square,
$x^2 + y^2 + (z-3)^2 = 9$. The graph of $x^2 + y^2 + (z-3)^2 = 9$ is a sphere with radius 3 and center at $(0, 0, 3)$.

PROBLEMS

Cylindrical coordinates

In problems 1 – 4, plot the points whose cylindrical coordinates are given.

1. $A(3, \pi/2, -2), B(2, \pi/6, 3), C(0, 30°, 3)$
2. $D(3, 3\pi/2, 2), E(4, 0, -2), F(2, \pi, 1)$
3. $P(1, 45°, -2), Q(3, 3\pi/2, 2), R(3, \pi/3, 0)$
4. $S(1, 2\pi/3, 2), T(0, 30°, -2), U(2, \pi/4, 3)$

In problems 5 – 10, values are given for two of the r, θ, z variables. Plot the graph as the third variable takes all possible values.

5. r = 3, θ = 0
6. r = 1, θ = π/6
7. r = 1, z = 3
8. r = 3, z = –2
9. θ = π/4, z = 2
10. θ = π, z = 3

In problems 11 – 16, a value is given for one of the r, θ, z variables. Plot the graph as the other two variables take all possible values.

11. r = 1
12. r = 3
13. θ = π/2
14. θ = π/6
15. z = 2
16. z = –1

In Problems 17 – 20, separate parametric graphs of the r, θ, z variables are given. Using the information in these graphs, sketch the graph of (r, θ, z) in the cylindrical coordinate system.

17. See Fig. 17.
18. See Fig. 18.
19. See Fig. 19.
20. See Fig. 20.

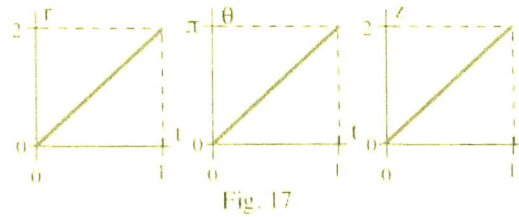

Fig. 17

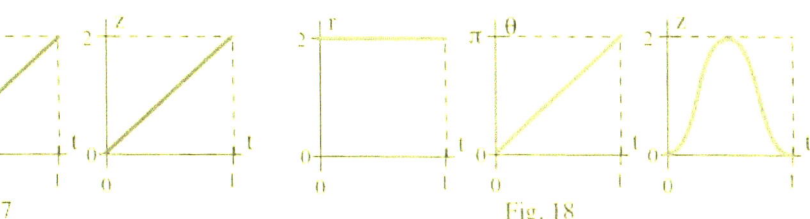

Fig. 18

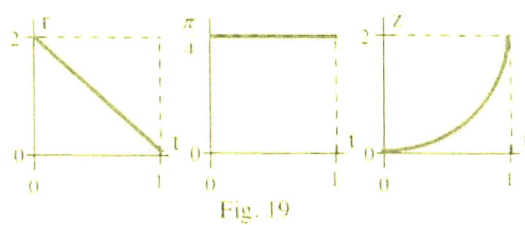

Fig. 19

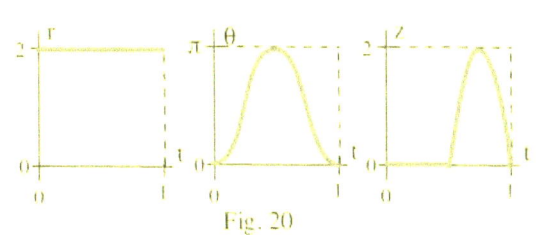

Fig. 20

In problems 21 – 24 sketch the set of points that satisfy the given conditions.

21. $1 \leq r \leq 3$, $0 \leq \theta \leq 90°$, and $z = 0$
22. $0 \leq r \leq 3$, $\theta = 90°$, and $0 \leq z \leq 2$
23. $r = 2$, $0 \leq \theta \leq 30°$, and $0 \leq z \leq 2$
24. $0 \leq r \leq 3$, $0 \leq \theta \leq 90°$, and $0 \leq z \leq 1$

In problems 25 – 28, convert the cylindrical coordinates of the given point to rectangular coordinates.

25. (5, π/6, 3)
26. (4, π/3, 5)
27. (3, 35°, –2)
28. (6, 75°, –4)

In problems 25 – 28, convert the rectangular coordinates of the given point to cylindrical coordinates.

29. (1, 2, 3)
30. (5, 2, –3)
31. (–4, 3, –1)
32. (7, –5, 3)

In problems 33 – 36, convert the cylindrical coordinate equations to equations in rectangular coordinates.

33. (a) $r^2 = 4r \cdot \sin(\theta) - 1$ (b) $r = 7$
34. (a) $r^2 = 8r \cdot \cos(\theta) + 3$ (b) $r = 1$
35. (a) $r = 5 \cdot \cos(\theta)$ (b) $z = r^2$
36. (a) $r = 5 \cdot \sec(\theta)$ (b) $z = r^2 \sin(\theta) \cos(\theta)$

In problems 37 – 40, convert the rectangular coordinate equations to equations in cylindrical coordinates.

37. (a) $z = x^2 + y^2 - 3x + 2y$ (b) $x = 3$

38. (a) $z = 3x^2 + 3y^2$ (b) $y = 2$

39. (a) $z = x^2 + 5y^2$ (b) $x + y + z = 5$

40. (a) $y = x^2$ (b) $x + 5y = z$

Spherical coordinates

In problems 41 – 44, plot the points whose spherical coordinates are given.

41. $A(3, \pi/2, \pi/4), B(2, 0, \pi/6), C(1, \pi, 90°)$

42. $D(3, \pi/6, 0), E(3, \pi/2, \pi/6), F(2, 60°, 20°)$

43. $P(2, 2\pi/3, 2\pi/3), Q(3, 1, 1), R(0, 71°, 7\pi/13)$

44. $S(3, \pi/3, \pi), T(0, \pi/3, \pi/7), U(2, 2, 1)$

In problems 45 – 50, values are given for two of the ρ, θ, φ variables. Plot the graph as the third variable takes all possible values.

45. $\rho = 3, \theta = 0$

46. $\rho = 1, \theta = \pi/6$

47. $\rho = 1, \varphi = \pi/4$

48. $\rho = 3, \varphi = \pi/2$

49. $\theta = \pi/4, \varphi = \pi/2$

50. $\theta = \pi, \varphi = \pi/6$

In problems 51 – 56, a value is given for one of the ρ, θ, φ variables. Plot the graph as the other two variables take all possible values.

51. $\rho = 1$

52. $\rho = 3$

53. $\theta = \pi/2$

54. $\theta = \pi/6$

55. $\varphi = 0$

56. $\varphi = \pi/2$

In problems 57 – 24 sketch the set of points that satisfy the given conditions.

57. $1 \le \rho \le 3, \theta = 90°$, and $10° \le \varphi \le 40°$

58. $\rho = 3, 0 \le \theta \le 90°$, and $10° \le \varphi \le 40°$

59. $\rho = 3, 0 \le \theta \le 30°$, and $0 \le \varphi \le 70°$

60. $0 \le \rho \le 3, 0 \le \theta \le 30°$, and $60° \le \varphi \le 90°$

In problems 61 – 64, convert the spherical coordinates of the given point to rectangular coordinates.

61. $(5, \pi/2, \pi/3)$

62. $(3, \pi/3, \pi/6)$

63. $(4, 45°, 30°)$

64. $(7, 90°, 45°)$

In problems 65 – 68, convert the rectangular coordinates of the given point to spherical coordinates.

65. $(1, 2, 3)$

66. $(5, 2, 7)$

67. $(-5, 3, 2)$

68. $(3, -4, -2)$

In problems 69 – 72, convert the spherical coordinate equations to equations in rectangular coordinates.

69. (a) $\rho = 5$ (b) $\theta = \pi/2$

70. (a) $\rho = 3$ (b) $\varphi = \pi/2$

71. (a) $\rho = 5 \cdot \sin(\varphi) \cdot \cos(\theta)$ (b) $\rho = 3 \cdot \sec(\varphi)$

72. $\rho \cdot \sin(\varphi) = 5\cos(\theta)$

In problems 73 – 76, convert the rectangular coordinate equation to an equation in spherical coordinates.

73. (a) $x^2 + y^2 + z^2 = 9$ (b) $x + z = 5$

74. (a) $x^2 + y^2 = 9$ (b) $y + z = 2$

75. (a) $z = 2x^2 + 2y^2$ (b) $z^2 = 25 - x^2$

76. (a) $z = 3$ (b) $z^2 = 25 - x^2 - y^2$

PRACTICE ANSWERS

Practice 1: The points $P(3, \pi/6, -1)$, $Q(3, \pi/2, 2)$, and $R(0, \pi, 3)$ are shown on Fig. 21.

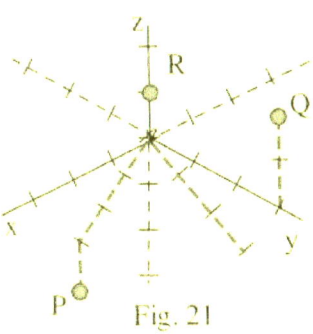

Fig. 21

Practice 2: Fig. 22 illustrates cylindrical coordinates as "rectangular coordinates on a door" by treating r and z as the rectangular coordinates (r along the bottom edge of the door and z as the amount above the bottom edge) and θ as the angular opening of the door.

Sometimes this is an excellent way to think of cylindrical coordinates.

Practice 3: The (approximate) graph of the path is shown in Fig. 23.

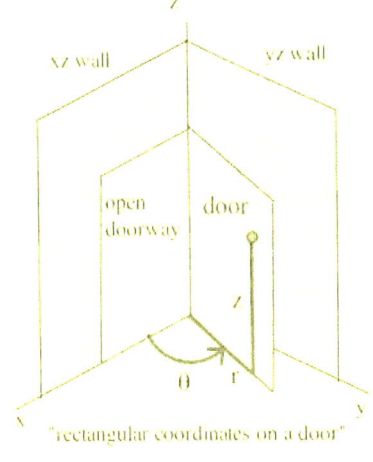

Fig. 22

Practice 4: (a) $(x,y,z) = (-5, 12, 1)$:
$r^2 = x^2 + y^2 = (-5)^2 + 12^2 = 13^2$. $\tan(\theta) = 12/(-5)$
so $\theta = \arctan(-12/5) \approx -1.176$ ($-67.4°$)
$\theta = \arctan(-12/5) \approx -1.176$ ($-67.4°$) but the point $(-5, 12)$ is in the second quadrant so the angle with the positive x–axis is $-1.176 + \pi \approx 1.966$ ($112.6°$). $z = 1$.

The cylindrical coordinates of P are approximately $(13, 1.966, 1)$.

(b) $x = r \cos(3\pi/4) = 4(-\frac{\sqrt{2}}{2}) = -2\sqrt{2}$. $y = r \sin(3\pi/4) = 4(\frac{\sqrt{2}}{2}) = 2\sqrt{2}$.

The rectangular coordinates of Q are approximately $(-2\sqrt{2}, 2\sqrt{2}, 3)$.

Practice 5: The points $P(3, 3\pi/4, \pi/6)$, $Q(3, 0, \pi/4)$, and $R(2, \pi/6, 0)$ are shown on Fig. 24.

Fig. 23

Practice 6: (a) $(x,y,z) = (2,9,6)$. $\rho^2 = x^2 + y^2 + z^2 = 2^2 + 9^2 + 6^2 = 11^2$.
$\tan(\theta) = 9/2 = 4.5$ so $\theta = \arctan(4.5) \approx 1.352$ ($77.5°$).
$\cos(\varphi) = z/\rho = 6/11$ so $\varphi = \arccos(6/11) \approx 0.994$ ($56.9°$).

The spherical coordinates of A are approximately $(11, 1.352, 0.994)$.

(b) $(\rho, \theta, \varphi) = (4, 3\pi/4, \pi/2)$. $x = 4 \sin(\pi/2) \cos(3\pi/4) = 4(1)(-\frac{\sqrt{2}}{2}) = -2\sqrt{2}$.

$y = 4 \sin(\pi/2) \sin(3\pi/4) = 4(1)(\frac{\sqrt{2}}{2}) = 2\sqrt{2}$. $z = 4 \cos(3\pi/4) = 4(0) = 0$.

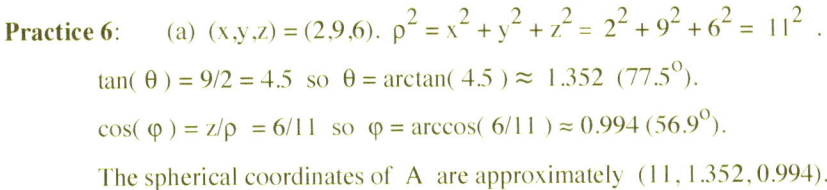

Fig. 24

The rectangular coordinates of B are approximately
$(-2\sqrt{2}, 2\sqrt{2}, 0) \approx (-1.414, 1.414, 0)$.

Appendix: MAPLE Commands for Graphs in Cylindrical & Spherical Coordinates

The computer language Maple has the ability to plot points whose locations are specified in cylindrical or spherical coordinates as well as paths and surfaces in cylindrical or spherical coordinates.

The following commands illustrate how to begin using Maple to graph paths and surfaces in cylindrical and spherical coordinates.

Begin your session with the following command to load the special commands we need:

    with(plots);   then press the *enter* key.

Now try the following, and some of your own too.

Cylindrical coordinate paths given parametrically:

    cylinderplot([3, theta, cos(theta)], theta=0..2*Pi, z = –1..1, axes = NORMAL);

    cylinderplot([3, theta, cos(theta)], theta=0..2*Pi, z = –1..1, grid = [30,30], axes = NORMAL);

Cylindrical coordinate surfaces:

    cylinderplot(1, theta = 0..2*Pi, z = –1..1, axes = NORMAL);   a cylinder

    cylinderplot((z+3*cos(2*theta), theta=0..Pi, z=0..3, axes = NORMAL);   strange, but nice

Spherical coordinate surfaces:

    sphereplot(1, theta=0..2*Pi, phi=0..Pi, grid = [30,30], axes = NORMAL);   a sphere

    sphereplot([t, 0.3, r], t=4..5, r=0..Pi/2, axes = NORMAL);

    sphereplot([r=0..5, t=0..2*Pi, Pi/4], axes = NORMAL);

    sphereplot([p*t, exp(t/10, p^2], t=0..Pi, p=–2..2, axes = NORMAL);   strange

    sphereplot((3*sin(x)^2–1)/2, x = –Pi..Pi, y = 0..Pi);

In Chapter 13 we will be working with surfaces $z = f(x,y)$. The Maple option for how we view these surfaces is "orientation = [theta, phi]" which specifies our viewing angle along the ray whose spherical coordinates have angles theta and phi (given in degrees). Try viewing the paraboloid of revolution $z = x^2 + y^2$ with several different orientations:

    plo3d(x^2 + y^2, x= –3..3, y= –3..3, orientation = [10,30]);   (try this command with some other orientations)

Now position the cursor on the graph, press (and hold) down the mouse button, and slowly move the mouse. This will rotate a "box" containing the graph (and display the orientation coordinates). When the button is released, the graph will be redrawn with the new orientation.

12.4 Cylindrical & Spherical Coordinates

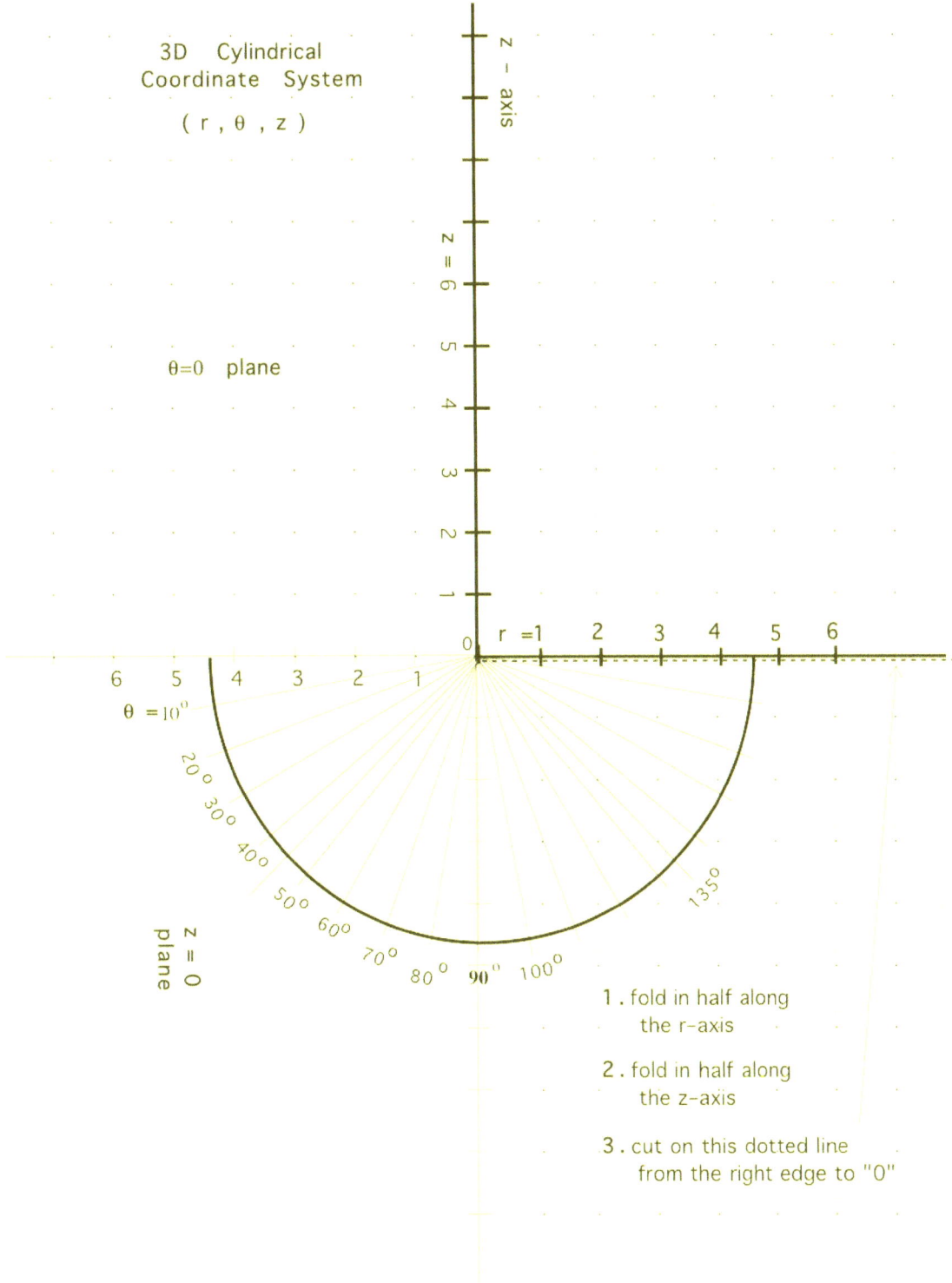

3D Cylindrical Coordinate System
(r, θ, z)

$\theta = 0$ plane

$z = 0$ plane

1. fold in half along the r-axis
2. fold in half along the z-axis
3. cut on this dotted line from the right edge to "0"

12.4 Cylindrical & Spherical Coordinates

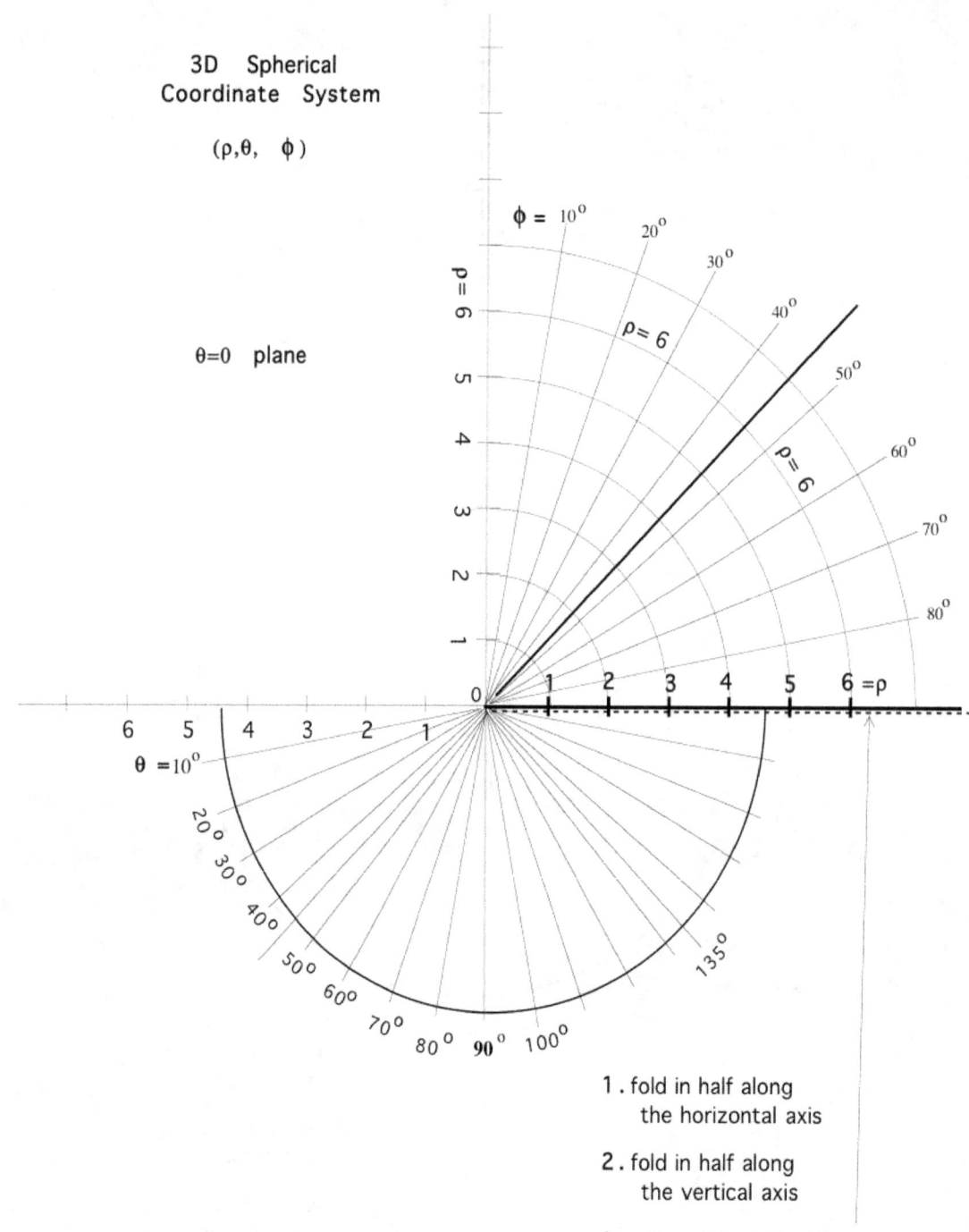

3D Spherical Coordinate System

(ρ, θ, ϕ)

$\theta = 0$ plane

1. fold in half along the horizontal axis
2. fold in half along the vertical axis
3. cut on this dotted line from the right edge to "0"

12.1 Selected Answers

The graphs for the odd problems 1 to 15 are given.

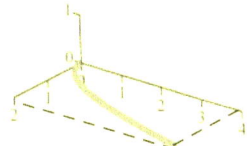
Fig. Problem 1

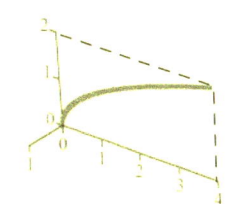
Fig. Problem 3

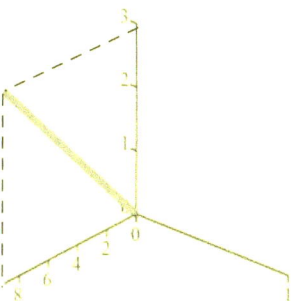
Fig. Problem 5

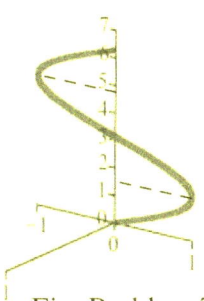
Fig. Problem 7

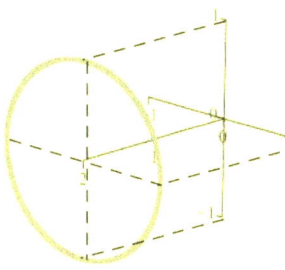
Fig. Problem 9

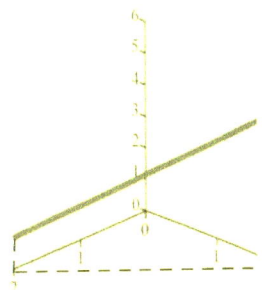
Fig. Problem 11

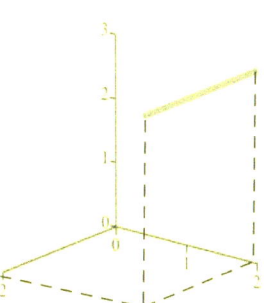
Fig. Problem 13

The graphs for problems 17 to 20 are given.

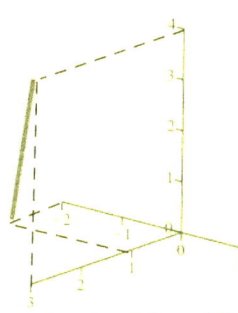
Fig. Problem 15

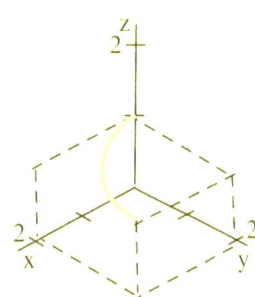
Fig. Problem 17

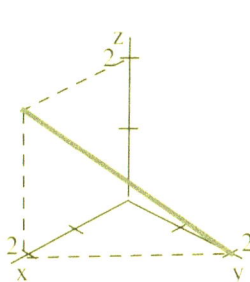
Fig. Problem 18

21. $\langle 3, 9/2, 1/2 \rangle$

22. $\langle 1, 0, 0 \rangle$

23. does not exist

24. $\langle 1, 8, 1 \rangle$

25. $\langle 0, 3, 0 \rangle$

26. does not exist

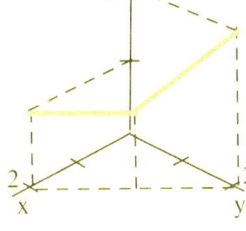
Fig. Problem 19

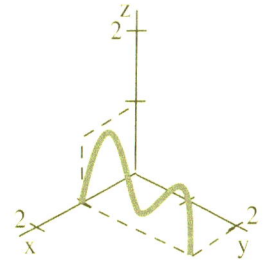
Fig. Problem 20

27. $\langle \pi/2, 0, 0 \rangle$

28. $\langle 1, 1, 1 \rangle$

29. continuous for all $t \neq 1$

30. continuous for all $t \neq 3$

31. continuous for all $t \geq -7$

32. continuous for all $t \geq 1$

33. continuous for all $t \neq -1, 0, 2$

34. continuous for all $t \neq -3, 2$

35. continuous for all $t > 0$

36. continuous for all $t \neq 0$

37. $x(t) = (1-t)^3(0) + 3(1-t)^2 \cdot t(1) + 3(1-t) \cdot t^2(1) + t^3(0) = -3t^2 + 3t$
    $y(t) = (1-t)^3(0) + 3(1-t)^2 \cdot t(0) + 3(1-t) \cdot t^2(2) + t^3(2) = -4t^3 + 6t^2$
    $z(t) = (1-t)^3(1) + 3(1-t)^2 \cdot t(1) + 3(1-t) \cdot t^2(0) + t^3(0) = 2t^3 - 3t^2 + 1$

38. $x(t) = (1-t)^3(0) + 3(1-t)^2 \cdot t(1) + 3(1-t) \cdot t^2(0) + t^3(1) = 4t^3 - 6t^2 + 3t$
    $y(t) = (1-t)^3(0) + 3(1-t)^2 \cdot t(0) + 3(1-t) \cdot t^2(2) + t^3(3) = -3t^3 + 6t^2$
    $z(t) = (1-t)^3(2) + 3(1-t)^2 \cdot t(1) + 3(1-t) \cdot t^2(0) + t^3(0) = t^3 - 3t + 2$

39. $x(t) = (1-t)^3(0) + 3(1-t)^2 \cdot t(0) + 3(1-t) \cdot t^2(2) + t^3(2) = -4t^3 + 6t^2$
    $y(t) = (1-t)^3(1) + 3(1-t)^2 \cdot t(0) + 3(1-t) \cdot t^2(1) + t^3(0) = -4t^3 + 6t^2 - 3t + 1$
    $z(t) = (1-t)^3(2) + 3(1-t)^2 \cdot t(2) + 3(1-t) \cdot t^2(0) + t^3(0) = 4t^3 - 6t^2 + 2$

40. $x(t) = (1-t)^3(0) + 3(1-t)^2 \cdot t(0) + 3(1-t) \cdot t^2(2) + t^3(1) = -5t^3 + 6t^2$
    $y(t) = (1-t)^3(1) + 3(1-t)^2 \cdot t(0) + 3(1-t) \cdot t^2(0) + t^3(1) = 3t^2 - 3t + 1$
    $z(t) = (1-t)^3(2) + 3(1-t)^2 \cdot t(2) + 3(1-t) \cdot t^2(0) + t^3(0) = 4t^3 - 6t^2 + 2$

The graphs for problems 41 to 44 are given.

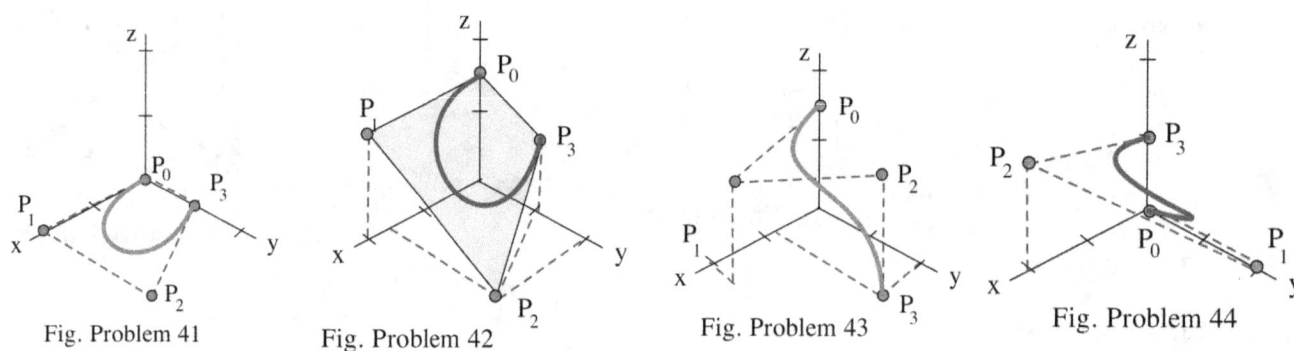

Fig. Problem 41   Fig. Problem 42   Fig. Problem 43   Fig. Problem 44

45. See Fig. 54
46. See Fig. 55
47. See Fig. 56

	A	B	C
x	D	D	D
y	I	I	I
z	D	D	D

Fig. 54

	A	B	C
x	D	D	D
y	I	I	I
z	D	I	D

Fig. 55

	A	B	C
x	D	I	D
y	I	I	I
z	D	D	D

Fig. 56

12.2 Selected Answers

1. $\mathbf{r}'(1) = \langle\, -\,,\, +\,,\, -\,\rangle$, $\mathbf{r}'(2) = \langle\, -\,,\, +\,,\, 0\,\rangle$, and $\mathbf{r}'(3) = \langle\, -\,,\, 0\,,\, +\,\rangle$.

3. $\mathbf{r}'(1) = \langle\, +\,,\, -\,,\, +\,\rangle$, $\mathbf{r}'(2) = \langle\, +\,,\, -\,,\, +\,\rangle$, and $\mathbf{r}'(3) = \langle\, +\,,\, -\,,\, +\,\rangle$.

5. $\mathbf{r}'(t) = \langle 3t^2, 2, 2t \rangle$, $\mathbf{r}''(t) = \langle 6t, 0, 2 \rangle$

   $\mathbf{v}(1) = \mathbf{r}'(1) = \langle 3, 2, 2 \rangle$, speed(1) $= |\mathbf{v}(1)| = \sqrt{17}$, direction(1) $= \dfrac{\mathbf{r}'(1)}{|\mathbf{r}'(1)|} = \dfrac{1}{\sqrt{17}}\langle 3, 2, 2 \rangle$

   $\mathbf{v}(2) = \mathbf{r}'(2) = \langle 12, 2, 4 \rangle$, speed(2) $= |\mathbf{v}(2)| = \sqrt{164}$, direction(2) $= \dfrac{\mathbf{r}'(2)}{|\mathbf{r}'(2)|} = \dfrac{1}{\sqrt{164}}\langle 12, 2, 4 \rangle$.

   $\mathbf{a}(1) = \mathbf{r}''(1) = \langle 6, 0, 2 \rangle$, $\mathbf{a}(2) = \mathbf{r}''(2) = \langle 12, 0, 2 \rangle$.

7. $\mathbf{r}'(t) = \langle -1, -4/t^2, 0 \rangle$, $\mathbf{r}''(t) = \langle 0, 8/t^3, 0 \rangle$

   $\mathbf{v}(1) = \mathbf{r}'(1) = \langle -1, -4, 0 \rangle$, speed(1) $= |\mathbf{v}(1)| = \sqrt{17}$, direction(1) $= \dfrac{\mathbf{r}'(1)}{|\mathbf{r}'(1)|} = \dfrac{1}{\sqrt{17}}\langle -1, -4, 0 \rangle$

   $\mathbf{v}(2) = \mathbf{r}'(2) = \langle -1, -1, 0 \rangle$, speed(2) $= |\mathbf{v}(2)| = \sqrt{2}$, direction(2) $= \dfrac{\mathbf{r}'(2)}{|\mathbf{r}'(2)|} = \dfrac{1}{\sqrt{2}}\langle -1, -1, 0 \rangle$.

   $\mathbf{a}(1) = \mathbf{r}''(1) = \langle 0, 8, 0 \rangle$, $\mathbf{a}(2) = \mathbf{r}''(2) = \langle 0, 1, 0 \rangle$.

9. $\mathbf{r}(t) = \langle t^3, 7, 1+5t \rangle$, $\mathbf{r}'(t) = \langle 3t^2, 0, 5 \rangle$.

   $\dfrac{d}{dt}\mathbf{r}(2t) = \mathbf{r}'(2t)\dfrac{d}{dt}(2t) = \langle 3(2t)^2, 0, 5 \rangle (2) = \langle 24t^2, 0, 10 \rangle$.

   Or, $\mathbf{r}(2t) = \langle (2t)^3, 7, 1+5(2t) \rangle = \langle 8t^3, 7, 1+10t \rangle$ so $\dfrac{d}{dt}\mathbf{r}(2t) = \langle 24t^2, 0, 10 \rangle$.

11. $\mathbf{r}(t) = \langle t, 2t^2, 3t^3 \rangle$, $\mathbf{r}'(t) = \langle 1, 4t, 9t^2 \rangle$.

    $\dfrac{d}{dt}\{\sin(t)\,\mathbf{r}(t)\} = \sin(t)\,\mathbf{r}'(t) + \cos(t)\,\mathbf{r}(t) = \sin(t)\langle 1, 4t, 9t^2 \rangle + \cos(t)\langle t, 2t^2, 3t^3 \rangle$.
    $= \langle \sin(t) + t\cos(t),\, 4t\sin(t) + 2t^2\cos(t),\, 9t^2\sin(t) + 3t^3\cos(t) \rangle$.

13. $\mathbf{r}(t) = (2 - 5t^3)\mathbf{i} + (7t)\mathbf{j} + (1+t)\mathbf{k} = \langle 2-5t^3, 7t, 1+t \rangle$, $\mathbf{r}'(t) = \langle -15t^2, 7, 1 \rangle$.

    $\dfrac{d}{dt}\mathbf{r}(3t) = \mathbf{r}'(3t)\dfrac{d}{dt}(3t) = \langle -15(3t)^2, 7, 1 \rangle (3) = \langle -405t^2, 21, 3 \rangle$.

    Or, $\dfrac{d}{dt}\mathbf{r}(3t) = \dfrac{d}{dt}\langle 2 - 5(3t)^3, 7(3t), 1+(3t) \rangle$
    $= \dfrac{d}{dt}\langle 2 - 135t^3, 21t, 1+3t \rangle = \langle -405t^2, 21, 3 \rangle$.

15. $\dfrac{d}{dt}\{\mathbf{u} + 2\mathbf{v}\} = \dfrac{d}{dt}\langle 2+10t, 8-t, 6+t^3 \rangle = \langle 10, -1, 3t^2 \rangle$

    $\dfrac{d}{dt}\{\mathbf{u}\cdot\mathbf{v}\} = \dfrac{d}{dt}\{4t - t^2 + 3t^3\} = 4 - 2t + 9t^2$

    $\dfrac{d}{dt}\{\mathbf{u}\times\mathbf{v}\} = \dfrac{d}{dt}\langle 3t - 4t^3 + t^4, t^3 + 5t^4, -t - 5t^2 \rangle = \langle 3 - 12t^2 + 4t^3, 3t^2 + 20t^3, -1 - 10t \rangle$.

17. $\frac{d}{dt}\{\mathbf{u}+2\mathbf{v}\} = \frac{d}{dt}\langle 5t^3-4t+2, 2-t, t+10\rangle = \langle 15t^2-4, -1, 1\rangle$

$\frac{d}{dt}\{\mathbf{u}\cdot\mathbf{v}\} = \frac{d}{dt}\{-10t^4+5t^3-21t^2+10t+8\} = -40t^3+15t^2-42t+10$

$\frac{d}{dt}\{\mathbf{u}\times\mathbf{v}\} = \frac{d}{dt}\langle -3t^2-34t+8, -20t^3-2t^2-3t+2, 15t^4-14t^2+11t-2\rangle$

$= \langle -6t-34, -60t^2-4t-3, 60t^3-28t+11\rangle$.

19. The curves intersect at the point (0,3,9) when t = 3: $\mathbf{u}(3) = \langle 0, 3, 9\rangle = \mathbf{v}(3)$.

$\mathbf{u}'(3) = \langle -1, 1, 6\rangle$ and $\mathbf{v}'(3) = \langle 0, 1, 0\rangle$.

$\cos(\theta) = \frac{\mathbf{u}'(3)\cdot\mathbf{v}'(3)}{|\mathbf{u}'(3)||\mathbf{v}'(3)|} = \frac{1}{\sqrt{38}} \approx 0.162$ so $\theta \approx 1.408$ ($\approx 80.7°$)

21. The curves intersect at the point (5,9,3) when t=−1 and s=3: $\mathbf{u}(-1) = \langle 5, 9, 3\rangle = \mathbf{v}(3)$.

$\mathbf{u}'(-1) = \langle -10, 0, -1\rangle$ and $\mathbf{v}'(3) = \langle 1, 3, -1\rangle$.

$\cos(\theta) = \frac{\mathbf{u}'(-1)\cdot\mathbf{v}'(3)}{|\mathbf{u}'(-1)||\mathbf{v}'(3)|} = \frac{-9}{\sqrt{101}\sqrt{11}} \approx -0.270$ so $\theta \approx 1.844$ ($\approx 105.7°$)

23. Area of the parallelogram $= |\mathbf{u}\times\mathbf{v}| = |\langle -2t^3, t^3, -t^2\rangle| = \sqrt{5t^6+t^4}$

Rate of change of area $= \frac{d}{dt}\sqrt{5t^6+t^4} = \frac{15t^5+2t^3}{\sqrt{5t^6+t^4}} = \frac{17}{\sqrt{6}}$ when t = 1 and $\frac{496}{\sqrt{336}}$ when t = 2.

25. Area of the triangle $= \frac{1}{2}|\mathbf{u}\times\mathbf{v}| = \frac{1}{2}|\langle 0, 6t, -2t^2\rangle| = \frac{1}{2}\sqrt{36t^2+4t^4} = \sqrt{9t^2+t^4}$

Rate of change of area $= \frac{d}{dt}\sqrt{9t^2+t^4} = \frac{9t+2t^3}{\sqrt{9t^2+t^4}} = \frac{11}{\sqrt{10}}$ when t = 1 and $\frac{34}{\sqrt{52}}$ when t = 2.

27. Volume of the tetrahedron $= \frac{1}{6}|\mathbf{u}\cdot(\mathbf{v}\times\mathbf{s})| = \frac{1}{6}|3t^2| = \frac{1}{2}t^2$.

Rate of change of volume $= \frac{d}{dt}\frac{1}{2}t^2 = t = 1$ when t = 1 and 2 when t = 2.

29. $\mathbf{r}(t) = \langle 6t^2+1, 4t^3+2, 6e^t-3\rangle$

30. $\mathbf{r}(t) = \langle 3t+2t^2+7, \sin(t)+2, t-3t^2+5\rangle$

31. $\mathbf{r}(t) = \langle 2t^3+4, 4t-2, 4t^2-5t-2\rangle$ 33 and 34. See Fig. 18   35 and 36. See Fig. 19

t	$\mathbf{r}''(t)$	$\mathbf{r}'(t)$	$\mathbf{r}(t)$
0	$\langle 0, 2, 5\rangle$	$\langle 1, 2, 3\rangle$	$\langle 0, 3, 1\rangle$
1	$\langle 4, 1, 3\rangle$	$\langle 5, 3, 6\rangle$	$\langle 5, 6, 7\rangle$
2	$\langle 6, 0, 1\rangle$	$\langle 11, 3, 7\rangle$	$\langle 16, 9, 14\rangle$
3	$\langle 4, -2, 0\rangle$	$\langle 15, 1, 7\rangle$	$\langle 31, 10, 21\rangle$
4	$\langle 2, 0, 2\rangle$	$\langle 17, 1, 9\rangle$	$\langle 48, 11, 30\rangle$
5	$\langle 8, 3, 4\rangle$	$\langle 25, 4, 13\rangle$	$\langle 73, 15, 43\rangle$

Fig. 18

t	$\mathbf{r}''(t)$	$\mathbf{r}'(t)$	$\mathbf{r}(t)$
0	$\langle 1, 2, 3\rangle$	$\langle 1, 6, 4\rangle$	$\langle 17, 3, 2\rangle$
1	$\langle 4, 2, 2\rangle$	$\langle 5, 8, 6\rangle$	$\langle 22, 11, 8\rangle$
2	$\langle 3, 1, 0\rangle$	$\langle 8, 9, 6\rangle$	$\langle 30, 20, 14\rangle$
3	$\langle 2, 3, 1\rangle$	$\langle 10, 12, 7\rangle$	$\langle 40, 32, 21\rangle$
4	$\langle 1, 4, 0\rangle$	$\langle 11, 16, 7\rangle$	$\langle 51, 48, 28\rangle$
5	$\langle 0, 1, 3\rangle$	$\langle 11, 17, 10\rangle$	$\langle 62, 65, 38\rangle$

Fig. 19

12.3 Selected Answers

1. $L = \int_{t=a}^{t=b} |\mathbf{r}'(t)| \, dt = \int_{t=0}^{t=2\pi} \sqrt{4\sin^2(t) + 4\cos^2(t) + 1} \, dt = \int_{t=0}^{t=2\pi} \sqrt{5} \, dt = 2\pi \cdot \sqrt{5} \approx 14.05$

2. $L = \int_{t=a}^{t=b} |\mathbf{r}'(t)| \, dt = \int_{t=0}^{t=2\pi} \sqrt{9\sin^2(t) + 9\cos^2(t) + 1} \, dt = \int_{t=0}^{t=2\pi} \sqrt{10} \, dt = 2\pi \cdot \sqrt{10} \approx 19.87$

3. $L = 2\pi\sqrt{17} \approx 25.91$

4. $L = 2\pi\sqrt{R^2 + 1}$

5. $L = \int_{t=a}^{t=b} |\mathbf{r}'(t)| \, dt = \int_{t=0}^{t=2\pi} \sqrt{4\sin^2(t) + 9\cos^2(t) + 1} \, dt \approx 17.08$

6. $L = \int_{t=a}^{t=b} |\mathbf{r}'(t)| \, dt = \int_{t=0}^{t=2\pi} \sqrt{4\sin^2(t) + 25\cos^2(t) + 1} \, dt \approx 23.93$

7. $L = \int_{t=a}^{t=b} |\mathbf{r}'(t)| \, dt = \int_{t=0}^{t=2\pi} \sqrt{A^2\sin^2(t) + B^2\cos^2(t) + 1} \, dt$

8. $L = \int_{t=a}^{t=b} |\mathbf{r}'(t)| \, dt = \int_{t=0}^{t=2\pi} \sqrt{4\sin^2(t) + 4\cos^2(t) + 1} \, dt = \int_{t=0}^{t=2\pi} \sqrt{5} \, dt = 2\pi \cdot \sqrt{5} \approx 14.05$

9. $L = \int_{t=a}^{t=b} |\mathbf{r}'(t)| \, dt = \int_{t=0}^{t=2\pi} \sqrt{t^2 + 2} \, dt \approx 22.43$

11. $L = \int_{t=a}^{t=b} |\mathbf{r}'(t)| \, dt = \int_{t=0}^{t=2\pi} \sqrt{(-2t\sin(t) + 2\cos(t))^2 + (t\cos(t) + \sin(t))^2 + 1} \, dt \approx 34.02$

13. $x'(t) = 3 - 6t, \; y'(t) = 18t - 21t^2, \; z'(t) = 3 - 18t + 5t^2$.

$L = \int_{t=a}^{t=b} |\mathbf{r}'(t)| \, dt = \int_{t=0}^{t=1} \sqrt{(x'(t))^2 + (y'(t))^2 + (z'(t))^2} \, dt$

15. $x'(t) = -6 + 24t - 24t^2, \; y'(t) = 3 - 6t + 9t^2, \; z'(t) = 3 - 6t$.

$L = \int_{t=a}^{t=b} |\mathbf{r}'(t)| \, dt = \int_{t=0}^{t=1} \sqrt{(x'(t))^2 + (y'(t))^2 + (z'(t))^2} \, dt$

17. $\mathbf{r}'(t) = \langle -\sin(t), \cos(t), 1 \rangle$, $\mathbf{r}''(t) = \langle -\cos(t), -\sin(t), 0 \rangle$, $\mathbf{r}'(t) \times \mathbf{r}''(t) = \langle \sin(t), -\cos(t), 1 \rangle$.

Then $|\mathbf{r}'(t)| = \sqrt{2}$ and $|\mathbf{r}'(t) \times \mathbf{r}''(t)| = \sqrt{2}$, so

$$\kappa = \frac{|\mathbf{r}' \times \mathbf{r}''|}{|\mathbf{r}'|^3} = \frac{\sqrt{2}}{(\sqrt{2})^3} = \frac{1}{2} \text{ for all values of } t.$$

19. $\mathbf{r}'(t) = \langle -R\sin(t), R\cos(t), 1 \rangle$, $\mathbf{r}''(t) = \langle -R\cos(t), -R\sin(t), 0 \rangle$, $\mathbf{r}'(t) \times \mathbf{r}''(t) = \langle R\sin(t), -R\cos(t), R^2 \rangle$.

Then $|\mathbf{r}'(t)| = \sqrt{R^2 + 1}$ and $|\mathbf{r}'(t) \times \mathbf{r}''(t)| = |R|\sqrt{R^2 + 1}$, so

$$\kappa = \frac{|\mathbf{r}' \times \mathbf{r}''|}{|\mathbf{r}'|^3} = \frac{|R|\sqrt{R^2+1}}{(\sqrt{R^2+1})^3} = \frac{|R|}{R^2+1} \text{ for all values of } t.$$

21. $x'(t) = -6 - 18t + 9t^2$, $y'(t) = 18t - 21t^2$, $z'(t) = 3 - 18t + 15t^2$.

$x''(t) = -18 + 18t$, $y''(t) = 18 - 42t$, $z''(t) = -18 + 30t$.

When $t = 0.2$, $x' = 2.76$, $y' = 2.76$, $z' = 0$, $x'' = -17.28$, $y'' = -26.4$, $z'' = -12$ so

$\mathbf{r}' \times \mathbf{r}'' = -33.12\mathbf{i} + 33.12\mathbf{j} - 25.1712\mathbf{k}$, $|\mathbf{r}' \times \mathbf{r}''| \approx 53.17$, $|\mathbf{r}'| \approx 3.90$

and $\kappa = \dfrac{|\mathbf{r}' \times \mathbf{r}''|}{|\mathbf{r}'|^3} = \dfrac{53.17}{(3.90)^3} = 0.896$. The radius of curvature is $\dfrac{1}{\kappa} \approx 1.12$.

When $t = 0.5$, $x' = -0.75$, $y' = 3.75$, $z' = -2.25$, $x'' = -9$, $y'' = -3$, $z'' = -3$

$\mathbf{r}' \times \mathbf{r}'' = -18\mathbf{i} + 18\mathbf{j} + 36\mathbf{k}$, $|\mathbf{r}' \times \mathbf{r}''| \approx 44.09$, $|\mathbf{r}'| \approx 4.44$

and $\kappa = \dfrac{|\mathbf{r}' \times \mathbf{r}''|}{|\mathbf{r}'|^3} = \dfrac{44.09}{(4.44)^3} = 0.504$. The radius of curvature is $\dfrac{1}{\kappa} \approx 1.98$.

23. $x'(t) = -3\sin(t)$, $y'(t) = 5\cos(t)$, $x''(t) = -3\cos(t)$, $y''(t) = -5\sin(t)$. Then

$$\kappa = \frac{|x'y'' - x''y'|}{((x')^2 + (y')^2)^{3/2}} = \frac{|15\sin^2(t) - -15\cos^2(t)|}{(9\sin^2(t) + 25\cos^2(t))^{3/2}}.$$

When $t = 0$, $\kappa = \dfrac{15}{(9\sin^2(0) + 25\cos^2(0))^{3/2}} = \dfrac{15}{125} = 0.12$. Radius of curvature $= \dfrac{25}{3} \approx 8.33$.

When $t = \dfrac{\pi}{4}$, $\kappa = \dfrac{15}{(9\sin^2(\pi/4) + 25\cos^2(\pi/4))^{3/2}} = \dfrac{15}{17^{3/2}} = 0.214$. Radius of curvature ≈ 4.67.

When $t = \dfrac{\pi}{2}$, $\kappa = \dfrac{15}{(9\sin^2(\pi/2) + 25\cos^2(\pi/2))^{3/2}} = \dfrac{15}{27} = 0.555$. Radius of curvature $= \dfrac{9}{5} = 1.8$.

25. $x'(t) = -A\sin(t)$, $y'(t) = B\cos(t)$, $x''(t) = -A\cos(t)$, $y''(t) = -B\sin(t)$. Then

$$\kappa = \frac{|x'y'' - x''y'|}{((x')^2 + (y')^2)^{3/2}} = \frac{|AB\sin^2(t) + AB\cos^2(t)|}{(A^2\sin^2(t) + B^2\cos^2(t))^{3/2}}.$$

When $t = 0$, $\kappa = \dfrac{|AB|}{(A^2\sin^2(0) + B^2\cos^2(0))^{3/2}} = \dfrac{|AB|}{|B|^3} = \dfrac{|A|}{|B|^2}$. Radius of curvature $= \dfrac{1}{\kappa} = \dfrac{|B|^2}{|A|}$.

When $t = \dfrac{\pi}{4}$, $\kappa = \dfrac{|AB|}{(A^2\sin^2(\pi/4) + B^2\cos^2(\pi/4))^{3/2}} = \dfrac{|AB|}{((A^2 + B^2)/2)^{3/2}} =$. Radius of curvature $= \dfrac{1}{\kappa}$.

When $t = \dfrac{\pi}{2}$, $\kappa = \dfrac{|AB|}{(A^2\sin^2(\pi/2) + B^2\cos^2(\pi/2))^{3/2}} = \dfrac{|AB|}{|A|^3} = \dfrac{|B|}{|A|^2}$. Radius of curvature $= \dfrac{1}{\kappa} = \dfrac{|A|^2}{|B|}$.

27. $y' = 3$ and $y'' = 0$ so $\kappa = \dfrac{|y''|}{(1 + (y')^2)^{3/2}} = \dfrac{0}{10^{3/2}} = 0$. (As we might expect, the curvature of the straight line $y = 3x + 5$ is 0.)

29. $y' = 2x$ and $y'' = 2$ so $\kappa = \dfrac{|y''|}{(1 + (y')^2)^{3/2}} = \dfrac{2}{(1 + 4x^2)^{3/2}}$.

When $x = 1$, $\kappa = \dfrac{2}{5^{3/2}} \approx 0.1789$. When $x = 2$, $\kappa = \dfrac{2}{17^{3/2}} \approx 0.0285$. When $x = 3$, $\kappa = \dfrac{2}{37^{3/2}} \approx 0.0089$.

12.4 Selected Answers

1 – 23 Odd: The answers are shown in the figures.

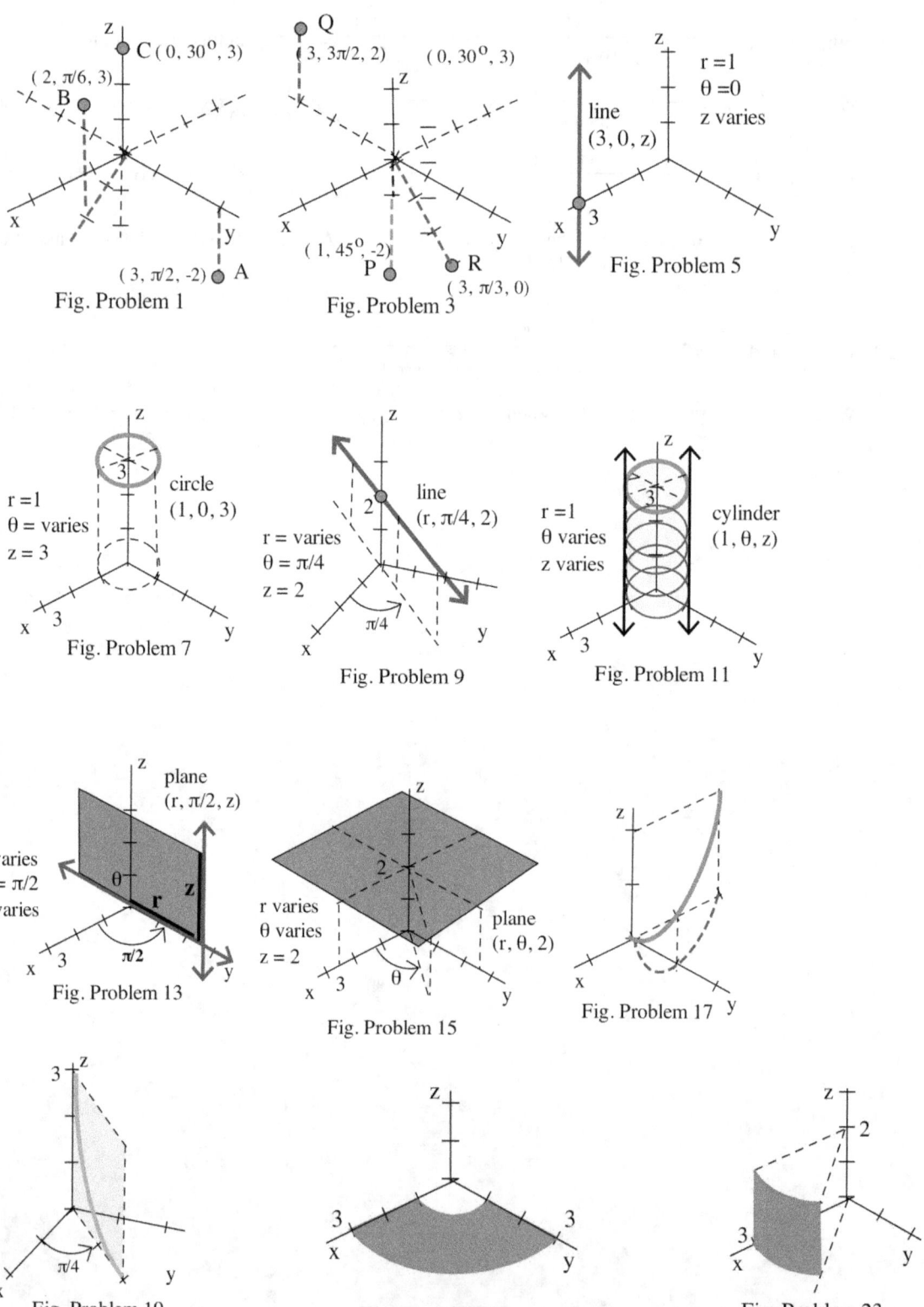

25. $x = 5\cos(\pi/6) \approx 4.33$, $y = 5\sin(\pi/6) = 2.5$, $z = 3$.

27. $x = 3\cos(35^\circ) \approx 2.46$, $y = 3\sin(35^\circ) \approx 1.72$, $z = -2$.

29. $r = \sqrt{x^2 + y^2} = \sqrt{1^2 + 2^2} = \sqrt{5}$,

    $\theta = \arctan(y/x) = \arctan(2/1) \approx 1.107$, $z = 3$.

31. $r^2 = x^2 + y^2 = 5^2 = 25$,

    $\theta = \arctan(y/x) = \arctan(-3/4) \approx -0.644$,

    $z = -1$. To get the correct location, we need to use $r = -5$.

33. (a) $x^2 + y^2 = 4y - 1$ or $x^2 + (y-2)^2 = 3$. (b) $\sqrt{x^2 + y^2} = 7$ or $x^2 + y^2 = 49$

35. (a) $x^2 + y^2 = 5x$ or $(x - 5/2)^2 + y^2 = 25/4$. (b) $z = x^2 + y^2$

37. (a) $z = r^2 - 3r\cos(\theta) + 2r\sin(\theta)$ (b) $r\cos(\theta) = 3$

39. $z = r^2\cos^2(\theta) + 5r^2\sin^2(\theta)$ (b) $r\cos(\theta) + r\sin(\theta) + z = 5$

41 – 59 Odd: The answers are shown in the figures.

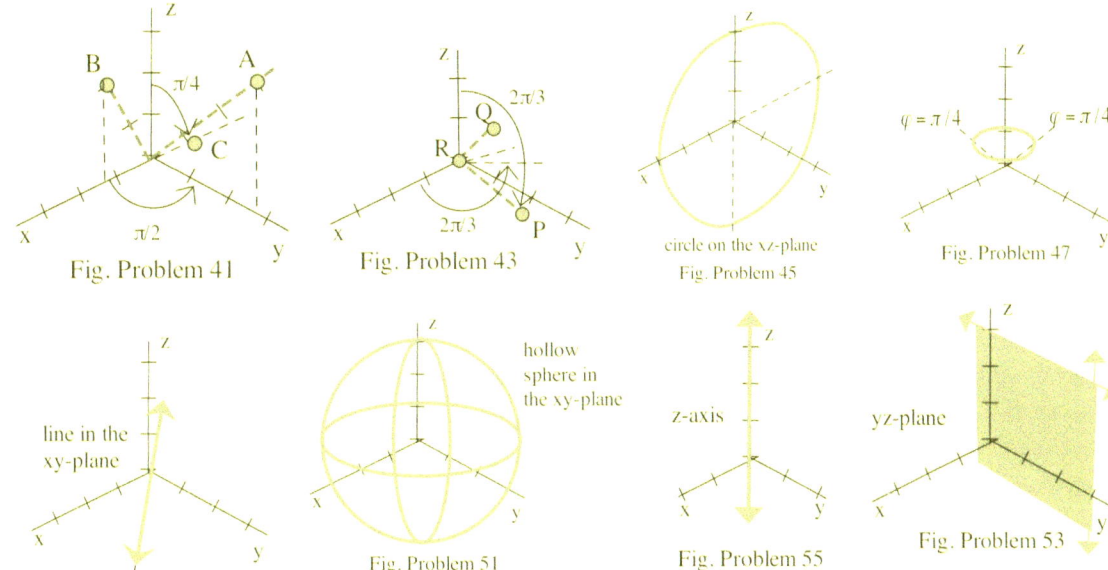

Fig. Problem 41

Fig. Problem 43

circle on the xz-plane
Fig. Problem 45

Fig. Problem 47

line in the xy-plane
Fig. Problem 49

hollow sphere in the xy-plane
Fig. Problem 51

z-axis
Fig. Problem 55

yz-plane
Fig. Problem 53

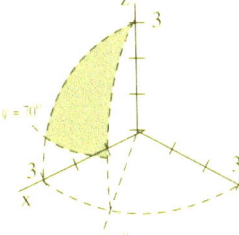

Fig. Problem 59

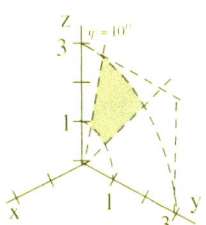

Fig. Problem 57

61. $x = 5 \cdot \sin(\pi/3) \cdot \cos(\pi/2) = 0$, $y = 5 \cdot \sin(\pi/3) \cdot \sin(\pi/2) = 5\sqrt{3}/2 \approx 4.330$, $z = 5 \cdot \cos(\pi/3) = 5/2$

63. $x = 4 \cdot \sin(30°) \cdot \cos(45°) = (4)(\frac{1}{2})(\frac{\sqrt{2}}{2}) = \sqrt{2} \approx 1.414$

$y = 4 \cdot \sin(30°) \cdot \sin(45°) = (4)(\frac{1}{2})(\frac{\sqrt{2}}{2}) = \sqrt{2} \approx 1.414$, $z = 4 \cdot \cos(30°) = (4)(\frac{\sqrt{3}}{2}) = 2\sqrt{3} \approx 3.468$

65. $\rho = \sqrt{x^2 + y^2 + z^2} = \sqrt{14} \approx 3.742$, $\theta = \arctan(y/x) = \arctan(2/1) \approx 1.107$ ($\approx 63.4°$),

$\varphi = \arccos(z/) = \arccos(3/\sqrt{14}) \approx 0.641$ ($\approx 36.7°$)

67. $\rho = \sqrt{x^2 + y^2 + z^2} = \sqrt{38} \approx 6.164$, $\theta = \arctan(y/x) = \arctan(-3/5) \approx -0.540$ ($\approx -30.9°$),

$\varphi = \arccos(z/\rho) = \arccos(2/\sqrt{38}) \approx 1.240$ ($\approx 71.0°$)

69. (a) $5 = \rho = \sqrt{x^2 + y^2 + z^2}$ or $x^2 + y^2 + z^2 = 25$,   (b) (Graphically) $x = 0$

71. (a) $\rho = 5 \cdot \sin(\varphi) \cdot \cos(\theta)$ (b) $\rho = 3 \cdot \sec(\varphi) = 3\dfrac{1}{\cos(\varphi)}$

$\rho^2 = 5\rho \cdot \sin(\varphi) \cdot \cos(\theta)$ $\rho \cdot \cos(\varphi) = 3$

$x^2 + y^2 + z^2 = 5x$ $z = 3$

73. (a) $\rho^2 = 9$ (b) $\rho \cdot \sin(\varphi) \cdot \cos(\theta) + \rho \cdot \cos(\varphi) = 5$

75. (a) $\rho \cdot \cos(\varphi) = 2\rho^2 \cdot \sin^2(\varphi)$ (b) $\rho^2 \cdot \cos^2(\varphi) = 25 - \rho^2 \cdot \sin^2(\varphi) \cdot \cos^2(\theta)$

13.1 FUNCTIONS OF TWO OR MORE VARIABLES

This section presents many of the precalculus concepts relating to functions of several variables, and the focus is on using formulas and tables of data to create and interpret graphical representations of functions of two variables.

Definition:

> A **function f of two variables** is a rule that assigns to each ordered pair (x, y) in the domain of the function a unique real number z. This can be written $z = f(x,y)$.

As with a function of one variable, a function of two variables is typically given by a table of values, a graph, or a formula.

Tables of Values

If we have data or perform measurements about the elevation of the ground (above sea level) at several locations, it is natural to record the elevation measurements using several columns of numbers as in Fig. 1.

Example 1: Describe the progression of elevations for the data in Fig. 1 if we start at the location (2,1), keep the x–values constant at $x = 2$, and proceed to increase the values of y: list the values of $f(2,1), f(2,2), f(2,3), f(2,4), ...$

Solution: $f(2,1) = 4.3, f(2,2) = 5.6, f(2,3) = 6.7$, $f(2,4) = 7.1, f(2,5) = 6.7$, and $f(2,6) = 5.6$. As y increases from 1 to 4, the z–values increase to a maximum of $z = 7.1$. As the y–values increase from 4 to 6, the z–values then decrease.

Practice 1: Describe the progression of elevations for the data in Fig. 1 if we start at the location (2,1), keep the y–values constant at $y = 1$, and proceed to increase the values of x: list the values of $f(2,1), f(3,1), f(4,1), f(5,1), ...$

x	y	z	x	y	z	x	y	z
0	0	1.6	3	0	3.8	6	0	1.6
0	1	1.8	3	1	5.3	6	1	1.8
0	2	2.0	3	2	7.1	6	2	2.0
0	3	2.1	3	3	9.1	6	3	2.1
0	4	2.2	3	4	10.0	6	4	2.2
0	5	2.1	3	5	9.1	6	5	2.1
0	6	2.0	3	6	7.1	6	6	2.0
1	0	2.4	4	0	3.3			
1	1	2.9	4	1	4.3			
1	2	3.3	4	2	5.6			
1	3	3.7	4	3	6.7			
1	4	5.9	4	4	7.1			
1	5	3.7	4	5	6.7			
1	6	3.3	4	6	5.6			
2	0	3.3	5	0	2.4			
2	1	4.3	5	1	2.9			
2	2	5.6	5	2	3.3			
2	3	6.7	5	3	3.7			
2	4	7.1	5	4	3.8			
2	5	6.7	5	5	3.7			
2	6	5.6	5	6	3.3			

Fig. 1: z is the elevation at location (x,y)

Many people find it difficult to recognize patterns and shapes from lists of numbers, and it is common to arrange the information in a table as in Fig. 2. With this table arrangement it is easier to answer the questions in Example 1 and Practice 1 as well as more complicated questions.

Example 2: Describe the progression of elevations for the data in Fig. 2 if we start at the location (2,1) and move "southeast" (x and y both increase at the same rate): list the values of f(2,1), f(3,2), f(4,3), f(5,4), ...

x \ y	0	1	2	3	4	5	6
0	1.6	1.8	2.0	2.1	2.2	2.1	2.0
1	2.4	2.9	3.3	3.7	5.9	3.7	3.3
2	3.3	4.3	5.6	6.7	7.1	6.7	5.6
3	3.8	5.3	7.1	9.1	9.9	9.1	7.1
4	3.3	4.3	5.6	6.7	7.1	6.7	5.6
5	2.4	2.9	3.3	3.7	3.8	3.7	3.3
6	1.6	1.8	2.0	2.1	2.2	2.1	2.0

Fig. 2: Elevations arranged in a table

Solution: f(2,1) = 4.3, f(3,2) = 7.1, f(4,3) = 6.7, f(5,4) = 3.8, and f(6,5) = 2.1 .

Graphs

Once the data is arranged in a table, it seems natural to create a partial graph of the function by plotting the elevations as points at the appropriate elevations above the xy–plane, but such a point graph is usually difficult to "read." Instead of a point graph, we could build a surface by plotting a small platform at each point (Fig. 3) or by connecting the points to nearby points (Fig. 4).

For a function f of two variables, the graph of the surface defined by f is the set of points z = f(x, y) where the ordered pairs (x,y) are in the domain of f.

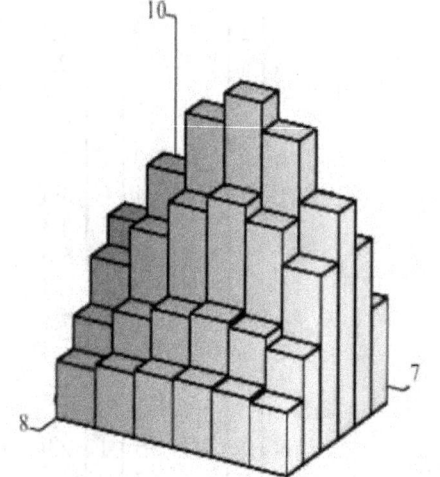

Fig. 3: A "graph" of the data from Fig. 2

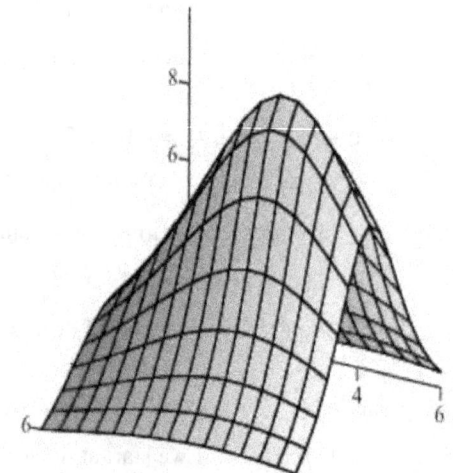

Fig. 4: A surface from the data in Fig. 2

Note: Figures 3 and 4 were drawn by computer using the language Maple. The Maple commands for drawing these and the other 3–dimensional surfaces in this section are given after the problem set.

If we "slice" the surface in Fig. 4 with the plane z = 8, the points where the plane cuts the surface are those points where the elevation of the surface is 8 units above the xy–plane. Fig. 5 shows the surface being sliced by the planes z = 8 and z = 4.

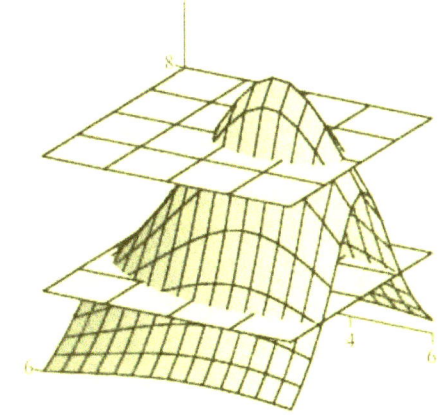

Fig. 5: The planes z=4 and z=8 "slicing" the surface

Fig. 6 shows the results of slicing the surface with several planes at different elevations. The thicker curves are those points where the planes z = 4 and z = 8 intersect the surface.

Definition:

    The **k level curve** of a function f of two variables is the set of points (x,y) that satisfy the equation f(x,y) = k, where k is a constant.

Fig. 6: Intersections of the surface and several planes

If we move all of the curves in Fig. 6 to the xy–plane (or, equivalently, view them from directly overhead), the result is a 2–dimensional graph of the level curves of the original surface. A graph of several level curves of the surface is shown in Fig. 7. We call such a graph a **level curve graph of f** or a **contour graph of f**.

Level curve graphs are an effective and efficient method of presenting a great deal of information about a surface or function

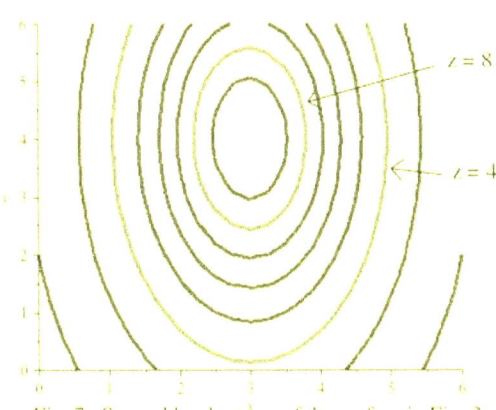

Fig. 7: Several level curves of the surface in Fig. 3

of two variables in a 2–dimensional way, and such graphs are very commonly used. Weather maps showing temperatures or barometric pressures over a region use a variation of contour graphs (Fig. 8). Hikers are familiar with topographic maps that show elevations of the terrain over which they plan to hike (Fig. 9). And almost every issue of scientific journals such as *Science* or *Nature* contains a variety of surface and contour graphs.

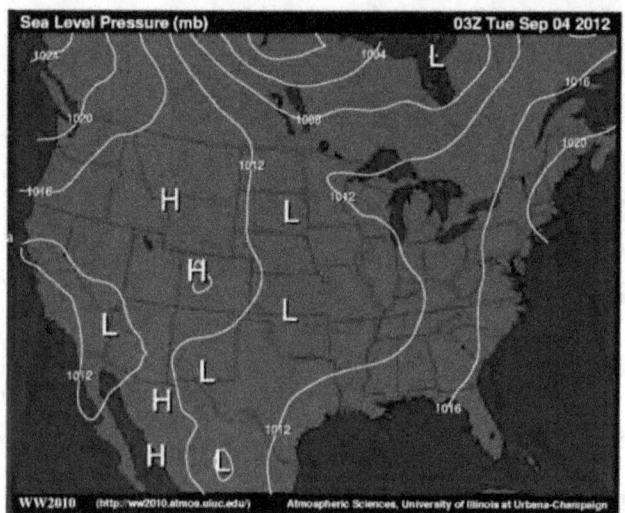

Fig. 8: Barometric pressure map showing isobars

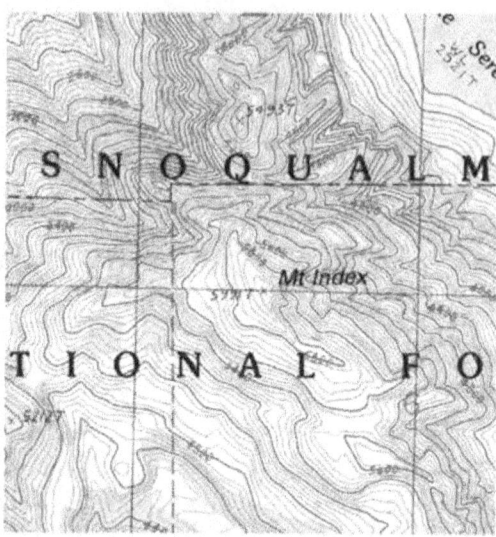

Fig. 9: Topographic map showing level contours

Practice 2: Fig. 10 shows a level curve graph of the temperature at each location on a table. On Fig. 11 sketch the a graph of the temperatures a bug experiences as it moves from point A to point B along a straight line. On Fig. 12 sketch the a graph of the temperatures a bug experiences as it moves from point C clockwise along the ellipse and returns to point C.

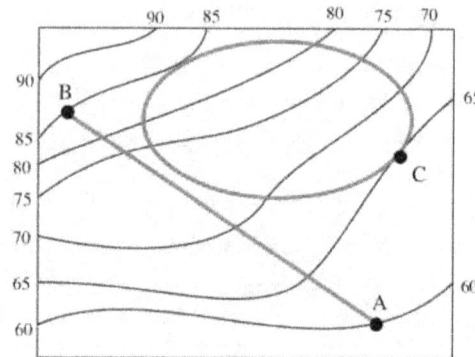

Fig. 10: Temperatures on a table

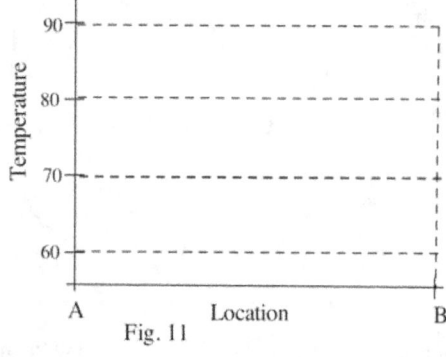

Fig. 11

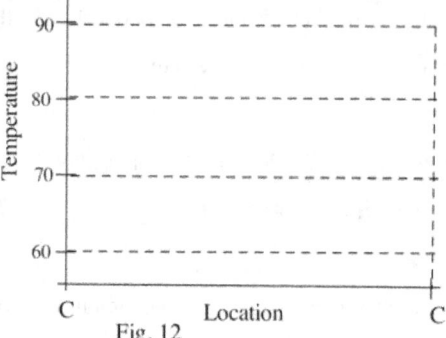

Fig. 12

Sketching Level Curves of a Surface From Data

In general, it is difficult to sketch good surface graphs (such as Fig. 4) from data or even from an equation for a surface, and such surface graphs are typically done by computers. It is much easier, however, to sketch level curves of a surface.

If our surface is described by data (by z–values at given locations (x,y)) and if we can assume that the surface does not have any holes or jumps, then there is a straightforward method for sketching crude level curves for the surface.

"Crude" Level Curve z = k Algorithm for Surface Data (Fig. 13)

Step 1: "Triangulate" the data locations by connecting adjacent locations with light (dotted) lines. The triangulation should be done so no dotted lines cross (but they can meet at the data locations).

Step 2: Plot a point (or small box) on each dotted line segment whose endpoints "surround" the value k (one endpoint is larger than k and one endpoint is smaller).

Step 3: In each triangle, connect the boxes with a line segment.

Step 4: Remove the light (dotted) lines.

Fig. 13: Steps for drawing a (crude) level curve z=8 from data

Fig. 14 shows the result when this algorithm is used in the data from Fig. 2 to approximate the level curves z = 4 and z = 8.

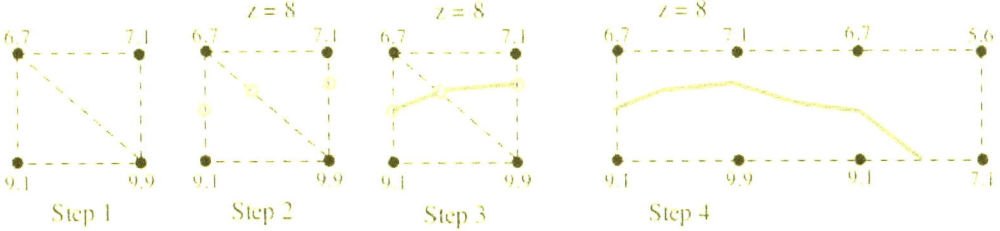

Fig. 14: Crude level curves z=4 and z=8 from data

If the data points are far apart, the resulting level curves may be only crude approximations of the actual level curves of the surface. However, if the data points are relatively close together and if the surface does not change elevation rapidly, then the algorithm results in a good approximation of the actual level curves of the surface. Many computer programs use variations of this algorithm to create level curves of surfaces.

Note: After a bit of practice most people don't bother actually sketching the dotted lines, but they still need to "think triangles."

It is possible to work with functions of three or more variables, but they are difficult (or impossible) to graph. The idea of a level curve for a function of 2 variables generalizes to a **level surface** for a function of 3 variables.

Sketching Level Curves of a Surface From A Formula

If the surface is described by a formula of the form $z = f(x,y)$, then we can often use algebra and our 2–dimensional graphing skills to create level curves for the surface.

Example 3: Sketch the level curve $z = 4$ for the surface $z = x^2 + 4y^2$ (Fig. 15).

Solution: The graph of the level curve $z = 4$ is the graph of $4 = x^2 + 4y^2$ or $\frac{x^2}{4} + \frac{y^2}{1} = 1$.

From Chapter 9, we know that the graph of $\frac{x^2}{4} + \frac{y^2}{1} = 1$ is an ellipse. The graph of the level curve $z = 4$ and other level curves are shown in Fig. 16.

Practice 3: Sketch the level curves $z = 4$ and $z = -8$ for the surface $z = xy$.

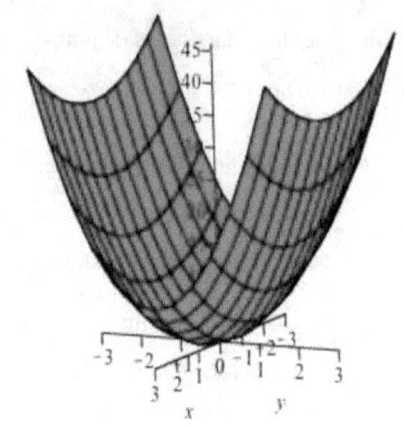

Fig. 15: $z = x^2 + 4y^2$ surface

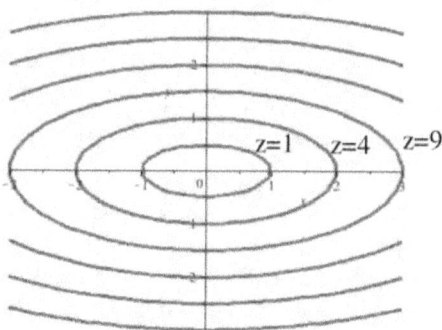

Fig. 16: $z = x^2 + 4y^2$ contours

Functions of Three Variables: $w = f(x, y, z)$

A function f of three variables is a rule that assigns to each ordered triple (x, y, z) in the domain of the function a unique real number w. This can be written $w = f(x,y,z)$. For example, the temperature w at each location (x,y,z) in a classroom is a function of the three variables x, y, and z that specify the location. The cost of a new car is a function of the make of the car, the options included in the car, and the time of year (and probably several more variables). Unfortunately, the graph of $w = f(x,y,z)$ requires four axes and four dimensions. However, a level surface of $w = f(x,y,z)$ is the set of points (x,y,z) such that $w = k$ is constant, and a level surface only requires three dimensions. The $w = 75°$ level surface in a classroom is the set of locations (x,y,z) at which the temperature is $75°$, and sometimes we can graph such level surfaces (Fig. 17).

We will examine how to differentiate and integrate functions of three or more variables, but we will do very little with their graphs.

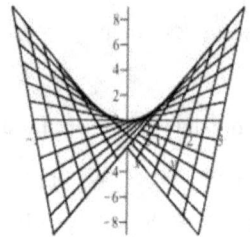

Fig. 17: $z = xy$ surface

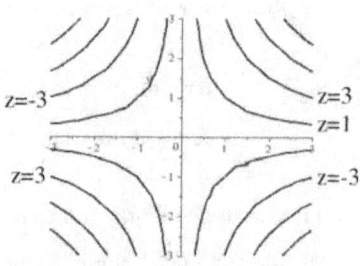

Fig. 18: $z = xy$ contours

13.1 Functions of Several Variables

Problems

1. The table of data in Fig. 19 shows the number thousands of gallons of drinks sold at a sports stadium as a function of the temperature at the beginning of the game and the number of people attending the game.
    (a) What are the minimum and maximum number of thousands of gallons sold?
    (b) When the attendance is 30,000 people, describe what happens to sales as the temperature increases from 50° to 90°.
    (c) When the temperature is 90°, describe what happens to sales as the attendance increases from 10,000 to 60,000.

	Temperature (°F)				
Attendance (1000s)	50	60	70	80	90
10	5	2	2	5	5
20	7	5	3	8	10
30	10	8	6	15	20
40	12	10	12	20	25
50	15	12	15	25	30
60	20	15	20	30	30

Fig. 19: Gallons (1000s) of drinks sold

2. The table of data in Fig. 20 shows the usual gas mileage (miles per gallon) for a truck hauling different loads and traveling at different speeds.
    (a) What are the minimum and maximum mileages for the truck?
    (b) When the truck has a load of 15,000 pounds, describe what happens to the mileage as the speed increases from 20 miles per hour (mph) to 60 mph.
    (c) When the truck is traveling at 50 mph, describe what happens to the mileage as the weight of the load varies from 20,000 pounds to 5,000 pounds.

	Speed (miles per hour)				
Weight of load (pounds)	20	30	40	50	60
5,000	14	18	20	22	20
10,000	11	14	17	19	17
15,000	9	12	16	17	13
20,000	7	11	15	14	10

Fig. 20: Gas Mileage of a Truck (miles per gallon)

3. Fig. 21 shows the depth of a lake at several locations.
    (a) What is the maximum depth of the lake in this region?
    (b) Describe the changing depth a fish swimming along the bottom would experience if the fish started at the location (2,1) and swam east.
    (c) Describe the changing depth a fish swimming along the bottom would experience if the fish started at the location (2,1) and swam south.

	East →				
x \ y	1	2	3	4	5
1	3	4	5	4	3
2	5	8	4	5	4
3	3	5	6	8	6
4	2	4	7	10	7
5	1	3	5	8	5

Fig. 21: Depth of a lake (meters)

4. Fig. 22 shows a map of several level elevation curves.
   (a) What is the maximum elevation in this region?
   (b) Sketch a graph of the elevation of a hiker moving along the straight path from A to B.
   (c) Sketch a graph of the elevation of a hiker moving along the curved path from C to D.
   (d) Sketch a path from E to F so that the path is never very steep.
   (e) At the point A, which directions can the hiker move (a short distance) without changing elevation?

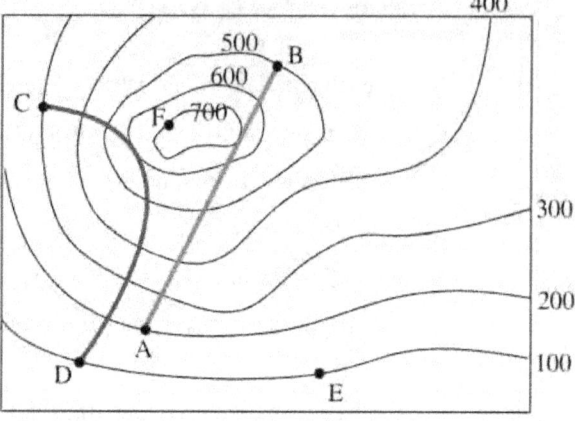

Fig. 22: Level elevation curves (topographic map)

5. Fig. 23 shows the nutrient concentration levels for a region at the bottom of the ocean.
   (a) Label the location of the highest nutrient concentration with an X.
   (b) Sketch a graph of the nutrient concentration level for an animal moving along the path from A to B.
   (c) Sketch a graph of the nutrient concentration level for an animal moving along the curved path from C to D.

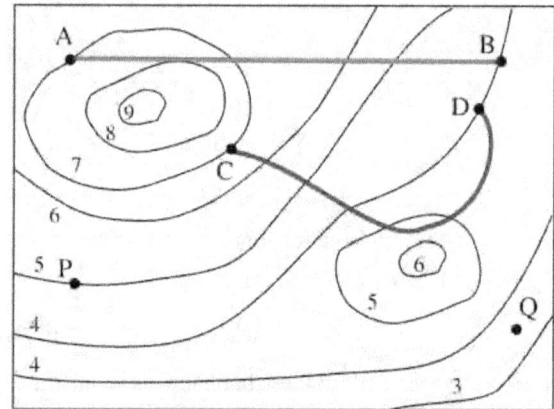

Fig. 23: Nutrient concentrations (parts per million)

   (d) Suppose an animal can sense the nutrient levels in the water near its location and always moves to increase the nutrient level. For an animal that starts at the point P, sketch several nutrient–increasing paths for the animal. Along which path does the nutrient concentration level seem to increase most rapidly?
   (e) Suppose the animal at location Q is at the nutrient concentration level that is best for it. Sketch an approximate path for the animal to move to stay at this nutrient concentration level.

6. Fig. 24 shows a level curve elevation graph for a piece of property with a river. Which direction (left–to–right or right–to–left) does the river flow? (The elevations of the level curves are intentionally unlabeled. You should be able to answer the question from the shape and location of the curves.)

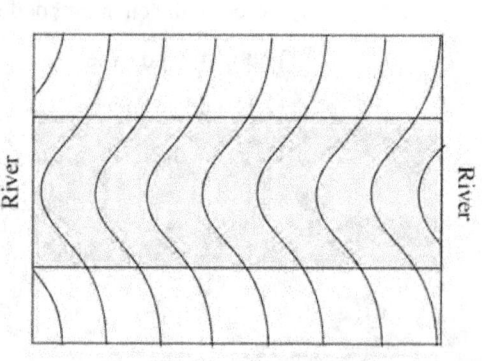

Fig. 24: River property

7. For the data in Fig. 19 sketch the approximate level curves for selling
   (a) 4,000 gallons of drinks. (b) 11,000 gallons of drinks (c) 15,000 gallons of drinks.

8. For the depth data in Fig. 23 sketch the approximate level curves
   (a) for the depth 9. (b) for the depth 6.5. (c) for the depth 4.

9. Find and sketch the level curves $z = 1, 4,$ and 9 for $z = x^2 + y^2$.

10. Find and sketch the level curves $z = 1, 4,$ and 9 for $z = x^2 - y^2$.

11. Find and sketch the level curves $z = 0, 7, 12$ and 16 for $z = 16 - x^2 - y^2$.

12. Find and sketch the level curves $z = 1, 4$ and 5 for $z = \dfrac{5}{1 + (x-3)^2 + (y-4)^2}$

13. Find and sketch the level curves $z = 0, 1, 4$ and 9 for $z = x^2 - y^2$.

14. Sketch several level curves for a surface with a hill of height 600 feet at the location $(4, 6)$

15. Sketch several level curves for a surface with a hill of height 600 feet at the location $(4, 6)$ so that if we move south or east from the peak the path is very steep, but if we move north or west from the peak the path is not steep.

16. Sketch several level curves for a surface with a hill of height 600 feet at the location $(4, 6)$ and a hill of height 400 feet at the location $(2,2)$.

In problems 17 – 23, a surface is given. Match the surface with one of the level curve graphs A – G.

17. The surface in Fig. 25.   18. The surface in Fig. 26.

19. The surface in Fig. 27.   20. The surface in Fig. 28.

21. The surface in Fig. 29.   22. The surface in Fig. 30.

23. The surface in Fig. 31.

24. A new manufacturing process uses a computer–controlled laser light and a vat of liquid plastic. The part of the plastic exposed to the laser light becomes solid to a depth of 1 cm. The process begins with a platform 1 cm below the liquid level of the plastic, and the laser light moves over a region of plastic which hardens. Then the platform is lowered 1 cm, and the laser light moves over another region of the plastic. This process of hardening a region of plastic and then lowering the platform is repeated until the solid plastic object is complete. How is this process related to the level curves of the object?

13.1 Functions of Several Variables Contemporary Calculus 174

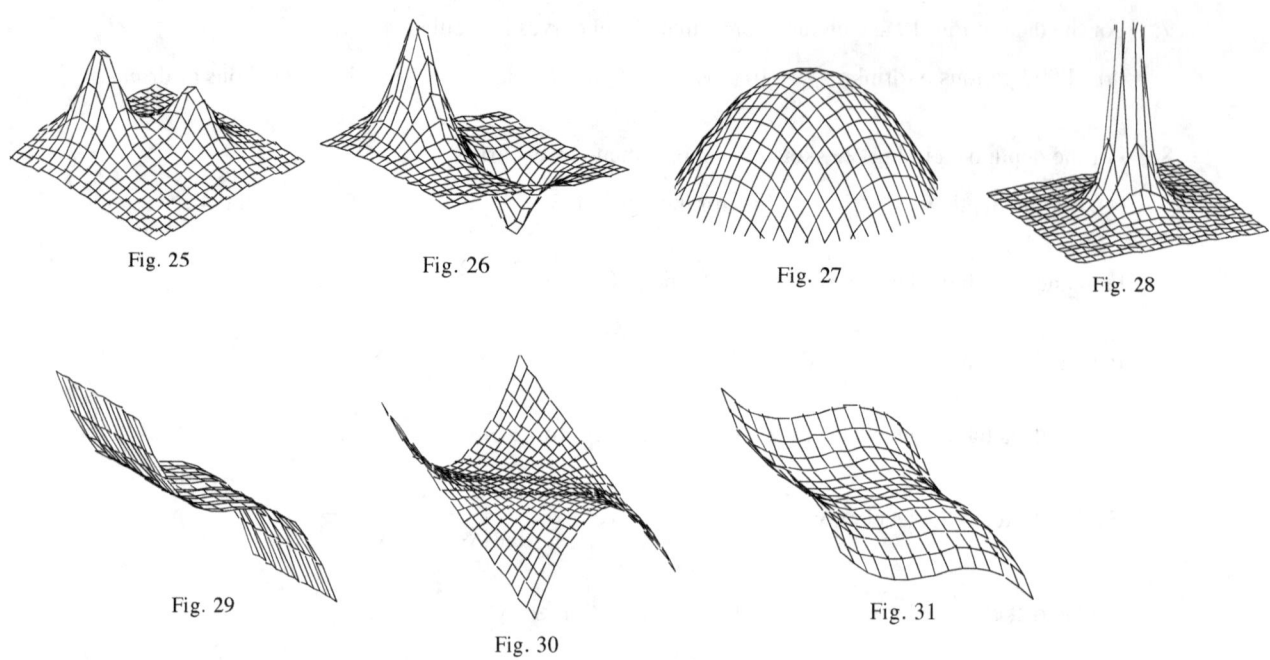

Fig. 25   Fig. 26   Fig. 27   Fig. 28

Fig. 29   Fig. 30   Fig. 31

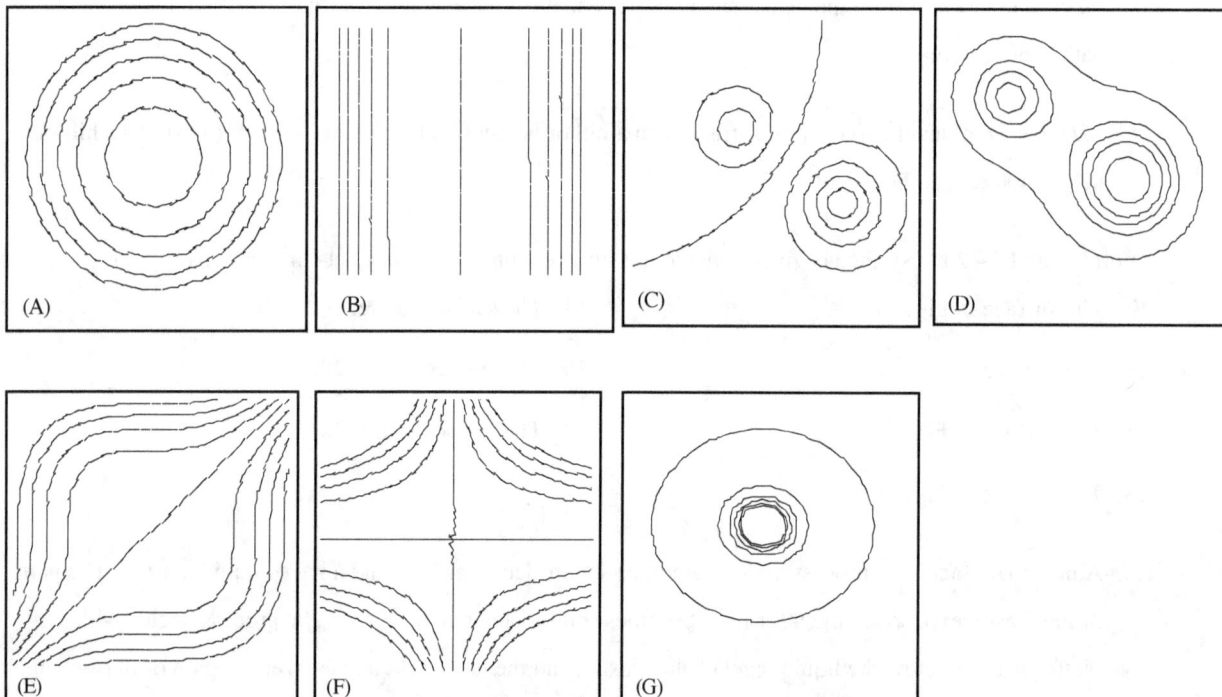

(A) (B) (C) (D)

(E) (F) (G)

Practice Answers

Practice 1: $f(2,1) = 4.3$, $f(3,1) = 5.3$, $f(4,1) = 4.3$, $f(5,1) = 2.9$, $f(6,1) = 1.8$

As y is held constant at 1 and the x values increase from 2 to 6, the values of f rise to a maximum of 5.3 (when $x = 3$) and then decrease to a minimum of 1.8 (when $x = 6$).

Practice 2: The graphs are given below.

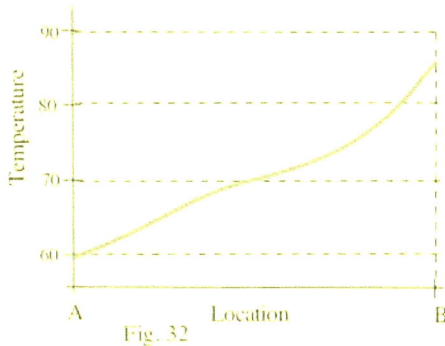

Fig. 32

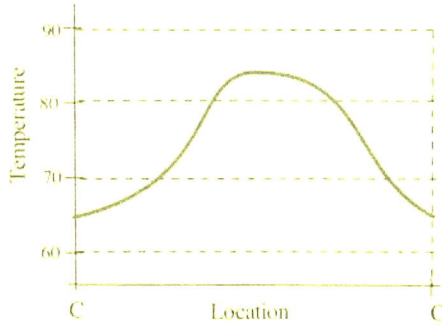

Practice 3: The level curves are given below.

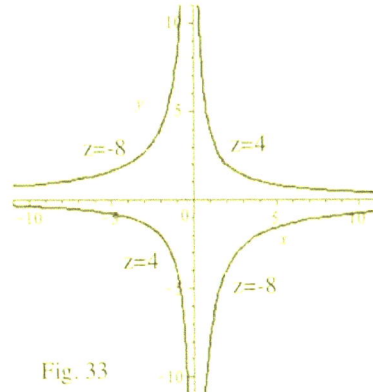

Fig. 33

Appendix: MAPLE commands to create 3D graphics for Section 13.1

> with(plots);   *loads additional graphics commands*
> with(linalg);   *loads additional matrix commands*

> f:=(i,j)->10/(1+0.4*(i-3)^2+0.1*(j-4)^2);   *defines the function f*
> M:=matrix(6,6,f);   *loads values of f into the matrix M*

> setoptions3d(axes=normal, tickmarks=[4,4,5], orientation=[30,70], thickness=0);
    sets options for the following graphs
    orientation=[30,70] puts viewing position at theta=30 and phi=70 (spherical coordinates)

> matrixplot(M);   *plots matrix M values as a surface*
> matrixplot(M, heights=histogram, gap=0);   *plots matrix M values as boxes* **(Fig. 3)**

> plot3d(10/(1+0.4*(x-3)^2+0.1*(y-4)^2), x=0..6, y=0..6, grid=[15,15]);
    plots surface **(Fig. 4)**

> plot3d({ 10/(1+0.4*(x-3)^2+0.1*(y-4)^2), 4, 8 }, x=0..6, y=0..6, grid=[15,15]);
    plots surface and planes z=8 and z=4 **(Fig. 5)**

> contourplot(10/(1+0.4*(x-3)^2+0.1*(y-4)^2), x=0..6, y=0..6, grid=[25,25],
    contours=[1,2,3,4,5,6,7,8,9], orientation=[30,70]);   *plots surface contours* **(Fig. 6)**

> contourplot(10/(1+0.4*(x-3)^2+0.1*(y-4)^2), x=0..6, y=0..6, grid=[25,25],
    contours=[1,2,3,4,5,6,7,8,9], orientation=[0,–1]);   *plots surface contours* **(Fig. 7)**

The following sequence of commands illustrates how to
    put data into a matrix C,
    convert the matrix values into a list of points (x,y,z) = (i , j, C(i,j)), and then
    graph the surface with the points (i , j, C(i,j)).

    > C:=matrix([[5,4,3,4,5,7,8], [6,3,2,1,3,5,4], [5,4,5,7,8,5,4], [2,3,2,3,2,1,1],
    [1,1,2,2,3,1,1]]);
    > Cdata:=[seq([seq([i,j,C[i,j]] , i=1..5)], j=1..6)] ;
    > surfdata(Cdata , axes=normal);

13.2 LIMITS AND CONTINUITY

Our development of the properties and the calculus of functions $z = f(x,y)$ of two (and more) variables parallels the development for functions $y = f(x)$ of a single variable, but the development for functions of two variables goes much quicker since you already understand the main ideas of limits, derivatives, and integrals. In this section we consider limits of functions of two variables and what it means for a function of two variables to be continuous. In many respects this development is similar to the discussions of limits and continuity in Chapter One and many of the results we state in this section are merely extensions of those results to a new setting. It may be a good idea to spend a little time now in Chapter One rereading the main ideas and results for limits of functions of one variable and reworking a few limit problems.

The main focus of this section is on functions of two variables since it is still possible to visualize these functions and to work geometrically, but the end of this section includes extensions to functions of three and more variables.

Limits of Functions of Two Variables

When we considered limits of functions of one variable, $\lim_{x \to a} f(x)$, we were interested in the values of $f(x)$ when x was close to the point a in the domain of f (Fig. 1), and we often read the symbols "$x \to a$" as "x approaches a."

For the limit of function of two variables, $\lim_{(x,y) \to (a,b)} f(x,y)$, we are

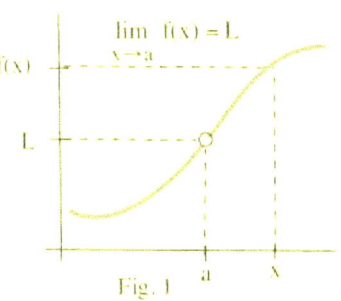

Fig. 1

interested in the values of $f(x,y)$ when the point (x,y) is close to the point (a,b) in the domain of f (Fig. 2).

Definition:

Let f be a function of two variables defined for all points "near" (a,b) but possibly not defined at the point (a,b). We say the

limit of $f(x,y)$ as (x,y) approaches (a,b) is L, written as

$$\lim_{(x,y) \to (a,b)} f(x,y) = L,$$

if the distance from $f(x,y)$ to L, $|f(x,y) - L|$, can be made arbitrarily small by taking (x,y) sufficiently close to (a,b),
(if $\sqrt{(x-a)^2 + (y-b)^2} = |\langle x-a, y-b \rangle|$ is sufficiently small).

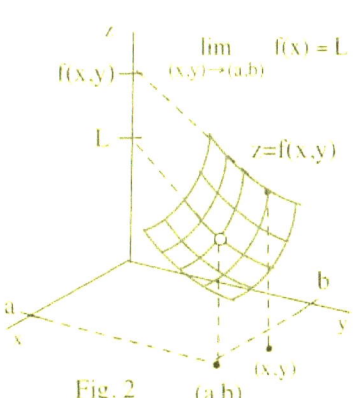

Fig. 2

13.2 Limits and Continuity

All of the limit properties in the Main Limit Theorem (Section 1.2) are also true for limits of functions of two variables, and many limits of functions of two variables are easy to calculate.

Example 1: Calculate the following limits:

(a) $\lim_{(x,y)\to(1,2)} \dfrac{xy}{x^2+y^2}$

(b) $\lim_{(x,y)\to(0,2)} \cos(xy^2) + \dfrac{x+6}{y}$

(c) $\lim_{(x,y)\to(5,3)} \sqrt{x^2-y^2}$

Solution:

(a) $\lim_{(x,y)\to(1,2)} \dfrac{xy}{x^2+y^2} = \dfrac{1\cdot 2}{1^2+2^2} = \dfrac{2}{5}$

(b) $\lim_{(x,y)\to(0,2)} \cos(xy^2) + \dfrac{x+6}{y} = \cos(0\cdot 2) + \dfrac{0+6}{2} = 4$

(c) $\lim_{(x,y)\to(5,3)} \sqrt{x^2-y^2} = \sqrt{5^2-3^2} = \sqrt{16} = 4$

Practice 1: Calculate the following limits:

(a) $\lim_{(x,y)\to(3,1)} \dfrac{xy}{x^2-y^2}$

(b) $\lim_{(x,y)\to(0,2)} \cos(x^2 y) + \dfrac{x+9}{y+1}$

(c) $\lim_{(x,y)\to(3,2)} \sqrt{x^2-y^3}$

At the end of this section we consider some examples of more complicated situations with functions whose limits do not exist. And we also extend the idea of limits to functions of three or more variables.

Continuity of Functions of Two Variables

A function of one variable is continuous at $x = a$ if $\lim_{x\to a} f(x) = f(a)$. Geometrically that means that the graph of f is connected at the point $(a, f(a))$ and does not have a hole or break there (Fig. 3). The definition and meaning of continous for functions of two variables is quite similar.

Definition:

A function of two variables defined at the point (a,b) and for all points near (a,b) is

continuous at (a,b) if $\lim_{(x,y)\to(a,b)} f(x,y) = f(a,b)$. (Fig. 4)

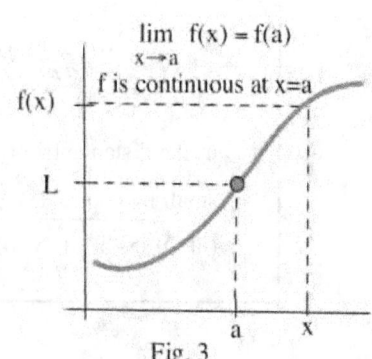

Fig. 3

And just as we talked about a function of one variable being continuous on an interval (or even on the entire real number line), we can talk about a function of two variables being continuous on a set D in the xy–plane or even on the entire xy–plane.

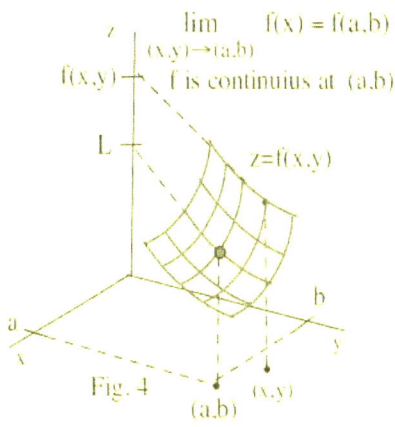

Fig. 4

Definition:

If a function of two variables is continuous at every point (a,b) in a set D, we say that the function is **continuous on D**.

Most of the functions we will work with are continuous either everywhere (at all points (x,y) in the plane) or continuous everywhere except at a "few" places.

- A **polynomial** function of two variables is **continuous everywhere**, at every point (x,y).
- A **rational** function of two variables is **continuous everywhere in its domain** (everywhere except where division by zero would occur).
- If $f(x,y)$ is continuous at (a,b), then $\sin(f(x,y))$, $\cos(f(x,y))$, and $e^{f(x,y)}$ are continuous at (a,b).
- More generally, if f (a function of two variables) is continuous at (a,b) and g (a function of one variable) is continuous at f(a,b), then $g(f(x,y)) = g \circ f(x,y)$ is continuous at (a,b).

Geometrically "f(x,y) is continuous at (a,b)" that means that the surface graph of f is connected at the point (a,b, f(a,b)) and does not have a hole or break there. Fig. 5 shows the surface graphs of several continuous functions of two variables.

Similar definitions and results are used for functions of three or more variables.

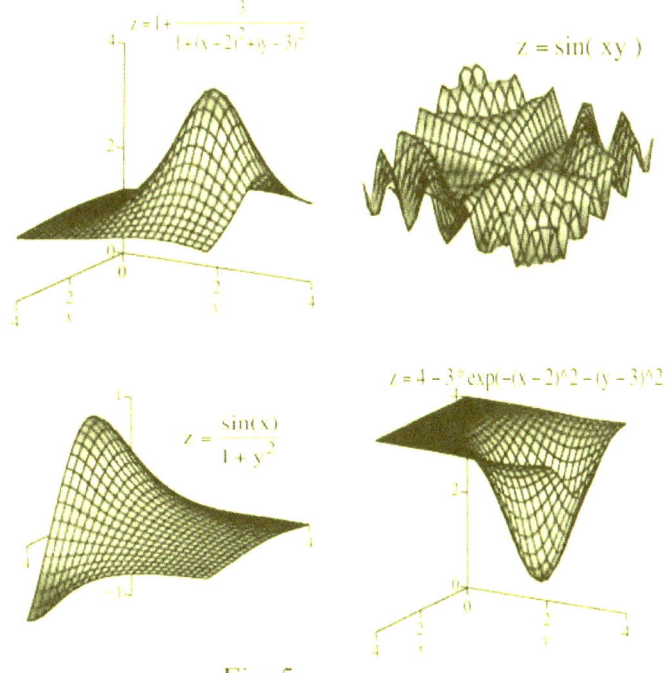

Fig. 5

Limits That Do Not Exist

Most of the functions we work with will have limits and will be continuous, but not all of them. A function of one variable did not have a limit if its left limit and its right limit had different values (Fig. 6). Similar situations can occur with functions of two variable as shown graphically in Fig. 7. For the function f(x,y) in Fig. 7, if (x,y) approaches the point (1,2) along path 1 (x = 1 and y→2⁻) then the values of f(x,y) approach 2. But if (x,y) approaches the point (1,2) along path 2 (x = 1 and y→2⁺) then the values of f(x,y) approach 1. Since two paths to the point (1,2) result in two different limiting values for f, we say that the limit of f(x,y) as (x,y) approaches (1,2) does not exist.

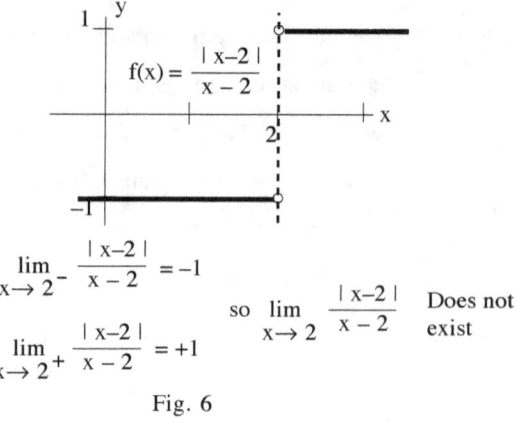

Fig. 6

Showing a limit does not exist:

If there are two paths so that f(x,y) → L_1 as (x,y)→(a,b) along path 1, and f(x,y) → L_2 as (x,y)→(a,b) along path 2, and $L_1 \neq L_2$,

then $\lim_{(x,y)\to(a,b)} f(x,y)$ does not exist.

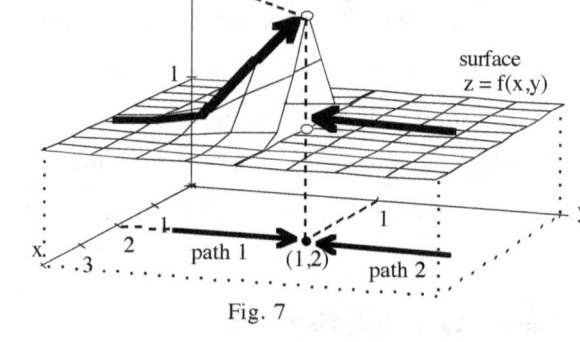

Fig. 7

Example 2: Find $\lim_{(x,y)\to(0,0)} \dfrac{xy}{x^2+y^2}$

Solution: Let path 1 be the x–axis, so y = 0.

Then $\dfrac{xy}{x^2+y^2} = \dfrac{0}{x^2+0} = 0$ and the limit of $\dfrac{xy}{x^2+y^2}$ as (x,y) → (0,0) along path 1 is 0.

However, if we take the path 2 to be the line y = x,

then $\dfrac{xy}{x^2+y^2} = \dfrac{x^2}{x^2+x^2} = \dfrac{1}{2}$ so the limit of $\dfrac{xy}{x^2+y^2}$ as (x,y) → (0,0) along path 2 is $\dfrac{1}{2}$.

Since the limits of f as (x,y) → (0,0) along two different paths is two different numbers, the limit of this f(x,y) as (x,y) → (0,0) **does not exist**.

Fig 8 shows these two paths and the different limits of f along them.

Practice 2: Find the limit of f(x,y) = $\dfrac{xy}{x^2+y^2}$ as (x,y)→(0,0) along the path y = 3x.

Since the limit of f(x,y) = $\dfrac{xy}{x^2+y^2}$ as (x,y)→(0,0) does not exist, this function is not continuous at (0,0).

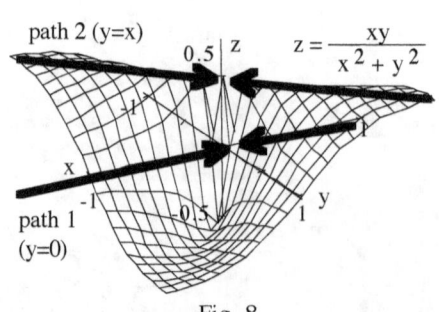

Fig. 8

Practice 3: Show that $\lim\limits_{(x,y)\to(0,0)} \dfrac{x^2-y^2}{x^2+y^2}$ does not exist.

Note: The "path method" only shows that a limit does not exist. Even if the limit of a function as $(x,y)\to(a,b)$ is the same value along two or three paths (or even along an infinite number of paths) we still cannot validly conclude that the limit exists.

Example 3: Evaluate $\lim\limits_{(x,y)\to(0,0)} \dfrac{x^2 y}{x^4+y^2}$ along the paths (a) the y–axis (x=0), (b) the x–axis (y=0), (c) the lines $y = mx$ for all values of $m\neq 0$, and (d) along the parabola $y = x^2$.

Solution: (a) Since $x=0$, $\lim\limits_{(x,y)\to(0,0)} \dfrac{x^2 y}{x^4+y^2} = \lim\limits_{y\to 0} \dfrac{0}{0+y^2} = 0$.

(b) Since $y=0$, $\lim\limits_{(x,y)\to(0,0)} \dfrac{x^2 y}{x^4+y^2} = \lim\limits_{x\to 0} \dfrac{0}{x^4+0} = 0$.

(c) Since $y = mx$ ($m\neq 0$), $\lim\limits_{(x,y)\to(0,0)} \dfrac{x^2 y}{x^4+y^2} = \lim\limits_{x\to 0} \dfrac{x^2(mx)}{x^4+(mx)^2} = \lim\limits_{x\to 0} \dfrac{x^2}{x^2}\cdot\dfrac{mx}{x^2+m^2} = 0$.

From parts (a), (b), and (c) we know that the limit of $\dfrac{x^2 y}{x^4+y^2}$ is 0 as $(x,y)\to(0,0)$ along **every straight line path**. But that is not enough to conclude that the limit along every path is 0.

(d) Along the parabolic path $y = x^2$, $\lim\limits_{(x,y)\to(0,0)} \dfrac{x^2 y}{x^4+y^2} = \lim\limits_{x\to 0} \dfrac{x^2(x^2)}{x^4+(x^2)^2} = \lim\limits_{x\to 0} \dfrac{x^4}{2x^4} = \dfrac{1}{2}$.

Part (d) together with any one of parts (a), (b), or (c) lets us conclude that

$\lim\limits_{(x,y)\to(0,0)} \dfrac{x^2 y}{x^4+y^2}$ does not exist.

Fig. 9 shows this surface.

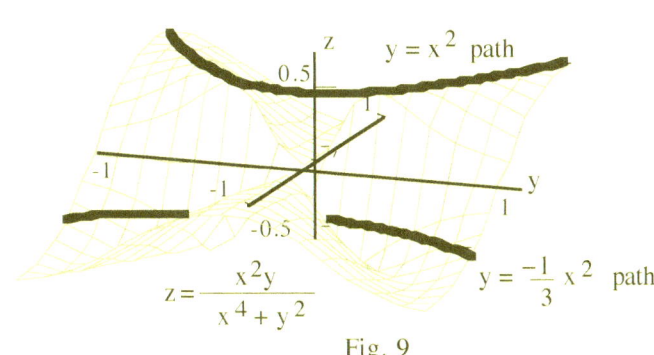

Fig. 9

Functions of More Than Two Variables: Limits and Continuity

Once we have made the adjustments to extend the ideas and definitions of limits and continuity to functions of two variables, it is straightforward to extend them to functions of three or more variables.

Definition:

Let f be a function of three variables defined for all points "near" (a,b,c) but possibly not defined at the point (a,b,c). We say the

limit of f(x,y,z) as (x,y,z) approaches (a,b,c) is L, written as

$$\lim_{(x,y,z) \to (a,b,c)} f(x,y,z) = L,$$

if the distance from f(x,y,z) to L, | f(x,y,z) – L |, can be made arbitrarily small by taking (x,y,z) sufficiently close to (a,b,c),

(if $\sqrt{(x-a)^2 + (y-b)^2 + (z-c)^2} = |\langle x-a, y-b, z-c \rangle|$ is sufficiently small).

Definition:

A function of three variables defined at the point (a,b,c) and for all points near (a,b,c) is **continuous at (a,b,c)** if $\lim_{(x,y,z) \to (a,b,c)} f(x,y,z) = f(a,b,c)$.

PROBLEMS

In Problems 1 – 4, the level curves of functions are given. Use the information from these level curves to determine the limits. (Since only a few level curves are shown, you need to make reasonable assumptions about the behavior of the functions.)

1. The level curves of z = f(x,y) are shown in Fig. 10.

   (a) $\lim_{(x,y) \to (1,2)} f(x,y)$ (b) $\lim_{(x,y) \to (1,1)} f(x,y)$

   (c) $\lim_{(x,y) \to (2,1)} f(x,y)$ (d) $\lim_{(x,y) \to (3,2)} f(x,y)$

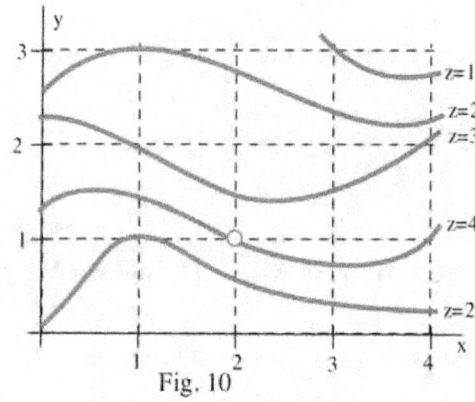

Fig. 10

13.2 Limits and Continuity

2. The level curves of $z = g(x,y)$ are shown in Fig. 11.

   (a) $\lim_{(x,y)\to(2,2)} g(x,y)$ (b) $\lim_{(x,y)\to(2,1)} g(x,y)$

   (c) $\lim_{(x,y)\to(1,2)} g(x,y)$ (d) $\lim_{(x,y)\to(3,2)} g(x,y)$

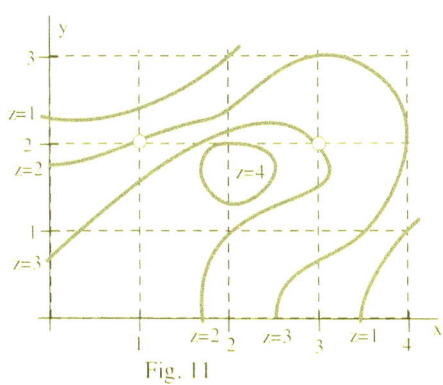
Fig. 11

3. The level curves of $z = S(x,y)$ are shown in Fig. 12.

   (a) $\lim_{(x,y)\to(1,2)} S(x,y)$ (b) $\lim_{(x,y)\to(2,1)} S(x,y)$

   (c) $\lim_{(x,y)\to(1,1)} S(x,y)$ (d) $\lim_{(x,y)\to(3,2)} S(x,y)$

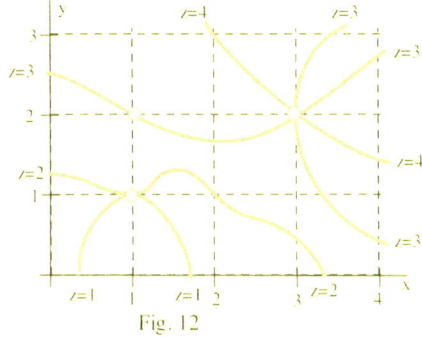
Fig. 12

4. The level curves of $z = T(x,y)$ are shown in Fig. 13.

   (a) $\lim_{(x,y)\to(3,3)} T(x,y)$ (b) $\lim_{(x,y)\to(2,2)} T(x,y)$

   (c) $\lim_{(x,y)\to(1,2)} T(x,y)$ (d) $\lim_{(x,y)\to(4,1)} T(x,y)$

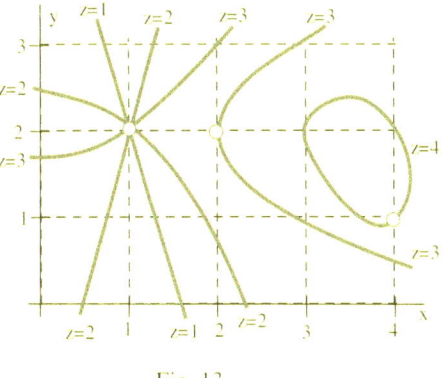
Fig. 13

In Problems 5 – 26, determine the limit if it exists or show that the limit does not exist.

5. $\lim_{(x,y)\to(2,3)} (x^2y^2 - 2xy^5 + 3y)$

6. $\lim_{(x,y)\to(-3,4)} (x^3 + 3x^2y^2 - 5y^3 + 1)$

7. $\lim_{(x,y)\to(0,0)} \dfrac{x^2y^3 + x^3y^2 - 5}{2 - xy}$

8. $\lim_{(x,y)\to(-2,1)} \dfrac{x^2 + xy + y^2}{x^2 - y^2}$

9. $\lim_{(x,y)\to(\pi,\pi)} x \cdot \sin\left(\dfrac{x+y}{4}\right)$

10. $\lim_{(x,y)\to(1,4)} e^{(\sqrt{x+2y})}$

11. $\lim_{(x,y)\to(0,0)} \dfrac{\sin(x^2+y^2)}{x^2+y^2}$

12. $\lim_{(x,y)\to(0,0)} \dfrac{x^2 - y^2}{x+y}$

13. $\lim_{(x,y)\to(0,0)} \dfrac{x-y}{x^2+y^2}$

14. $\lim_{(x,y)\to(0,0)} \dfrac{x^2}{x^2+y^2}$

15. $\lim_{(x,y)\to(0,0)} \dfrac{8x^2y^2}{x^4+y^4}$

16. $\lim_{(x,y)\to(0,0)} \dfrac{x^3+xy^2}{x^2+y^2}$

17. $\lim_{(x,y)\to(0,0)} \dfrac{2xy}{x^2+y^2}$

18. $\lim_{(x,y)\to(0,0)} \dfrac{(x+y)^2}{x^2+y^2}$

19. $\lim_{(x,y)\to(0,0)} \dfrac{\sqrt{xy}}{\sqrt{x^2+y^2}}$

20. $\lim_{(x,y)\to(0,0)} \dfrac{2x^2+3xy+4y^2}{3x^2+5y^2}$

21. $\lim_{(x,y)\to(0,0)} \dfrac{xy+1}{x^2+y^2+1}$

22. $\lim_{(x,y)\to(0,0)} \dfrac{xy^3}{x^2+y^6}$

23. $\lim_{(x,y)\to(0,0)} \dfrac{2x^2y}{x^4+y^2}$

24. $\lim_{(x,y)\to(0,0)} \dfrac{x^3y^2}{x^2+y^2}$

25. $\lim_{(x,y)\to(0,0)} \dfrac{x^2+y^2}{\sqrt{x^2+y^2+1}-1}$

26. $\lim_{(x,y)\to(0,0)} \dfrac{\sqrt{x^2+y^2+1}-1}{x^2+y^2}$

27. $\lim_{(x,y)\to(0,1)} \dfrac{xy-x}{x^2+y^2-2x+2y+2}$

28. $\lim_{(x,y)\to(1,-1)} \dfrac{x^2+y^2-2x-2y}{x^2+y^2-2x+2y+2}$

29. $\lim_{(x,y,z)\to(1,2,3)} \dfrac{xz^2-y^2z}{xyz-1}$

30. $\lim_{(x,y,z)\to(2,3,0)} \{ xe^x + \ln(2x-y) \}$

31. $\lim_{(x,y,z)\to(0,0,0)} \dfrac{x^2-y^2-z^2}{x^2+y^2+z^2}$

32. $\lim_{(x,y,z)\to(0,0,0)} \dfrac{xy+yz+zx}{x^2+y^2+z^2}$

33. $\lim_{(x,y,z)\to(0,0,0)} \dfrac{xy+yz^2+xz^2}{x^2+y^2+z^2}$

34. $\lim_{(x,y,z)\to(0,0,0)} \dfrac{x^2y^2z^2}{x^2+y^2+z^2}$

35. The function f whose level curves are shown in Fig. 10 is not defined at (2,1). Define a value for f(2,1) so f is continuous at (2,1).

36. The function g whose level curves are shown in Fig. 11 is not defined at (1,2) and (3,2). Can we define values for g(1,2) and g(3,2) so g is continuous at (1,2) and (3,2)?

37. The function S whose level curves are shown in Fig. 12 is not defined at (1,1), (1,2) and (3,2). Can we define values for S(1,1), S(1,2) and S(3,2) so S is continuous at each of those points?

38. The function T whose level curves are shown in Fig. 13 is not defined at (1,2), (2,2) and (4,1). Can we define values for T(1,2), T(2,2) and T(4,1) so T is continuous at eaach of those points?

In Problems 39 – 51, determine where the given function is not continuous.

39. $f(x,y) = \dfrac{x^2 + y^2 + 1}{x^2 + y^2 - 1}$

40. $f(x,y) = \dfrac{x^6 + x^3 y^3 + y^6}{x^3 + y^3}$

41. $g(x,y) = \ln(2x + 3y)$

42. $S(x,y) = e^{xy} \sin(x + y)$

43. $T(x,y) = \sqrt{x + y} - \sqrt{x - y}$

44. $T(x,y) = 2^{x \tan(y)}$

45. $F(x,y) = x \ln(yz)$

46. $F(x,y) = x + y\sqrt{x + z}$

Practice Answers

Practice 1: (a) $\displaystyle\lim_{(x,y)\to(3,1)} \dfrac{xy}{x^2 - y^2} = \dfrac{3\cdot 1}{3^2 - 1^2} = \dfrac{3}{8}$

(b) $\displaystyle\lim_{(x,y)\to(0,2)} \cos(x^2 y) + \dfrac{x+9}{y+1} = \cos(0^2\cdot 2) + \dfrac{0+9}{2+1} = 1 + \dfrac{9}{3} = 4$

(c) $\displaystyle\lim_{(x,y)\to(3,2)} \sqrt{x^2 - y^3} = \sqrt{3^2 - 2^3} = \sqrt{1} = 1$

Practice 2: Along the path $y = 3x$, $\dfrac{xy}{x^2 + y^2} = \dfrac{x(3x)}{x^2 + (3x)^2} = \dfrac{3x^2}{10x^2} = \dfrac{3}{10}$ for $x \neq 0$.

Then the limit of $\dfrac{xy}{x^2 + y^2}$ as $(x,y) \to (0,0)$ along the path $y = 3x$ is $\dfrac{3}{10}$.

Practice 3:

Along the path $y = x$, $\dfrac{x^2 - y^2}{x^2 + y^2} = \dfrac{x^2 - x^2}{x^2 + x^2} = \dfrac{0}{2x^2} = 0$ for $x \neq 0$, so the limit along this path is 0.

Along the x–axis $y = 0$, so $\dfrac{x^2 - y^2}{x^2 + y^2} = \dfrac{x^2}{x^2} = 1$ for $x \neq 0$, so the limit along this path is 1.

(Also, along the y–axis $x = 0$, so $\dfrac{x^2 - y^2}{x^2 + y^2} = \dfrac{-y^2}{y^2} = -1$ for $y \neq 0$, so the limit along this path is –1.)

13.3 PARTIAL DERIVATIVES

For a function $y = f(x)$ of one variable, the derivative $\frac{dy}{dx}$ measured the rate of change of the variable y with respect to the variable x. For a function $z = f(x,y)$ of two variables we can ask about the rate of change of z with respect to the variable x or the variable y: how do changes in x effect z, and how do changes in y effect z? In scientific and economic settings with many variables, it is common to try to determine the effect of each variable by holding all of the other variables constant and then measuring the outcomes as that single variable in allowed to vary.

Example 1: The table of data in Fig. 1 shows the number of thousands of gallons of drinks sold at a sports stadium as a function of the temperature at the beginning of the game and the number of people attending the game. At a game with 30,000 people on a $70°$ day,

(a) what is the average rate of change of drink sales as the temperature rises to $80°$?

(b) what is the average rate of change of drink sales as the attendance increases to 40,000 people?

Attendance (1000s)	Temperature (°F)				
	50	60	70	80	90
10	5	2	2	5	5
20	7	5	3	8	10
30	10	8	6	15	20
40	12	10	12	20	25
50	15	12	15	25	30
60	20	15	20	30	30

Fig. 1: Gallons (1000s) of drinks sold

Solution: (a) In this situation the attendance is constant at 30,000 people, and the temperature changes from $70°$ to $80°$. The average rate of change is

$$\frac{f(30,80°) - f(30,70°)}{80° - 70°} = \frac{15000 - 6000 \text{ gallons}}{10°} = 900 \text{ gallons per degree rise in temperature.}$$

(b) In this case the temperature is constant at $70°$, and the attendance changes from 30,000 people to 40,000 people. The average rate of change is

$$\frac{f(40,70) - f(30,70)}{40000 - 30000} = \frac{12000 - 6000 \text{ gallons}}{10000 \text{ people}} = 0.6 \text{ gallons per additional person in attendance.}$$

Note that these rates of change depend on the starting attendance and temperature as well as on the variable that is allowed to change. You should also notice that the units of the two answers are different — one is "gallons/degree" and the other is "gallons/person."

Practice 1: Using the data in Fig. 1 and at a game with 20,000 people on a $80°$ day,

(a) what is the average rate of change of drink sales as the temperature rises to $90°$?

(b) what is the average rate of change of drink sales as the attendance increases to 30,000 people?

13.3 Partial Derivatives

The definition of a partial derivative follows from this idea of holding one variable constant and measuring the rate of change as the other variable changes.

Definition:

The **partial derivative of $f(x,y)$ with respect to x at the point (a,b)** is

$$f_x(a,b) = \lim_{h \to 0} \frac{f(a+h,b) - f(a,b)}{h} \quad \text{(if the limit exists and is finite).}$$

Meaning: $f_x(x,y)$ measures the *instantaneous rate of change* of f at the point (x,y) in the direction of increasing x values.

To calculate $f_x(x,y)$ when $z = f(x,y)$ is given by a formula,

treat y as a constant and differentiate with respect to x.

Example 2: (a) For $f(x,y) = 3x^2 + 7y^2 - 10xy$, find $f_x(x,y)$, $f_x(1,2)$ and $f_x(3,1)$.

(b) For $g(x,y) = \sin(3xy) + \ln(5y) + x^3 y^5$, find $g_x(x,y)$ and $g_x(1,2)$.

Solution: (a) $f_x(x,y) = 6x + 0 - 10y = 6x - 10y$ and $f_x(1,2) = 6(1) - 10(2) = -14$. $f_x(3,1) = 6(3) - 10(1) = 8$.

(b) $g_x(x,y) = 3y \cdot \cos(3xy) + 0 + 3x^2 y^5$ and

$g_x(1,2) = 3(2)\cos(3(1)(2)) + 3(1)^2(2)^5 = 6\cos(6) + 96 \approx 101.8$.

Practice 2: (a) For $f(x,y) = x^3 + 4y^2 + 5x^2 y$, find $f_x(x,y)$ and $f_x(2,5)$.

(b) For $g(x,y) = e^{xy} + \frac{x}{y}$, find $g_x(x,y)$ and $g_x(0,2)$.

We can also interpret the partial derivatives graphically. The graph of $z = f(x,y)$ is typically a surface (Fig. 2) and the graph of "y = a constant" is a plane, so the graph of "$z = f(x,y)$ with y held constant" is the curve resulting from the intersection of the surface and the plane. Fig. 3 shows such a surface and plane and their curve of intersection when $y = 2$. $f_x(1,2)$ is the slope of the line tangent to this curve at the point $(1,2)$ as shown in Fig. 4.

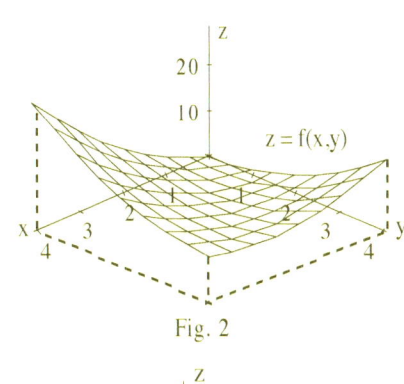

Fig. 2

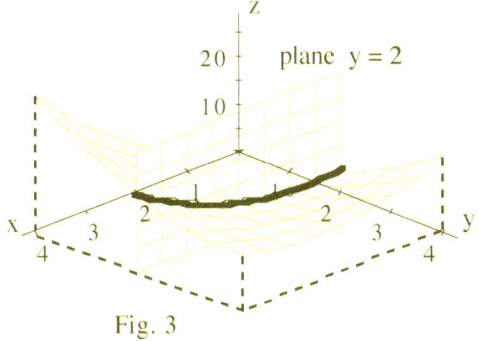

Fig. 3

Example 3: Use the information in Fig. 4 to estimate the value of $f_x(1,2)$.

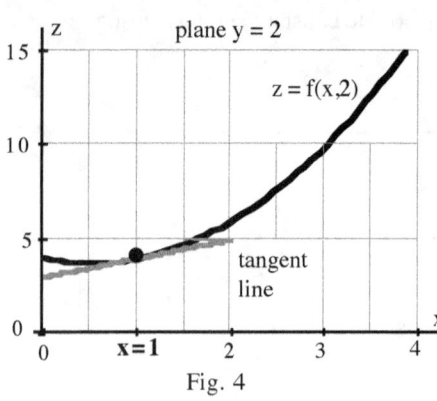

Fig. 4

Solution: We can estimate the slope of the tangent line in Fig. 4 by picking two points on the line and calculating the slope of the line connecting the two points. It looks like (0,3) and (2,5) are points on the tangent line, and the slope of the segment between those two points is $\frac{5-3}{2-0} = 1$. Then we estimate $f_x(1,2) \approx 1$.

Practice 3: Use the information in Fig. 4 to estimate the values of $f(3,2)$ and $f_x(3,2)$.

The partial derivative with respect to y is similar, but now we treat x as the constant.

Definition:

The **partial derivative of f(x,y) with respect to y at the point (a,b)** is

$$f_y(a,b) = \lim_{h \to 0} \frac{f(a,b+h) - f(a,b)}{h}$$ (if the limit exists and is finite).

Meaning: $f_y(x,y)$ measures the *instantaneous rate of change* of f at the point (x,y) in the direction of increasing y values.

To calculate $f_y(x,y)$ when z = f(x,y) is given by a formula,

treat x as a constant and differentiate with respect to y.

Example 4: (a) For $f(x,y) = 3x^2 + 7y^2 - 10xy$, find $f_y(x,y)$, $f_y(1,2)$, and $f_y(3,1)$.

(b) For $g(x,y) = \sin(3xy) + \ln(5y) + x^3 y^5$, find $g_y(x,y)$ and $g_y(1,2)$.

Solution: (a) $f_y(x,y) = 0 + 14y - 10x = 14y - 10x$. Then $f_y(1,2) = 14(2) - 10(1) = 18$ and $f_y(3,1) = -16$.

(b) $g_y(x,y) = 3x \cdot \cos(3xy) + \frac{5}{5y} + 5x^3 y^4$. Then

$g_y(1,2) = 3(1) \cdot \cos(3(1)(2)) + \frac{5}{5(2)} + 5(1)^3(2)^4 \approx 2.88 + 0.5 + 80 = 83.38$.

Practice 4: (a) For $f(x,y) = x^3 + 4y^2 + 5x^2 y$, find $f_y(x,y)$ and $f_y(2,5)$.

(b) For $g(x,y) = e^{xy} + \frac{x}{y}$, find $g_y(x,y)$ and $g_y(0,2)$.

Notations: The following notations are all commonly used to represent partial derivatives of z = f(x,y)

$f_x(x,y) = f_x = \frac{\partial f}{\partial x} = \frac{\partial}{\partial x} f(x,y) = \frac{\partial z}{\partial x} = D_x f$ Partial derivative of f with respect to x

$f_y(x,y) = f_y = \frac{\partial f}{\partial y} = \frac{\partial}{\partial y} f(x,y) = \frac{\partial z}{\partial y} = D_y f$ Partial derivative of f with respect to y

Example 5: Use the information in Figures 5 and 6 to estimate the value of $f_y(1,2)$.

Solution: Fig. 5 shows the surface $z = f(x,y)$ and the plane $x = 1$, but it is difficult to estimate the value of $f_y(1,2)$ from it. Fig. 6 shows the intersection of the surface graph with the plane, and the tangent line at the point $(1,2)$ is included. We can estimate the value of $f_y(1,2)$ by picking two points on the tangent line and calculating the slope between them. It looks like $(1,2)$ and $(3,6)$ are points on the tangent line, and the slope of the segment is $\frac{6-2}{3-1} = 2$. Then we estimate $f_y(1,2) \approx 2$.

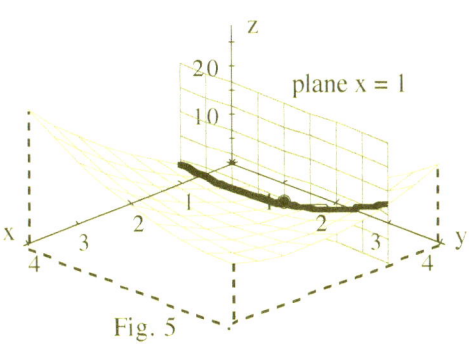

Fig. 5

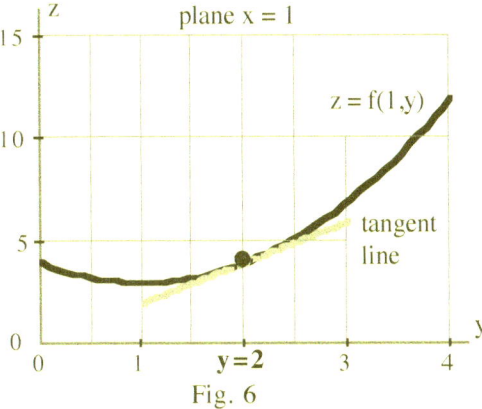

Fig. 6

Practice 5: Use the information in Figures 5 and 6 to estimate the signs (positive, negative or zero) of
(a) $f_y(1,3)$, (b) $f_y(1,1)$, and (c) $f_y(1,4)$.

Partial Derivatives in Context

Of course it is very important to be able to calculate partial derivatives, but it also important to understand and to be able to communicate what they mean and measure. And you need to be able to attach the correct units to your answers.

Example 6: The surface area A (square inches) of a small child is a function of the length L (inches) and the weight W (pounds) of the child: $A = A(L, W)$. Explain (in clear English sentences) the meaning of the following. Be sure to include units.

(a) $A(26, 46) = 164$ (b) $\frac{\partial A(26,46)}{\partial W} = 7$ (c) $\frac{\partial A(26,46)}{\partial L} = 5$

Solution: (a) $A(26, 46) = 164$ square inches. A child who is 26 inches long and weighs 46 pounds will have a surface area of 164 square inches.

(b) $\frac{\partial A(26,46)}{\partial W} = 7$ square inches per pound. The surface area of this child (length 26 inches, weight 46 pounds, area 164 square inches) is **increasing at an INSTANTANEOUS RATE of** 7 square inches per each additional pound of weight if the length stays constant. Units: (square inches)/pound

(c) $\frac{\partial A(26,46)}{\partial L} = 5$ square inches per inch. The surface area of this child (length 26 inches, weight 46 pounds, area 164 square inches) is **increasing at an INSTANTANEOUS RATE of** 5 square inches per each additional inch of length if the weight stays constant. Units: (square inches)/inch

Practice 6: A certain biotech process using bacteria to produce a vaccine V (in grams) depends on the number B of bacteria and the temperature T (in $^\circ C$) of the laboratory: V = V(B, T). Explain (in clear English sentences) the meaning of the following. Be sure to include units.

(a) $V(2000,40) = 8.9$ (b) $\dfrac{\partial V(2000,40)}{\partial B} = 0.003$ (c) $\dfrac{\partial V(2000,40)}{\partial T} = -1.4$

Partial Derivatives and Level Curves

Level curves of a surface $z = f(x,y)$ give us information about f and also about the rate of change of f as x and y increase, the partial derivatives f_x and f_y.

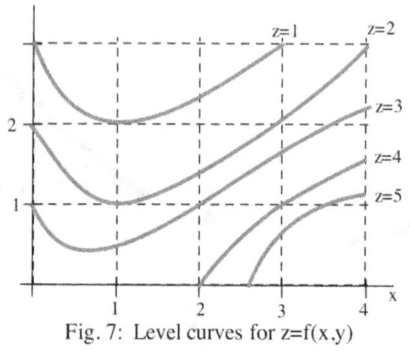

Fig. 7: Level curves for z=f(x,y)

Example 7: Use the information in Fig. 7 to estimate the signs (positive, negative or zero) of

(a) $f_x(3,2)$ and (b) $f_y(3,2)$.

Solution: (a) As we move through the point (3,2) in the increasing x direction (Fig. 8a), the level curves are increasing in value so $f_x(3,2)$ is positive. Fig. 8b shows the graph of z along the line segment of increasing x–values (y is constantly 2), and the slope of the tangent line to this graph is positive when $x = 3$ so $f_x(3,2)$ is positive.

(b) As we move through the point (3,2) in the increasing y direction (Fig. 9a), the level curves are decreasing in value so $f_y(3,2)$ is negative. Fig. 9b shows the graph of z along the line segment of increasing y–values (x is constantly 3), and the slope of the tangent line to this graph is negative when $y = 2$ so $f_y(3,2)$ is negative.

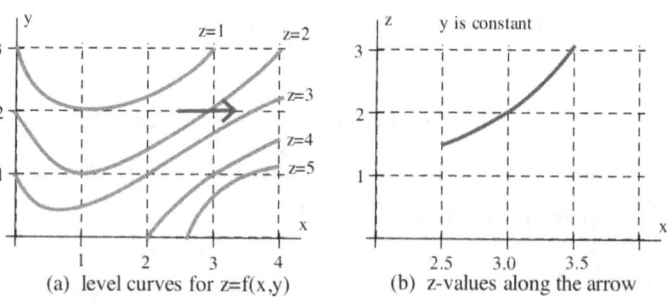

(a) level curves for z=f(x,y)   (b) z-values along the arrow

Fig. 8: the sign of $f_x(3,2)$

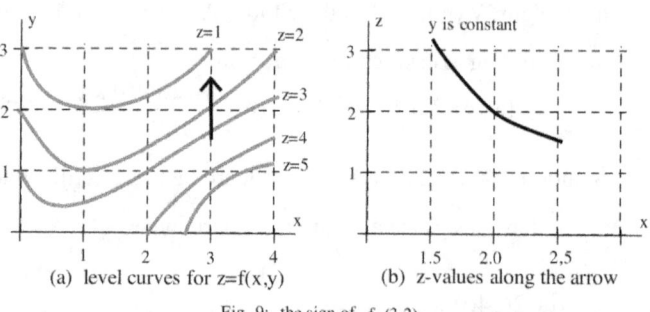

(a) level curves for z=f(x,y)   (b) z-values along the arrow

Fig. 9: the sign of $f_y(3,2)$

Note: If the level curves are close together in our direction of movement, then the z–values are changing rapidly in that direction and the magnitude (the absolute value of the magnitude) of the rate of change in that direction is large. For the function described by the level curves in Fig. 7, $|f_x(3,2)| > |f_y(3,2)|$ because the level curve lines are closer together as when we move from (3,2) in the x–direction than when we move in the y–direction.

Practice 7: Use the information in Fig. 7 to estimate the signs (positive, negative or zero) of $f_x(3,1)$, $f_y(3,1)$, $f_x(1,1)$ and $f_y(1,1)$. Which of these partial derivatives has the largest absolute value?

Second Partial Derivatives

For a function $y = f(x)$ of one variable, the second derivative $f''(x) = \frac{d}{dx}\left(\frac{df}{dx}\right) = \frac{d^2f}{dx^2}$ is the rate of change of the rate of change of f, and it measures the concavity of the graph of f. The situation for $z = f(x,y)$ is similar. The second derivative of a function of one variable was used to determine whether a critical point was a local maximum or minimum, and the second partial derivatives will be used to help determine whether critical points of functions of two variables are local maximums or minimums.

Definition: Second Partial Derivatives of $z = f(x,y)$:

$$f_{xx}(x,y) = f_{xx} = \frac{\partial}{\partial x}\left(\frac{\partial f}{\partial x}\right) = \frac{\partial^2 f}{\partial x^2} = \frac{\partial^2 z}{\partial x^2} \qquad \text{differentiate twice with respect to } x$$

$$f_{yy}(x,y) = f_{yy} = \frac{\partial}{\partial y}\left(\frac{\partial f}{\partial y}\right) = \frac{\partial^2 f}{\partial y^2} = \frac{\partial^2 z}{\partial y^2} \qquad \text{differentiate twice with respect to } y$$

$f_{xx}(x,y)$ measures the concavity of the graph of f in the x–direction. $f_{yy}(x,y)$ measures the concavity in the y–direction.

We can also differentiate first with respect to one variable and then differentiate the result with respect to the other variable.

Definition: Second Mixed Partial Derivatives of $z = f(x,y)$:

$$f_{xy} = (f_x)_y = \frac{\partial}{\partial y}\left(\frac{\partial f}{\partial x}\right) = \frac{\partial^2 f}{\partial y \partial x} \qquad \text{differentiate first with respect to } x, \text{ then with respect to } y$$

$$f_{yx} = (f_y)_x = \frac{\partial}{\partial x}\left(\frac{\partial f}{\partial y}\right) = \frac{\partial^2 f}{\partial x \partial y} \qquad \text{differentiate first with respect to } y, \text{ then with respect to } x$$

$f_{xy}(x,y)$ measures the rate of change in the y–direction of the rate of change in the x–direction. This is more difficult to interpret graphically.

Note how order of the x and y changes depending on the notation: $f_{xy} = \frac{\partial}{\partial y}\left(\frac{\partial f}{\partial x}\right) = \frac{\partial^2 f}{\partial y \partial x}$ and $f_{yx} = \frac{\partial^2 f}{\partial x \partial y}$.

Example 8: For $f(x,y) = 3x^3 + 7y^4 - 10x^2 y$, calculate f_{xx}, f_{yy}, f_{xy} and f_{yx}.

Solution: $f_x = 9x^2 - 20xy$ and $f_y = 28y^3 - 10x^2$. Then

$$f_{xx} = \frac{\partial}{\partial x}(9x^2 - 20xy) = 18x - 20y. \qquad f_{yy} = \frac{\partial}{\partial y}(28y^3 - 10x^2) = 84y^2.$$

$$f_{xy} = \frac{\partial}{\partial y}(9x^2 - 20xy) = -20x. \qquad f_{yx} = \frac{\partial}{\partial x}(28y^3 - 10x^2) = -20x.$$

Practice 8: For $g(x,y) = e^{xy} + \frac{x}{y}$, calculate g_{xx}, g_{yy}, g_{xy} and g_{yx}.

In the previous Example and Practice it turned out that the mixed partials were equal: $f_{xy} = f_{yx}$ and $g_{xy} = g_{yx}$. The next theorem says this is always the case for "nice" (sufficiently smooth) surfaces.

Clairaut's Theorem:

If $f(x,y)$ is defined and continuous at (a,b) and for all points near (a,b)
and f_{xy} and f_{yx} are both continuous at all points near (a,b),

then $f_{xy}(a,b) = f_{yx}(a,b)$.

We can also define higher partial derivatives in a natural way such as $f_{xyy} = (f_{xy})_y = \frac{\partial}{\partial y}\left(\frac{\partial^2 f}{\partial y \partial x}\right) = \frac{\partial^3 f}{\partial y \partial y \partial x}$.

These higher partial derivatives are sometimes useful in physics and other areas, but we will not use them.

Partial Derivatives Implicitly

In all of the previous examples we knew z explicitly as a function of x and y. But sometimes it is not possible to algebraically isolate z in order to calculate a partial derivative. In that case we can still determine the partial derivatives, but we need to do so implicitly.

Example 8: $xy + yz = xz$. Determine $\frac{\partial z}{\partial x}$ and $\frac{\partial z}{\partial y}$ in general and at the point $(3, 2, 6)$.

Solution: In this case we can calculate the partial derivatives both explicitly and implicitly.

Explicitly: Solving for z we get $xy = xz - yz$ so $z = \frac{xy}{x-y}$. Then, using the quotient rule,

$$\frac{\partial z}{\partial x} = \frac{(x-y)\cdot \frac{\partial(xy)}{\partial x} - xy\cdot \frac{\partial(x-y)}{\partial x}}{(x-y)^2} = \frac{(x-y)\cdot y - xy\cdot 1}{(x-y)^2} = \frac{-y^2}{(x-y)^2}.$$

At $(3, 2, 6)$, $\frac{\partial z}{\partial x} = -4$. Similarly, $\frac{\partial z}{\partial y} = \frac{x^2}{(x-y)^2}$ which equals 9 at $(3, 2, 6)$.

Implicitly: Taking the partial derivative of each side, $\frac{\partial}{\partial x}(xy + yz) = \frac{\partial}{\partial x}(xz)$, we get

$$\left[x \cdot \frac{\partial y}{\partial x} + y \cdot \frac{\partial x}{\partial x} \right] + \left[y \cdot \frac{\partial z}{\partial x} + z \cdot \frac{\partial y}{\partial x} \right] = x \cdot \frac{\partial z}{\partial x} + z \cdot \frac{\partial x}{\partial x}.$$ But $\frac{\partial y}{\partial x} = 0$ (why?) and $\frac{\partial x}{\partial x} = 1$ so the previous equation simplifies to $[0 + y] + \left[y \cdot \frac{\partial z}{\partial x} + 0 \right] = x \cdot \frac{\partial z}{\partial x} + z$.

Then $\frac{\partial z}{\partial x} = \frac{z - y}{y - x}$ which equals -4 at $(3, 2, 6)$, the same result we got differentiating explicitly. Similarly, $\frac{\partial z}{\partial u} = \frac{x + z}{x - y} = \frac{9}{1}$ at $(3, 2, 6)$.

Practice 9: $xy^2 + \sin(z) + 3 = 2x + 3z$. Determine $\frac{\partial z}{\partial x}$ in general and at the point $(3, 1, 0)$.

A Final Comment!

Partial derivatives are used extensively in the remaining sections on multivariate calculus, and it is vital that you understand what they measure and that you become able to calculate partial derivatives quickly and accurately. Extra practice now will save you time (and points) in the rest of the course.

PROBLEMS

1. For $f(x,y) = 16 - 4x^2 - y^2$, find $f_x(1,2)$ and $f_y(1,2)$ and interpret these numbers as slopes. Illustrate with sketches.

2. For $f(x,y) = \sqrt{4 - x^2 - 4y^2}$, find $f_x(1,0)$ and $f_y(1,0)$ and interpret these numbers as slopes. Illustrate with sketches.

In problems 3 – 11, find the indicated partial derivatives.

3. $f(x,y) = x^3 y^5$; $f_x(3, -1)$

4. $f(x,y) = xe^{-y} + 3y$; $\frac{\partial}{\partial y}(1, 0)$

5. $z = \frac{x^3 + y^3}{x^2 + y^2}$; $\frac{\partial z}{\partial x}$, $\frac{\partial z}{\partial y}$

6. $z = \frac{x}{y} + \frac{y}{x}$; $\frac{\partial z}{\partial x}$

7. $xy + yz = xz$; $\frac{\partial z}{\partial x}$, $\frac{\partial z}{\partial y}$

8. $\sin(x) + y \cdot e^z = z$; $\frac{\partial z}{\partial x}$, $\frac{\partial z}{\partial y}$

9. $y^2 + yz^2 = zx^2$; $\frac{\partial z}{\partial x}$, $\frac{\partial z}{\partial y}$

10. $x^2 + y^2 - z^2 = 2x(y + z)$; $\frac{\partial z}{\partial x}$, $\frac{\partial z}{\partial y}$

11. $u = xy \sec(xy)$; $\frac{\partial u}{\partial x}$

12. $f(x,y,z) = xyz$; $f_y(0, 1, 2)$

13. $u = xy + yz + zx$; u_x, u_y, u_z

In problems 14 – 29, find the first partial derivatives of the given functions.

14. $f(x,y) = x^3 y^5 - 2x^2 y + x$

15. $f(x,y) = x^4 + x^2 y^2 + y^4$

16. $f(x,y) = \frac{x-y}{x+y}$

17. $f(x,y) = e^x \tan(x-y)$

18. $f(s,t) = \sqrt{2 - 3s^2 - 5t^2}$

19. $f(u,v) = \arctan(u/v)$

20. $g(x,y) = y \tan(x^2 y^3)$

21. $z = \ln(x + \sqrt{x^2 + y^2})$

22. $z = \sinh(\sqrt{3x + 4y})$

23. $f(x,y) = \int_x^y e^{(t^2)} \, dt$

24. $f(x,y,z) = x^2 y z^3 + xy - z$

25. $f(x,y,z) = x^{yz}$

26. $u = z \sin\left(\frac{y}{x+z}\right)$

27. $u = xy^2 z^3 \ln(x + 2y + 3z)$

28. $f(x,y,z,t) = \frac{x-y}{z-t}$

29. $u = \sqrt{x_1^2 + x_2^2 + \ldots + x_n^2}$

30. Use the definition of partial derivatives as limits to find $f_x(x,y)$ and $f_y(x,y)$ when $f(x,y) = x^2 - xy + 2y^2$.

In problems 31 – 33, find $\partial z/\partial x$ and $\partial z/\partial y$.

31. $z = f(x) + g(y)$

32. $z = f(x + y)$

33. $z = f(x/y)$

In problems 34 – 36, find all of the second partial derivatives.

34. $f(x,y) = x^2 y + x\sqrt{y}$

35. $z = (x^2 + y^2)^{3/2}$

36. $z = t \cdot \arcsin(\sqrt{x})$

In problems 37 and 38, verify that the conclusion of Clairaut's Theorem holds, that is, $u_{xy} = u_{yx}$.

37. $u = x^5 y^4 - 3x^2 y^3 + 2x^2$

38. $u = \arcsin(xy^2)$

39. Verify that the function $u = e^{-a^2 k^2 t} \sin(kx)$ is a solution of the heat equation $u_t = a^2 u_{xx}$.

40. The total resistance R produced by three conductors with resistances R_1, R_2, R_3 connected in a parallel electrical circuit is given by the formula $\frac{1}{R} = \frac{1}{R_1} + \frac{1}{R_2} + \frac{1}{R_3}$. Find $\partial R/\partial R_1$.

Practice Answers

Practice 1: (a) (2000 gallons)/(10 degrees) = 200 gallons/degree

(b) (7000 gallons)/(10,000 people) = 0.7 gallons/person

Practice 2: (a) $f_x(x,y) = 3x^2 + 10xy$, $f_x(2,5) = 112$

(b) $g_x(x,y) = y \cdot e^{xy} + \frac{1}{y}$, $g_x(0,2) = 2.5$

Practice 3: $f(3,2) \approx 9$, $f_x(3,2) \approx 9$

Practice 4: (a) $f_y(x,y) = 8y + 5x^2$, $f_y(2,5) = 60$

(b) $g_y(x,y) = x\, e^{xy} - x/y^2$, $g_y(0,2) = 0$

Practice 5: (a) $f_y(1,3)$ is positive (b) $f_y(1,1)$ is approximately 0 (c) $f_y(1,4)$ is positive

Practice 6: (a) This process will produce 8.9 grams of vaccine when we have 2000 bacteria and the laboratory temperature is 40 °C.

(b) The amount of vaccine produced (when we have 2000 bacteria at a temperature of 40 °C) will **increase at an INSTANTANEOUS RATE of** 0.003 grams for each additional bacteria when the temperature stays constant. Units: grams/bacteria

(c) The amount of vaccine produced (when we have 2000 bacteria at a temperature of 40 °C) will **decrease at an INSTANTANEOUS RATE of** 1.4 grams for each degree increase in temperature when the number of bacteria stays constant. Units: grams/°C

Practice 7: $f_x(3,1)$ is positive, $f_y(3,1)$ is negative,

$f_x(1,1)$ is zero (z has a local max for increasing x values), and $f_y(1,1)$ is negative.

I estimate that $f_y(3,1)$ has the largest absolute value (contour lines are closest together).

Practice 8: If $g(x,y) = e^{xy} + \frac{x}{y}$, then

$$g_x = y \cdot e^{xy} + \frac{1}{y} \qquad\qquad g_y = x \cdot e^{xy} - \frac{x}{y^2}.$$

$$g_{xx} = \frac{\partial}{\partial x}(y \cdot e^{xy} + \frac{1}{y}) = y^2 \cdot e^{xy} \qquad g_{yy} = \frac{\partial}{\partial y}(x \cdot e^{xy} - \frac{x}{y^2}) = x^2 \cdot e^{xy} + \frac{2x}{y^3}$$

$$g_{xy} = \frac{\partial}{\partial y}(y \cdot e^{xy} + \frac{1}{y}) = xy \cdot e^{xy} + e^{xy} - \frac{1}{y^2}$$

$$g_{yx} = \frac{\partial}{\partial x}(x \cdot e^{xy} - \frac{x}{y^2}) = xy \cdot e^{xy} + e^{xy} - \frac{1}{y^2}$$

Practice 9: $\frac{\partial}{\partial x}(x \cdot y^2 + \sin(z) + 3) = \frac{\partial}{\partial x}(2x + 3z)$ so

$\left[x \cdot 2y \cdot \frac{\partial y}{\partial x} + y^2 \cdot \frac{\partial x}{\partial x} \right] + \cos(z) \cdot \frac{\partial z}{\partial x} + 0 = 2 \cdot \frac{\partial x}{\partial x} + 3 \cdot \frac{\partial z}{\partial x}$ which simplifies to

$\left[y^2 \right] + \cos(z) \cdot \frac{\partial z}{\partial x} = 2 + 3 \cdot \frac{\partial z}{\partial x}$ so $\frac{\partial z}{\partial x} = \frac{2 - y^2}{\cos(z) - 3}$ which equals $\frac{1}{-2}$ at $(3, 1, 0)$.

13.4 TANGENT PLANES and DIFFERENTIALS

In Section 2.8 we were able to use the derivative f' of a function $y = f(x)$ of one variable to find the equation of the line tangent to the graph of f at a point $(a, f(a))$ (Fig. 1): $y = f(a) + f'(a) \cdot (x - a)$. And then we used this tangent line to approximate values of f near the point $(a, f(a))$, and we introduced the idea if the differential $df = f'(a) \cdot dx$ of the function f. In this section we extend these ideas to functions $z = f(x,y)$ of two variables. But here we will find tangent planes (Fig. 2) rather than tangent lines, and we will use the tangent plane to approximate values of $f(x,y)$. Finally, we will extend the concept of a differential to functions of two variables.

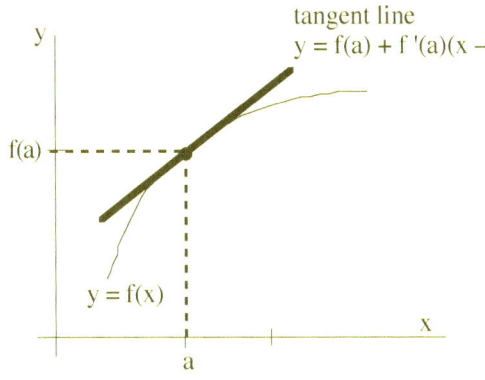

Fig. 1: Tangent line to $y = f(x)$ at $x = a$

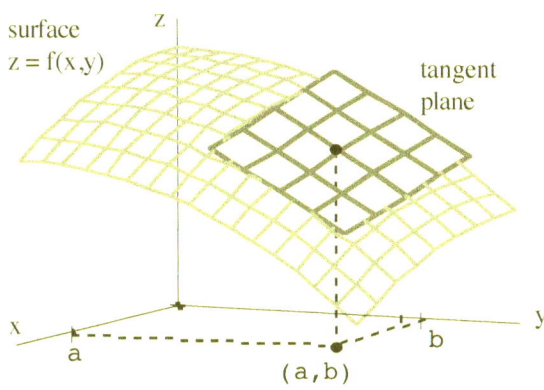

Fig. 2: Tangent plane to $z = f(x,y)$ at (a,b)

Tangent Planes

In Section 11.6 we saw how to use a point (a, b, c) and two (nonparallel) vectors to determine the equation of the plane through the point and containing lines parallel to the given vectors (Fig. 3):

(1) we used the cross product of the two given vectors to find a normal vector $\mathbf{N} = \langle n_1, n_2, n_3 \rangle$ to the plane, and then

(2) we used the normal vector $\mathbf{N}$ and the point to write the equation of the plane as $n_1(x - a) + n_1(y - b) + n_1(z - c) = 0$.

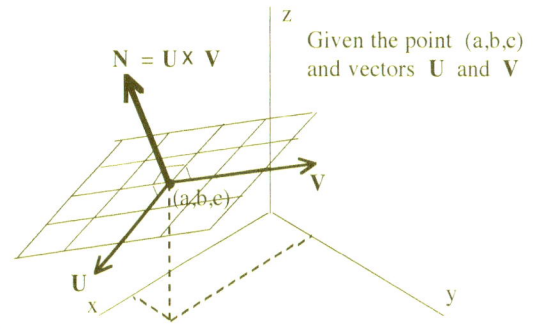

Given the point (a,b,c) and vectors $\mathbf{U}$ and $\mathbf{V}$

Equation of the tangent plane is
$n_1(x-a) + n_2(y-b) + n_3(z-c) = 0$

Fig. 3

This approach also works when we need the equation of a plane tangent to a surface $z = f(x,y)$, but we will use the formula for the surface to find the two needed vectors.

Example 1: Find the equation of the plane tangent to the surface $f(x,y) = 2x^3 + y^2 + 3$ at the point $P = (1, 2, f(1,2)) = (1, 2, 9)$ on the surface (Fig. 4).

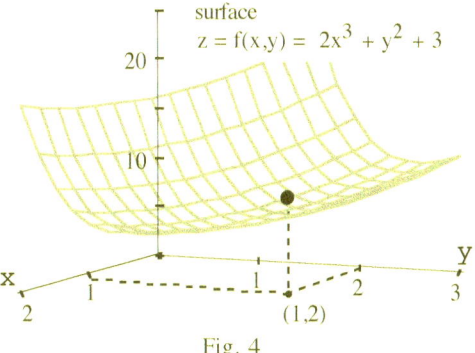

Fig. 4

Solution: We are given a point $(1, 2, 9)$ on the plane, but we need two vectors. These vectors are the rates of change of the surface $f(x,y)$ in the x and y directions. The rate of change of $f(x,y)$ in the x–direction is $f_x(x,y) = 6x^2$, and at the point $(1,2,9)$ we have $f_x(1,2) = 6(1)^2 = 6$. Similarly, the rate of change of $f(x,y)$ in the y–direction is $f_y(x,y) = 2y$, and at the point $(1,2,9)$ we have $f_y(1,2) = 2(2) = 4$.

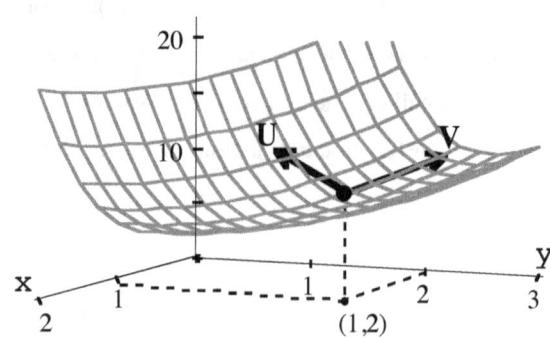

Then a "rate of change vector in the x–direction" is $\mathbf{U} = \langle 1, 0, 6 \rangle$ formed by taking 1 "step" in the x–direction, taking 0 "steps" in the y–direction (y is constant), and then taking 6 "steps" in the z–direction ($6 = f_x(1,2)$ = rate of change of z with respect to increasing x–values). Similarly, a "rate of change" vector in the y–direction is $\mathbf{V} = \langle 0, 1, 4 \rangle$. These vectors are shown in Fig. 5.

Fig. 5: Surface and tangent vectors $\mathbf{U}$ and $\mathbf{V}$

Now a normal vector $\mathbf{N}$ to the plane we want is formed by taking

$$\mathbf{N} = \mathbf{V} \times \mathbf{U} = \begin{vmatrix} \mathbf{i} & \mathbf{j} & \mathbf{k} \\ 0 & 1 & 4 \\ 1 & 0 & 6 \end{vmatrix} = (6)\mathbf{i} - (-4)\mathbf{j} + (-1)\mathbf{k} = 6\mathbf{i} + 4\mathbf{j} - 1\mathbf{k}$$

(Note: taking $\mathbf{N} = \mathbf{U} \times \mathbf{V} = -6\mathbf{i} - 4\mathbf{j} + 1\mathbf{k}$ also works.)

Finally, using the point $P = (1, 2, 9)$ and the normal vector $\mathbf{N} = \mathbf{V} \times \mathbf{U} = 6\mathbf{i} + 4\mathbf{j} - 1\mathbf{k}$, we know that the equation of the plane is

$6(x - 1) + 4(y - 2) - 1(z - 9) = 0$ or $z = 9 + 6(x - 1) + 4(y - 2)$

Looking at the equation $z = 9 + 6(x - 1) + 4(y - 2)$ of the plane, you should notice that the 9 is the z–coordinate of our original point, that the coefficient of the x variable, 6, is $f_x(1,2)$, and that the coefficient of the y variable, 4, is $f_y(1,2)$. Fig. 6 shows the surface and the tangent plane.

Fortunately we do not need to go through all of those calculations every time we need the equation of a plane tangent to a surface at a point: the pattern that we noted about the coefficients of the variables in the tangent plane equation and the values of the partial derivatives holds for every differentiable function.

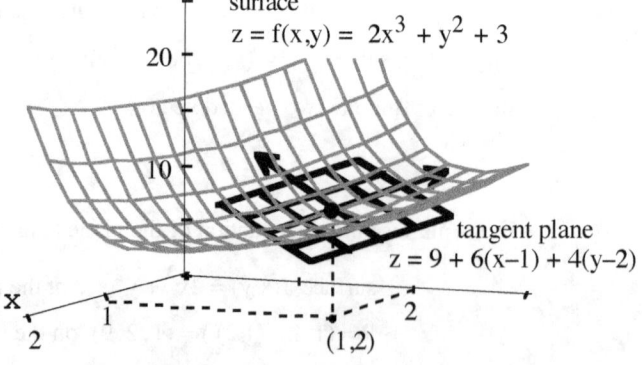

Fig. 6: Surface and tangent plane

Equation for a Tangent Plane

If f(x,y) is differentiable at the point (a, b, f(a,b)),

then the equation of the plane tangent to the surface $z = f(x, y)$ at the point P(a, b, f(a,b))

is $z = f(a,b) + f_x(a,b)(x - a) + f_y(a,b)(y - b)$.

Proof: The proof simply involves the steps we went through for Example 1. $\mathbf{U} = \langle 1, 0, f_x(a,b) \rangle$ is formed by taking 1 "step" in the x–direction, taking 0 "steps" in the y–direction (y is constant), and then taking $f_x(a,b)$ "steps" in the z–direction. Similarly, $\mathbf{V} = \langle 0, 1, f_y(a,b) \rangle$. Then

$$\mathbf{N} = \mathbf{V} \times \mathbf{U} = \begin{vmatrix} \mathbf{i} & \mathbf{j} & \mathbf{k} \\ 0 & 1 & f_y(a,b) \\ 1 & 0 & f_x(a,b) \end{vmatrix} = (f_x(a,b))\mathbf{i} - (-f_y(a,b))\mathbf{j} + (-1)\mathbf{k} = f_x(a,b)\mathbf{i} + f_y(a,b)\mathbf{j} - 1\mathbf{k}.$$

Finally, using the point (a, b, f(a,b)) and $\mathbf{N} = \mathbf{V} \times \mathbf{U} = f_x(a,b)\mathbf{i} + f_y(a,b)\mathbf{j} - 1\mathbf{k}$, we have that the equation of the plane is

$f_x(a,b)(x - a) + f_y(a,b)(y - b) - 1(z - f(a,b)) = 0$ or $z = f(a,b) + f_x(a,b)(x - a) + f_y(a,b)(y - b)$.

Example 2: Find the plane tangent to the surface $z = 2x^2y^3 + \ln(xy) + 7$ at the point (1, 1, 9).

Solution: $f_x(x, y) = 4xy^3 + \frac{1}{xy}(y) = 4xy^3 + \frac{1}{x}$ and $f_y(x, y) = 6x^2y^2 + \frac{1}{xy}(x) = 6x^2y^2 + \frac{1}{y}$

so $f_x(1, 1) = 5$ and $f_y(1, 1) = 7$.

Then the equation of the tangent plane is $z = 9 + 5(x - 1) + 7(y - 1)$ or $\mathbf{z = 5x + 7y - 3}$.

(This is much quicker than the "from scratch" method of Example 1.)

Fig. 7 shows two views of this surface and the tangent plane — notice that in this case the tangent plane cuts through the surface.

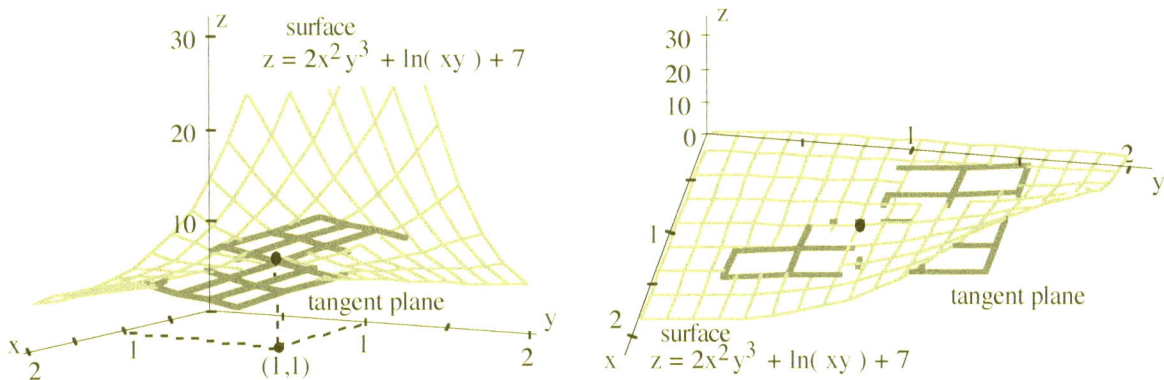

Fig. 7: Two views of the surface and tangent plane

Practice 1: Find the plane tangent to the surface $z = 5xy^2 + 7y + \sin(xy) - 2$ at the point (0, 1, 5).

Differentials

The following "boxed" material summarizes, from Section 2.8, the definition and results about the differential dy of a function $y = f(x)$ of one variable.

Definition for $y = f(x)$: The **differential** of $y = f(x)$ is $\mathbf{dy = f'(x)\,dx} = \frac{df}{dx}\,dx$.

Meaning of dy: **dy** is the change in the y–value, **along the tangent line** to f obtained by a step of dx in the x–value.

Result: If f is differentiable at $x = a$ and dx is "small"

then $f(a + dx) - f(a) \approx \mathbf{dy}$ or $f(a + dx) \approx f(a) + \mathbf{dy}$.

Meaning of the Result: For a small step dx, the actual change in f is approximately equal to the change along the tangent line: $f(a + dx) \approx f(a) + f'(a)\,dx$.

The extension to functions of two variables is given in the next "box."

Definition for $z = f(x,y)$: The **differential** (or **total differential**) of $z = f(x,y)$ is

$$\mathbf{dz = f_x(x,y)\,dx + f_y(x,y)\,dy} = \frac{\partial f}{\partial x}\,dx + \frac{\partial f}{\partial y}\,dy .$$

Meaning of dz: **dz** is the change in the z–value, **along the tangent plane** to f, obtained by a step of dx in the x direction and a step of dy in the y direction. (Fig. 8)

Result: If $z = f(x,y)$ is differentiable at the point (a, b),

then $f(a + dx, b + dy) - f(a,b) \approx \mathbf{dz} = f_x(a, b)\,dx + f_y(a, b)\,dy = \frac{\partial f}{\partial x}\,dx + \frac{\partial f}{\partial y}\,dy$.

Meaning of the Result: For a small step of dx in the x direction and a small step dy in the y direction, the change in f is approximately equal to the change along the tangent plane:

$f(a + dx, b + dy) \approx f(a,b) + dz = f(a,b) + f_x(a, b)\,dx + f_y(a,b)\,dy$

Example 3: Find the differential of $z = 5 + 3x^2y^3$ (a) in general, and (b) at the point $(x,y) = (2, 1)$.

Solution: (a) $\mathbf{dz} = f_x(a, b)\,dx + f_y(a, b)\,dy = \{\,6xy^3\,\}\,dx + \{\,9x^2y^2\,\}\,dy$

(b) at $(2, 1)$, $\mathbf{dz} = \{\,6(2)(1)^3\,\}\,dx + \{\,9(2)^2(1)^2\,\}\,dy = (12)\,dx + (36)\,dy$.

Example 4: For $z = 5 + 3x^2y^3$ and the point $(2,1)$, use the result of Example 3 that $dz = (12)\,dx + (36)\,dy$ to approximate $f(2.02, 1.01)$ and $f(2.01, 0.97)$.

Compare these approximate values with the exact values of $f(2.02, 1.01)$ and $f(2.01, 0.97)$.

Solution: For f(2.02, 1.01), dx = 0.02 and dy = 0.01 so dz = (12)(0.02) + (36)(0.01) = 0.6.

Then f(2.02, 1.01) ≈ f(2, 1) + dz = 17 + 0.6 = 17.6.

Actually, f(2.02, 1.01) = 17.6121206012 so the approximation "error" using the differential is 0.012.

For f(2.01, 0.97), dx = 0.01 and dy = –0.03, so dz = (12)(0.01) + (36)(–0.03) = –0.96.

Then f(2.01, 0.97) ≈ f(2,1) + dz = 17 + (–0.96) = 16.04.

Actually, f(2.01, 0.97) = 16.0618705619 so the approximation "error" using the differential is 0.022.

Practice 2: Find the differential of $z = 3 + x \cdot \sin(2xy)$ (a) in general, and (b) at the point $(x,y) = (1, \pi/2)$.

(c) Use the result of part (b) to approximate $f(1.3, \frac{\pi}{2} - 0.1)$ and $f(0.99, \frac{\pi}{2} + 0.2)$.

(d) Compare the results of (c) with the exact values of $f(1.3, \frac{\pi}{2} - 0.1)$ and $f(0.99, \frac{\pi}{2} + 0.2)$.

Examples 4 and Practice 2(c and d) compare the value of f found by moving **along the tangent plane** to the actual value of f found **on the surface**. When the sideways movement is "small" (when dx and dy are both small), then the "along the tangent plane" value of z is close to the "on the surface" value of z, the actual value of f.

PROBLEMS

In problems 1 – 8, find an equation for the tangent plane to the given surface at the given point.

1. $z = x^2 + 4y^2$ at $(2, 1, 8)$
2. $z = x^2 - y^2$ at $(3, -2, 5)$
3. $z = 5 + (x - 1)^2 + (y + 2)^2$ at $(2, 0, 10)$
4. $z = \sin(x + y)$ at $(1, -1, 0)$
5. $z = \ln(2x + y)$ at $(-1, 3, 0)$
6. $z = e^x \cdot \ln(y)$ at $(3, 1, 0)$
7. $z = xy$ at $(-1, 2, -2)$
8. $z = \sqrt{x - y}$ at $(5, 1, 2)$

In problems 9 – 18, find the differential of the given function.

9. $z = x^2 y^3$
10. $z = x^4 - 5x^2 y + 6xy^3 + 10$
11. $z = \dfrac{1}{x^2 + y^2}$
12. $z = y \cdot e^{xy}$
13. $u = e^x \cdot \cos(xy)$
14. $v = \ln(2x - 3y)$
15. $w = x^2 y + y^2 z$
16. $w = x \sin(yz)$
17. $w = \ln(\sqrt{x^2 + y^2 + z^2})$
18. $w = \dfrac{x + y}{y + z}$

19. If $z = 5x^2 + y^2$ and (x, y) changes from $(1, 2)$ to $(1.05, 2.1)$, compare the values of Δz and dz.

20. If $z = x^2 - xy + 3y^2$ and (x, y) changes from $(3, -1)$ to $(2.96, -0.95)$, compare the values of Δz and dz.

In problems 21 – 24, use differentials to approximate the value of f at the given point.

21. $f(x, y) = \sqrt{x^2 - y^2}$ at (5.01, 4.02)

22. $f(x, y) = \sqrt{20 - x^2 - 7y^2}$ at (1.95, 1.08)

23. $f(x, y) = \ln(x - 3y)$ at (6.9, 2.06)

24. $f(x, y) = y \cdot e^{xy}$ at (0.2, 1.96)

25. The length and width of a rectangle are measured as 30 cm and 24 cm, respectively, with an error in measurement of at most 0.1 cm in each. Use differentials to estimate the maximum error in the calculated area of the rectangle.

26. The dimensions of a closed rectangular box are measured as 80 cm, 60 cm, and 50 cm, respectively, with a possible error of 0.2 cm in each dimension. Use differentials to estimate the maximum error in calculating the surface area of the box.

27. Use differentials to estimate the amount of tin in a closed tin can with diameter 8 cm and height 12 cm if the tin is 0.04 cm thick.

28. Use differentials to estimate the amount of metal in a closed cylindrical can that is 10 cm high and 4 cm in diameter if the metal in the wall is 0.05 cm thick and the metal in the top and bottom is 0.1 cm thick.

29. A boundary stripe 3 in. wide is painted around a rectangle whose dimensions are 100 ft. by 200 ft. Use differentials to approximate the number of square feet of paint in the stripe.

30. The pressure, volume, and temperature of a mole of an ideal gas are related by the equation $PV = 8.31T$, where P is measured in kilopascals, V in liters, and T in °K (= °C + 273). Use differentials to find the approximate change in pressure if the volume increases from 12 L to 12.3 L and the temperature decreases from 310°K to 305°K.

PRACTICE ANSWERS

Practice 1: $z = f(a,b) + f_x(a,b)(x - a) + f_y(a,b)(y - b)$

with $f(x,y) = 5xy^2 + 7y + \sin(xy) - 2$, $a = 0$, $b = 1$, and $f(0,1) = 5$.

$f_x(x,y) = 5y^2 + y \cdot \cos(xy)$ so $f_x(0,1) = 5 + 1 = 6$.

$f_y(x,y) = 10xy + 7 + x \cdot \cos(xy)$ so $f_y(0,1) = 7$.

Then the equation of the tangent plane to f at (0,1,5) is $z = 5 + 6(x - 0) + 7(y - 1) = 6x + 7y - 2$.

Practice 2: (a) $dz = f_x(a,b)\,dx + f_y(a,b)\,dy = \{x\cdot\cos(2xy)\cdot 2y + \sin(2xy)\}\,dx + \{x\cdot\cos(2xy)\cdot 2x\}\,dy$

(b) at $(1, \pi/2)$,

$dz = \{1\cdot\cos(2\cdot 1\cdot\pi/2)\cdot 2\cdot\pi/2 + \sin(2\cdot 1\cdot\pi/2)\}\,dx + \{1\cdot\cos(2\cdot 1\cdot\pi/2)\cdot 2\cdot 1\}\,dy$ so

$dz = (-\pi)\,dx + (-2)\,dy$.

(c)&(d) For $f(1.3, \frac{\pi}{2} - 0.1)$, $dx = 0.3$ and $dy = -0.1$ so $dz = (-\pi)(0.3) + (-2)(-0.1) \approx -0.742$.

Then $f(1.3, \frac{\pi}{2} - 0.1) \approx f(1, \frac{\pi}{2}) + dz = 3 + (-0.742) = 2.258$.

Actually, $f(1.3, \frac{\pi}{2} - 0.1) = 2.18006691401$ so the approximation "error" is 0.078,

an "error" of less than 4%.

For $f(0.99, \frac{\pi}{2} + 0.2)$, $dx = -0.01$ and $dy = 0.2$ so $dz = (-\pi)(-0.01) + (-2)(0.2) = -0.369$.

Then $f(0.99, \frac{\pi}{2} + 0.2) \approx f(1, \frac{\pi}{2}) + dz = 3 + (-0.369) = 2.631$.

Actually, $f(0.99, \frac{\pi}{2} + 0.2) = 2.64700487061$ so the approximation "error" is 0.016,

an "error" of less than 1%.

13.5 DIRECTIONAL DERIVATIVES and the GRADIENT VECTOR

Directional Derivatives

In Section 13.3 the partial derivatives f_x and f_y were defined as

$$f_x(x,y) = \lim_{h \to 0} \frac{f(x+h,y) - f(x,y)}{h} \quad \text{(if the limit exists and is finite)} \quad \text{and}$$

$$f_y(x,y) = \lim_{h \to 0} \frac{f(x,y+h) - f(x,y)}{h} \quad \text{(if the limit exists and is finite)}.$$

These partial derivative measured the instantaneous rate of change of $z = f(x,y)$ as we moved in the increasing x–direction (while holding y constant) and in the increasing y–direction (while holding x constant). Sometimes, however, we are interested in the rate of change of $z = f(x,y)$ as we move in some other direction, and that leads to the idea of a **directional derivative** to measure the instantaneous rate of change of $z = f(x,y)$ as we move in any direction.

Definition:

The **directional derivative** of $z = f(x,y)$ in the direction of a unit vector $\mathbf{u} = \langle a, b \rangle$ is

$$D_{\mathbf{u}}f(x,y) = \lim_{h \to 0} \frac{f(x+ah, y+bh) - f(x,y)}{h} \quad \text{(if the limit exists and is finite)}.$$

Figures 1a and 1b illustrate the slope of the line tangent to the curve where the plane above the vector u intersects the surface $z = f(x,y)$.

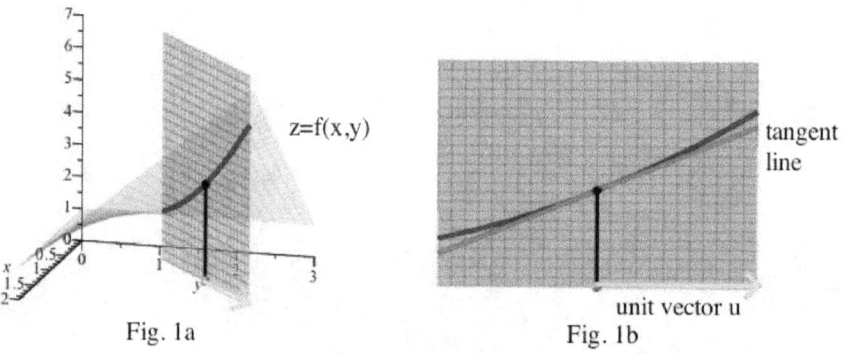

Fig. 1a Fig. 1b

Using this definition can be tedious and algebraically messy, but it is worth doing once. Fortunately we will soon see a much more efficient method.

13.5 Directional Derivatives and the Gradient Vector

Example 1: Use this definition to calculate $D_{\mathbf{u}}f(1, 2)$ for $f(x,y) = 1 + xy$ and $\mathbf{u} = \langle 0.6, 0.8 \rangle$

Solution: $D_{\mathbf{u}}f(1,2) = \lim\limits_{h \to 0} \dfrac{f(1+0.6h, 2+0.8h) - f(1,2)}{h} = \lim\limits_{h \to 0} \dfrac{[1 + (1+0.6h)(2+0.8h)] - 3}{h}$

$\lim\limits_{h \to 0} \dfrac{[1 + 2 + + 2(0.6)h + 0.8h] - 3}{h} = 2(0.6) + 0.8 = 2$

Practice 1: Calculate $D_{\mathbf{u}}f(2, 1)$ for $f(x,y) = x^2 + 2x + 3y + 1$ and $\mathbf{u} = \left\langle \dfrac{5}{13}, \dfrac{12}{13} \right\rangle$.

For more complicated functions it becomes extremely difficult to use the definition to calculate directional derivatives. Fortunately there is a much easier way given by the next theorem.

Theorem:
If $f(x,y)$ is differentiable,
then f has a directional derivative in the direction of every <u>unit vector</u> $\mathbf{u} = \langle a, b \rangle$, and
$$D_{\mathbf{u}}f(x,y) = f_x(x,y) \, a + f_y(x,y) \, b \, .$$

A proof of this result uses a multivariable "Chain Rule" that we will discuss in Section 13.8.

Example 2: Verify that this theorem gives the same answer for $D_{\mathbf{u}}f(x,y)$ as our use of the directional derivative definition in Example 1.

Solution: $f_x(x,y) = y$ and $f_y(x,y) = x$ so $f_x(1, 2) = 2$ and $f_y(1, 2) = 1$. $\mathbf{u} = \langle 0.6, 0.8 \rangle$ so

$D_{\mathbf{u}}f(1, 2) = (2)(0.6) + (1)(0.8) = 2$, the same result as in Example 1.

Practice 2: Verify that this theorem gives the same answer for $D_{\mathbf{u}}f(x,y)$ as your use of the directional derivative definition in Practice 1.

Example 3: Calculate the directional derivatives of $z = f(x,y) = x + 5x^2y^3$ at the point $(2,1)$ in the directions of the unit vectors (a) $\mathbf{u} = \langle 0.6, 0.8 \rangle$, (b) $\mathbf{u} = \langle -0.6, -0.8 \rangle$, (c) $\mathbf{u} = \langle 0.8, 0.6 \rangle$, (d) $\mathbf{u} = \mathbf{i} = \langle 1, 0 \rangle$, and (e) $\mathbf{u} = \mathbf{j} = \langle 0, 1 \rangle$.

Solution: $f_x(x,y) = 1 + 10xy^3$ and $f_y(x,y) = 15x^2y^2$ so $f_x(2,1) = 21$ and $f_y(2,1) = 60$. Then, by the previous Theorem,

(a) for $\mathbf{u} = \langle 0.6, 0.8 \rangle$, $D_{\mathbf{u}}f(2,1) = f_x(2,1) \, a + f_y(2,1) \, b = (21)(0.6) + (60)(0.8) = 60.6$.

(b) for $\mathbf{u} = \langle -0.6, -0.8 \rangle$, $D_\mathbf{u}f(2,1) = (21)(-0.6) + (60)(-0.8) = -60.6$.

(c) for $\mathbf{u} = \langle 0.8, 0.6 \rangle$, $D_\mathbf{u}f(2,1) = (21)(0.8) + (60)(0.6) = 52.8$.

(d) for $\mathbf{u} = \mathbf{i} = \langle 1, 0 \rangle$, $D_\mathbf{u}f(2,1) = (21)(1) + (60)(0) = 21$ ($= f_x(2,1)$).

(e) for $\mathbf{u} = \mathbf{j} = \langle 0, 1 \rangle$, $D_\mathbf{u}f(2,1) = (21)(0) + (60)(1) = 60$ ($= f_y(2,1)$).

The Gradient Vector

You might have noticed that the pattern $D_\mathbf{u}f(x,y) = f_x(x,y) a + f_y(x,y) b$ for calculating the directional derivative can be viewed as the dot product of the vector $\langle f_x(x,y), f_y(x,y) \rangle$ with the unit direction vector $\mathbf{u} = \langle a, b \rangle$. This vector $\langle f_x(x,y), f_y(x,y) \rangle$ shows up in a variety of contexts and is called the **gradient** of f.

Definition of the Gradient Vector:

The **gradient vector** of $f(x,y)$ is $\nabla f(x,y) = \langle f_x(x,y), f_y(x,y) \rangle = \frac{\partial f}{\partial x} \mathbf{i} + \frac{\partial f}{\partial y} \mathbf{j}$.

The symbol "∇f" is read as "grad f" or "del f."

Example 4: Calculate $\nabla f(x,y)$ and $\nabla f(0,1)$ for (a) $f(x,y) = x + 5x^2y^3$ and (b) $f(x,y) = y^2 + \sin(x)$.

Solution: (a) For $f(x,y) = x + 5x^2y^3$, $f_x(x,y) = 1 + 10xy^3$ and $f_y(x,y) = 15x^2y^2$ so
$\nabla f(x,y) = \langle 1 + 10xy^3, 15x^2y^2 \rangle$ and $\nabla f(0,1) = \langle 1, 0 \rangle$.

(b) For $f(x,y) = y^2 + \sin(x)$, $f_x(x,y) = \cos(x)$ and $f_y(x,y) = 2y$ so
$\nabla f(x,y) = \langle \cos(x), 2y \rangle$ and $\nabla f(0,1) = \langle 1, 2 \rangle$.

Practice 3: Calculate $\nabla f(x,y)$ and $\nabla f(1,2)$ for (a) $f(x,y) = e^{xy} + 2x^3y + y^2$ and (b) $f(x,y) = \cos(2x+3y)$.

The directional derivative can now be written simply as the dot product of the gradient and the unit direction vector.

$$D_\mathbf{u}f(x,y) = \nabla f(x,y) \cdot \mathbf{u}$$

The gradient vector $\nabla f(x,y)$ is useful for much more than a compact notation for directional derivatives. $\nabla f(x,y)$ has a number of special features that make it useful for investigating the behavior of surfaces.

> **Three very important properties of the gradient vector $\nabla f(x,y)$:**
>
> (1) At a point (x,y), the maximum value of the directional derivative $D_{\mathbf{u}}f(x,y)$ is $|\nabla f(x,y)|$.
>
> (2) At a point (x,y), the maximum value of $D_{\mathbf{u}}f(x,y)$ occurs when $\mathbf{u}$ has the same direction as the gradient vector $\nabla f(x,y)$. (At each point (x,y), the gradient vector $\nabla f(x,y)$ "points" in the direction of maximum increase for $f(x,y)$.)
>
> (3) At a point (x,y), the gradient vector $\nabla f(x,y)$ is normal (perpendicular) to the level curve that goes through the point (x,y).

One of the beauties of mathematics is that sometimes a result like the powerful and non-obvious properties of the gradient can be proven in rather simple ways.

Proof of (1) and (2): $\quad D_{\mathbf{u}}f(x,y) = \nabla f(x,y) \cdot \mathbf{u}$

$\qquad\qquad\qquad\qquad = |\nabla f(x,y)| |\mathbf{u}| \cos(\theta) \qquad \theta$ is the angle between the vectors $\nabla f(x,y)$ and $\mathbf{u}$

$\qquad\qquad\qquad\qquad = |\nabla f(x,y)| \cos(\theta)$

The maximum value of $\cos(\theta)$ is 1 (when $\theta = 0$), so the maximum value of $D_{\mathbf{u}}f(x,y)$ is $|\nabla f(x,y)|$ and this maximum occurs when $\theta = 0$, when $\nabla f(x,y)$ and $\mathbf{u}$ have the same direction.

Proof of (3): f is constant along the level curve at the point (x,y) so $D_{\mathbf{u}}f(x,y) = 0$ when $\mathbf{u}$ is the direction of the level curve. But $D_{\mathbf{u}}f(x,y) = \nabla f(x,y) \cdot \mathbf{u}$ so $\nabla f(x,y)$ is perpendicular to $\mathbf{u}$, the direction of the level curve.

Example 5: Find (a) the maximum rate of change of $f(x,y) = xe^y$ at the point $(2,0,)$ and (b) the direction in which this maximum rate of change occurs.

Solution: $f_x(x,y) = e^y$ and $f_y(x,y) = xe^y$ so $f_x(2,0) = e^0 = 1$ and $f_y(2,0) = 2e^0 = 2$.

(a) The maximum value of the rate of change of f is $D_{\mathbf{u}}f(x,y) = |\nabla f(1,2)| = |\langle 1, 2 \rangle| = \sqrt{5}$.

(b) This maximum value occurs when $\mathbf{u}$ is in the direction of $\nabla f(1,2)$: $\mathbf{u} = \langle 1/\sqrt{5}, 2/\sqrt{5} \rangle$.

Practice 4: Find (a) the maximum rate of change of $f(x,y) = \sqrt{2x+3y}$ at the point $(5, 2)$ and (b) the direction in which this maximum rate of change occurs.

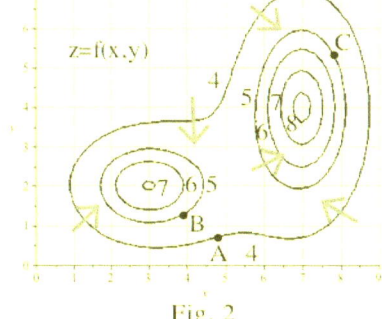

Fig. 2

Fig. 2 shows several level curves for a function $z = f(x,y)$ and the gradient vector at several locations. (Note: the lengths of these gradient vectors are exaggerated.)

Practice 5:   Sketch the gradient vector $\nabla f(x,y)$ for the function f in Fig. 2 at A, B and C.

A ball placed at (x,y) will begin to roll in the direction $u = -\nabla f(x,y)$.

Climbing to a (local) maximum

Property (2) is the foundation for using the gradient vector $\nabla f(x,y)$ in iterative methods for finding local maximums of functions of several variables:

(i) At any point (x,y) we take a short "step" in the direction of $\nabla f(x,y)$ — this takes us "uphill" along the steepest route at that point.

(ii) Repeat step (i) until a (local) maximum is reached.

Property (3) provides an easy way to geometrically determine the direction of the gradient from the level curves of a surface.

Fig. 3 shows level curves for a function $z = f(x,y)$ and the "uphill gradient" paths for several starting points.

Practice 6: Sketch the "uphill gradient" path for the function f in Fig. 3 at starting points A, B and C.

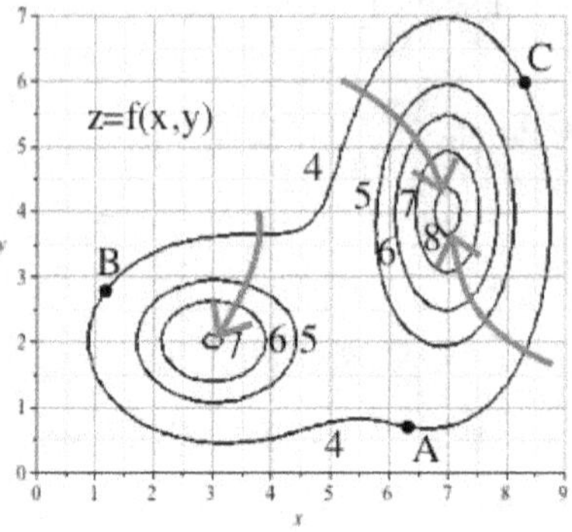

Beyond $z = f(x,y)$

Fig. 3

So far all of the examples have dealt with functions of two variables, $z = f(x,y)$, but that was just for convenience. The definitions and ideas of gradient vectors and directional derivatives and their properties extend in a natural way to functions of three (or more) variables.

Extensions to $w = f(x,y,z)$

Definition: $\nabla f(x,y,z) = \langle f_x(x,y,z), f_y(x,y,z), f_z(x,y,z) \rangle = \frac{\partial f}{\partial x}\mathbf{i} + \frac{\partial f}{\partial y}\mathbf{j} + \frac{\partial f}{\partial z}\mathbf{k}$.

Theorem: For a differentiable function $f(x,y,z)$ and a unit direction vector $\mathbf{u} = \langle a, b, c \rangle$,

$$D_\mathbf{u} f(x,y,z) = \nabla f(x,y,z) \cdot \mathbf{u}$$

Features: (1) The maximum value of the directional derivative $D_\mathbf{u} f(x,y,z)$ is $|\nabla f(x,y,z)|$.

(2) The maximum value of $D_\mathbf{u} f(x,y,z)$ occurs when $\mathbf{u}$ has the same direction as $\nabla f(x,y,z)$.

(2) The gradient vector $\nabla f(x,y,z)$ is normal (perpendicular) to the level surface through the point (x,y,z).

These same ideas also extend very naturally to functions of more than three variables.

For example, if x, y and z (all in meters) give the location in a room then w=f(x,y,z) could be the temperature (°C) at that location. Then instead of a level curve (in 2D), the points (x,y,z) where w=70 would be a level **surface** in 3D. $D_{\mathbf{u}}f(x,y,z)$ would be the instantaneous rate of change of temperature at location (x,y,z) in the direction **u**, and the units of $D_{\mathbf{u}}f(x,y,z)$ would be °C/m. The gradient vector $\nabla f(x,y,z)$ would still give the maximum value of the directional derivative and would point in the direction of maximum rate of temperature increase. A heat-seeking flying bug in the room would follow a path in the direction of the gradient at each point.

PROBLEMS

In problems 1 – 4, find the directional derivative of f at the given point in the direction indicated by the given angle θ. (Note: θ is the angle the direction vector **u** makes with the positive x–axis, so the components of **u** are $\langle \cos(\theta), \sin(\theta) \rangle$.)

1. $f(x,y) = x^2y^2 + 2x^4y$ at $(1,-2)$ with $\theta = \pi/3$ 2. $f(x,y) = (x^2 - y)^3$ at $(3,1)$ with $\theta = 3\pi/4$

3. $f(x,y) = y^x$ at $(1,2)$ with $\theta = \pi/2$ 4. $f(x,y) = \sin(x + 2y)$ at $(4,-2)$ with $\theta = -2\pi/3$

In problems 5 – 8, (a) find the gradient of f, (b) evaluate the gradient at the given point P, and (c) find the rate of change of f at P in the direction of the given vector **u**.

5. $f(x,y) = x^3 - 4x^2y + y^2$ at $P = (0,-1)$ with $\mathbf{u} = \langle 3/5, 4/5 \rangle$.

6. $f(x,y) = e^x \cdot \sin(y)$ at $P = (1, \pi/4)$ with $\mathbf{u} = \langle -1/\sqrt{5}, 2/\sqrt{5} \rangle$.

7. $f(x,y,z) = xy^2z^3$ at $P = (1,-2,1)$ with $\mathbf{u} = \langle 1/\sqrt{3}, -1/\sqrt{3}, 1/\sqrt{3} \rangle$.

8. $f(x,y,z) = xy + yz^2 + xz^3$ at $P = (2,0,3)$ with $\mathbf{u} = \langle -2/3, -1/3, 2/3 \rangle$.

In problems 9 – 14, find the directional derivative of the given function at the given point in the direction of the given vector **v**.

9. $f(x,y) = \sqrt{x-y}$ at $(5,1)$ with $\mathbf{v} = \langle 12, 5 \rangle$.   10. $f(x,y) = x/y$ at $(6,-2)$ with $\mathbf{v} = \langle -1, 3 \rangle$.

11. $g(x,y) = x \cdot e^{xy}$ at $(-3,0)$ with $\mathbf{v} = 2\mathbf{i} + 3\mathbf{j}$.   12. $g(x,y) = e^x \cos(y)$ at $(1, \pi/6)$ with $\mathbf{v} = \mathbf{i} - \mathbf{j}$.

13. $f(x,y,z) = \sqrt{xyz}$ at $(2,4,2)$ with $\mathbf{v} = \langle 4, 2, -4 \rangle$.

14. $f(x,y,z) = z^3 - x^2y$ at $(1,6,2)$ with $\mathbf{v} = \langle 3, 4, 12 \rangle$.

In problems 15 – 20, find the maximum rate of change of f at the given point and the direction in which it occurs.

15. $f(x, y) = x \cdot e^{-y} + 3y$ at the point $(1, 0)$

16. $f(x, y) = \ln(x^2 + y^2)$ at the point $(1, 2)$

17. $f(x, y) = \sqrt{x^2 + 2y}$ at the point $(4, 10)$

18. $f(x, y, z) = x + y/z$ at the point $(4, 3, -1)$

19. $f(x, y) = \cos(3x + 2y)$ at the point $(\pi/6, -\pi/8)$

20. $f(x, y, z) = \dfrac{x}{y} + \dfrac{y}{z}$ at the point $(4, 2, 1)$

21. At each dot in Fig. 4 sketch the gradient vector.

22. At each dot in Fig. 5 sketch the gradient vector.

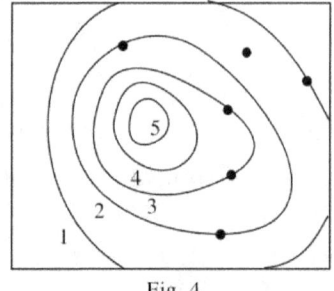

Fig. 4

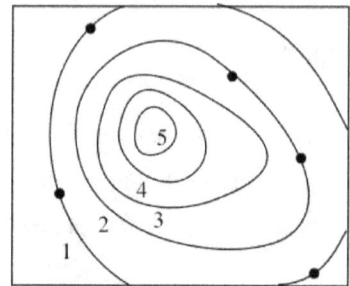

Fig. 5

23. At each dot in Fig. 6 sketch the "uphill gradient" path.

24. At each dot in Fig. 7 sketch the "uphill gradient" path.

25. Show that a differentiable function f decreases most rapidly at (x,y) in the direction opposite to the gradient vector, that is, in the direction $-\nabla f(x,y)$.

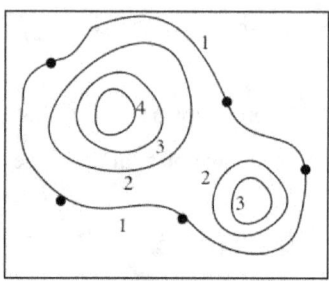

Fig. 6

26. Use the result of Problem 23 to find the direction in which the function $f(x,y) = x^4 y - x^2 y^3$ decreases fastest at the point $(2, -3)$.

27. The temperature T in a metal ball is inversely proportional to the distance from the center of the ball, which we take to be the origin. The temperature at the point $(1, 2, 2)$ is $120°$.

(a) Find the rate of change of T at $(1, 2, 2)$ in the direction toward the point $(2, 1, 3)$.

(b) Show that at any given point in the ball the direction of greatest increase in temperature is given by a vector that points toward the origin.

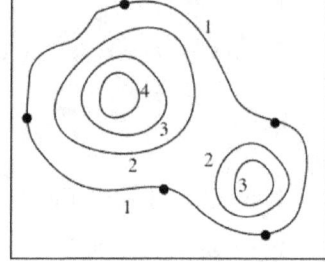

Fig. 7

28. The temperature at a point (x, y, z) is given by $T(x,y,z) = 200 \cdot e^{(-x^2 - 3y^2 - 9z^2)}$ where T is measured in $°C$ and x, y, and z in meters.

(a) Find the rate of change of temperature at the point $P(2, -1, 2)$ in the direction toward the point $(3, -3, 3)$.

(b) In which direction does the temperature increase fastest at P?

13.5 Directional Derivatives and the Gradient Vector   Contemporary Calculus

(c) Find the maximum rate of increase at P.

29. Suppose that over a certain region of space the electrical potential V is given by $V(x,y,z) = 5x^2 - 3xy + xyz$.
    (a) Find the rate of change of the potential at $P(3, 4, 5)$ in the direction of the vector $\mathbf{v} = \mathbf{i} + \mathbf{j} - \mathbf{k}$.
    (b) In which direction does V change most rapidly at P?
    (c) What is the maximum rate of change at P?

30. Suppose that you are climbing a hill whose shape is given by the equation $z = 1000 - 0.01x^2 - 0.02y^2$ and you are standing at the point with coordinates $(60, 100, 764)$.
    (a) In which direction should you proceed initially in order to reach the top of the hill fastest?
    (b) If you climb in that direction, at what angle above the horizontal will you be climbing initially?

31. Let F be a function of two variables that has continuous partial derivatives and consider the points $A(1,3)$, $B(3,3)$, $C(1,7)$, and $D(6,15)$. The directional derivative of A in the direction of the vector AB is 3, and the direction derivative at A in the direction of AC is 26. Find the direction derivative of f at A in the direction of the vector AD.

Practice Answers

Practice 1: $f(x,y) = x^2 + 2x + 3y + 1$ and $\mathbf{u} = \left\langle \frac{5}{13}, \frac{12}{13} \right\rangle$

$$D_\mathbf{u} f(2,1) = \lim_{h \to 0} \frac{f\left(2 + \frac{5}{13}h, 2 + \frac{12}{13}h\right) - f(2,1)}{h} = \lim_{h \to 0} \frac{\left[\left(2 + \frac{5}{13}h\right)^2 + 2\left(2 + \frac{5}{13}h\right) + 3\left(1 + \frac{12}{13}h\right) + 1\right] - 12}{h}$$

$$= \lim_{h \to 0} \frac{\left[\left(4 + \frac{20}{13}h + \frac{25}{13}h^2\right) + \left(4 + \frac{10}{13}h\right) + \left(3 + \frac{36}{13}h\right) + 1\right] - 12}{h} = \lim_{h \to 0} \frac{\frac{66}{13}h + \frac{25}{13}h^2}{h} = \frac{66}{13}$$

Practice 2: $f_x(x,y) = 2x + 2$ and $f_y(x,y) = 3$ so $f_x(2, 1) = 6$ and $f_y(2, 1) = 3$. $\mathbf{u} = \left\langle \frac{5}{13}, \frac{12}{13} \right\rangle$ so

$$D_\mathbf{u} f(2, 1) = (6)\left(\frac{5}{13}\right) + (3)\left(\frac{12}{13}\right) = \frac{66}{13}$$, the same result as in Example 2 but much easier.

Practice 3: (a) For $f(x,y) = e^{xy} + 2x^3 y + y^2$, $f_x(x,y) = y \cdot e^{xy} + 6x^2 y$ and $f_y(x,y) = x \cdot e^{xy} + 2x^3 + 2y$ so

$$\nabla f(x,y) = \left\langle y \cdot e^{xy} + 6x^2 y, \; x \cdot e^{xy} + 2x^3 + 2y \right\rangle \text{ and } \nabla f(1,2) = \left\langle 2e^2 + 12, \; e^2 + 6 \right\rangle.$$

(b) For $f(x,y) = \cos(2x+3y)$, $f_x(x,y) = -2\sin(2x+3y)$ and $f_y(x,y) = -3\sin(2x+3y)$ so

$$\nabla f(1, 2) = \left\langle -2\sin(2x+3y), \; -3\sin(2x+3y) \right\rangle \text{ and } \nabla f(1, 2) = \left\langle -2\sin(8), \; -3\sin(8) \right\rangle.$$

Practice 4: $f_x(x,y) = \dfrac{1}{\sqrt{2x+3y}}$ and $f_y(x,y) = \dfrac{3}{2\sqrt{2x+3y}}$ so $f_x(5,2) = \dfrac{1}{4}$ and $f_y(5,2) = \dfrac{3}{8}$.

(a) The maximum value of the rate of change of f is
$D_{\mathbf{u}} f(5,2) = |\nabla f(5,2)| = \left\langle \dfrac{1}{4}, \dfrac{3}{8} \right\rangle = \dfrac{\sqrt{13}}{8} \approx 0.45$.

(b) This maximum value occurs when **u** is in the direction of $\nabla f(5,2)$: $\mathbf{u} = \dfrac{\nabla f(5,2)}{|\nabla f(5,2)|} = \left\langle \dfrac{2}{\sqrt{13}}, \dfrac{3}{\sqrt{13}} \right\rangle$.

Practice 5: See Fig. 9. Note that each gradient vector is perpendicular to the level curve and points uphill.

Practice 6: See Fig. 8. Note that "uphill gradient" path is always perpendicular to the level curves.

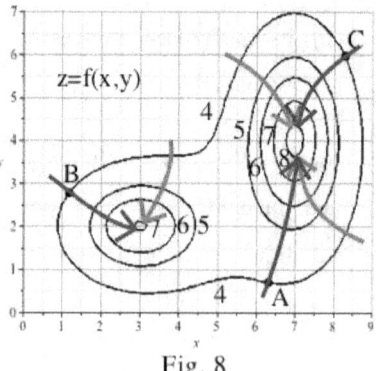

Fig. 8

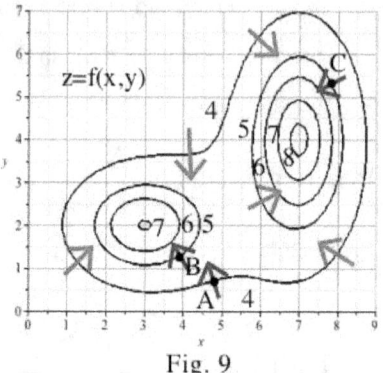

Fig. 9

13.6 MAXIMUMS AND MINIMUMS

One important use of derivatives in beginning calculus was to find maximums and minimums of functions of a single variable. Similarly, an important use of partial derivatives is to find maximums and minimums of functions of two (or more) variables.

We are going to consider three situations, and each situation will require a different method.

Three max/min situations: (a) domain = entire xy-plane,

(b) domain = bounded region of the xy-plane, and

(c) domain = a path in the xy-plane.

Situation (a) might ask for the highest elevation anywhere on earth (= at the summit of Mt. Everest = 29,029 ft), (b) might ask for the highest elevation in the state of Washington (= at the summit of Mt. Rainier = 14,4011 ft), and (c) might ask for the highest elevation we achieved during a hike on Mt. Rainier even if we did not reach the summit.

We consider situations (a) and (b) in this section and situation (c) in the next section.

(a) Domain of our max/min search of f is the ENTIRE xy-plane

Definition: A function of two variables $f(x,y)$ has a **local maximum** at (a,b) if $f(x,y) \leq f(a,b)$ for all points (x,y) in some disk with center (a,b). The value $f(a,b)$ is called a **local maximum value** of f.

The next theorem tells us where we should look for maximums and minimums.

Theorem: If f is differentiable and has a local maximum or minimum at (a,b),

then $f_x(a,b) = 0$ and $f_y(a,b) = 0$.

Note: It is **possible** that $f_x(a,b) = 0$ and $f_y(a,b) = 0$ and that $f(a,b)$ is **not** a local maximum or minimum. (See Example 2 below.)

Proof: The proof is simply a process of elimination. If one (or both) of $f_x(a,b)$ or $f_y(a,b)$ is positive, then moving a small distance Δ in the direction of that variable will increase the value of f so either $f(a+\Delta,b) > f(a,b)$ or $f(a,b+\Delta) > f(a,b)$ and $f(a,b)$ is not a local maximum. A similar argument also shows that $f(a,b)$ can not be a local minimum.

13.6 Maximums and Minimums

Example 1: Find all local maximums and minimums of $f(x,y) = x^2 + y^2 - 2x - 6y + 7$.

Solution: $f_x(x,y) = 2x - 2$ so $f_x(x,y) = 0$ when $x = 1$. $f_y(x,y) = 2y - 6$ so $f_y(x,y) = 0$ when $y = 3$.

The only possible location of a maximum or minimum of $f(x,y)$ is at the point $(1,3)$, but we do not know if we have a maximum or minimum or neither at that point. (The graph of $z = f(x,y)$ in Fig. 1 indicates that $f(1,3)$ is a maximum.)

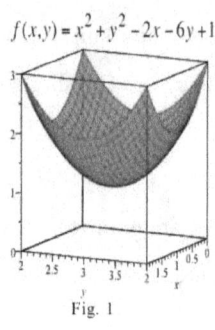
Fig. 1

Example 2: Find all local maximums and minimums of $f(x,y) = x^2 - y^2$.

Solution: $f_x(x,y) = 2x$ so $f_x(x,y) = 0$ when $x = 0$. $f_y(x,y) = -2y$ so $f_y(x,y) = 0$ when $y = 0$.

The only possible location of a maximum or minimum of $f(x,y)$ is at the point $(0,0)$, but $f(0,0) = 0$ is neither a local maximum nor a minimum of f: for any $a \neq 0$, $f(a,0) = a^2 > f(0,0)$ so $f(0,0)$ is not a maximum; for any $b \neq 0$, $f(0,b) = -b^2 < f(0,0)$ so $f(0,0)$ is not a minimum. (The surface $z = x^2 - y^2$ in Fig. 2 is called a "saddle.")

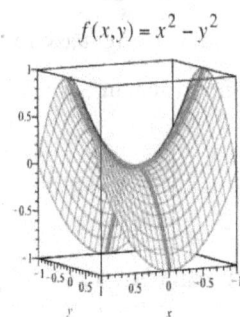 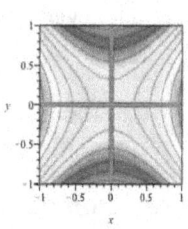
Fig. 2

Example 3: Find all critical points of $f(x,y) = x^2y - 2x^2 - y^3 + 3y + 7$.

Solution: $f_x(x,y) = 2xy - 4x$ and $f_y(x,y) = x^2 - 3y^2 + 3$ so we need to solve the system $\{2xy - 4x = 0$ and $x^2 - 3y^2 + 3 = 0\}$. In order for $0 = 2xy - 4x = 2x(y-2)$ we know that either $x=0$ or $y=2$.

x=0 case: Putting x=0 into f_y we have $f_y(0,y) = -3y^2 + 3 = 0$ so $y^2 = 1$ and $y = \pm 1$. This gives us two critical points: $(0, 1)$ and $(0, -1)$.

y=2 case: Putting y=02 into f_y we have $f_y(x,2) = x^2 - 12 + 3 = 0$ so $x^2 = 9$ and $x = \pm 3$. This gives us two new critical points: $(3, 2)$ and $(-3, 2)$.

This function has 4 critical points, and any local maximums or minimums can only occur at one of those 4 locations. Unfortunately it is not easy to decide whether each critical point gives us a local maximum, a local minimum or a saddle. For that we need a graph or the Second Derivative Test.

Practice 1: Find all critical points of $f(x,y) = 2x^3 + xy^2 + 5x^2 + y^2 + 3$.

In beginning calculus we had a Second Derivative Test to help determine whether a critical point was a local maximum or a local minimum. There is also a Second (Mixed Partial) Derivative Test to help us determine whether a critical point of a function of two variables is a local maximum or minimum or saddle.

Second Derivative Test for Maximums and Minimums

Suppose the second partial derivatives f_{xx}, f_{xy}, f_{yx}, and f_{yy} are continuous in a disk with center (a,b) and $f_x(a,b) = 0$ and $f_y(a,b) = 0$. Let $D = D(a,b) = f_{xx}(a,b)f_{yy}(a,b) - \{ f_{xy}(a,b) \}^2$.

(i) If $D > 0$ and $f_{xx}(a,b) > 0$, then $f(a,b)$ is a local minimum.

(ii) If $D > 0$ and $f_{xx}(a,b) < 0$, then $f(a,b)$ is a local maximum.

(iii) If $D < 0$, then $f(a,b)$ is not a local minimum or a local maximum. (It is a saddle point.)

(iv) If $D = 0$, then the test is "indeterminate": $f(a,b)$ could be a local maximum or a local minimum or neither.

The proof of this theorem is given in an appendix after the Practice solutions.

Example 4: In Example 3 we found 4 critical points of $f(x,y) = x^2 y - 2x^2 - y^3 + 3y + 7$:

(0, 1), (0, -1), (3, 2) and (-3, 2). Use the Second Derivative Test to determine whether each of these gives a local maximum of f, a local minimum of f, or is a saddle point.

Solution: $f_{xx}(x,y) = 2y - 4$, $f_{yy}(x,y) = -6y$, and $f_{xy}(x,y) = 2x$. Many people find it easiest (and safest) to arrange the numerical information in a table. Fig. 3 shows the surface and level curves for this $z = f(x,y)$.

point	(0, 1)	(0, -1)	(3, 2)	(-3, 2)
f_{xx}	-2	-6	0	0
f_{yy}	-6	6	-12	-12
f_{xy}	0	0	6	-6
D	12	-36	-36	-36
result	max	saddle	saddle	saddle

Practice 2: Use the Second Derivative Test to determine whether each critical point of $f(x,y) = 2x^3 + xy^2 + 5x^2 + y^2 + 3$ (from Practice 1) gives a local maximum of f, a local minimum of f, or is a saddle point.

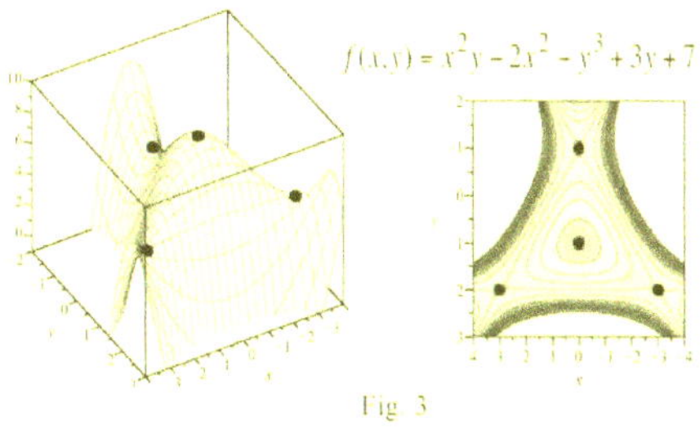

Example 5: Use the ideas of this section to find the shortest distance from the point $(1, 0, -2)$ to the plane $x + 2y + z = 4$.
(Suggestion: Minimize $f(x,y)$ = the square of the distance.)

Fig. 3

Solution: The square of the distance of (x,y,z) to the point $(1,0,-2)$ is $(x-1)^2 + (y-0)^2 + (z+2)^2$, is a function of three variables. However, we know that (x,y,z) is on the plane $x + 2y + z = 4$ so $z = 4 - x - 2y$. Replacing z with $4 - x - 2y$ in the distance (squared) formula, we want to minimize $f(x,y) = (x-1)^2 + (y-0)^2 + (6 - x - 2y)^2$.

$f_x = 2(x-1) + 2(6 - x - 2y)(-1) = 4x + 4y - 14$ and $f_y = 2y + 2(6 - x - 2y)(-2) = 4x + 10y - 24$. We need to find the values of x and y that make f_x and f_y both equal to zero, and the only place that occurs is at the point $(a,b) = (11/6, 5/3)$.

At the point $(a,b) = (11/6, 5/3)$ we have $f_{xx}(a,b) = 4, f_{xy} = 4$, and $f_{yy} = 10$ so $D(a,b) = f_{xx}f_{yy} - (f_{xy})^2 = 24 > 0$ and $f_{xx} > 0$. Then by part (i) of the Second Derivative Test we can conclude that $f(x,y)$ has a local minimum at $(11/6, 5/3)$: the shortest distance is

$$\sqrt{f(11/6, 5/3)} = \sqrt{(5/6)^2 + (5/3)^2 + (5/6)^2} = \sqrt{\frac{5}{6}} = \frac{5\sqrt{6}}{6}.$$

(Note: This problem probably would be easier using the ideas from section 11.6.)

(b) Domain of our max/min of f search is a BOUNDED REGION of the xy-plane

In beginning calculus we sometimes needed to find the maximum or minimum value of a function $f(x)$ for x in an interval [a, b]. In that situation we found critical points in [a, b] where $f'(x) = 0$ or was undefined and then we also needed to check the values of f at the ENDPOINTS when x=a and x=b. When looking for max/mins on a bounded region R we still need to find critical points (x,y) in R, but we also need to consider values f on the BOUNDARY of R.

Method for finding Maximums and Minimums on Bounded Domains:

To find the maximum and minimum values of a differential function on a closed bounded region R:

(1)   Find the values of f at the critical points of f in R.

(2)   Find the extreme values of f on the boundary of R.

(3)   The largest value of f from steps (1) and (2) is the absolute maximum value of f on R; the smallest value of f is the absolute minimum value of f on R.

Note: The Second Derivative Test is not used in this situation.

Example 6: Find the absolute maximum and minimum values of $f(x,y) = x^2 - 2xy + 2y + 3$ on the rectangle $R = \{(x,y) : 0 \leq x \leq 3$ and $0 \leq y \leq 2\}$.

Solution: Step (1): Find the critical points of f in R. These occur when $f_x(x,y) = 2x - 2y = 0$ and $f_y(x,y) = -2x + 2 = 0$, so (solving algebraically) we have $x = 1$ and $y = 1$. The only critical point from step (1) is $(x,y) = (1,1)$ and $f(1,1) = 4$.

Step (2): Find the critical points and extreme values of f on the boundary of R.

The boundary of the rectangle R consists of four line segments L1, L2, L3, and L4 where L1 = segment from (0,0) to (3,0), L2 = segment from (3,0) to (3,2), L3 = segment from (3,2) to (0,2), and L4 = segment from (0,2) to (0,0).

On L1, $0 \le x \le 3$ and $y = 0$ so $f(x,y) = f(x,0) = x^2 + 3$ which has minimum value $f(0,0) = 3$ and maximum value $f(3,0) = 12$.

On L2, $x = 3$ and $0 \le y \le 2$ so $f(x,y) = f(3,y) = 9 - 6y + 2y + 3 = 12 - 4y$ which has minimum value $f(3,2) = 4$ and maximum value $f(3,0) = 12$.

On L3, $0 \le x \le 3$ and $y = 2$ so $f(x,y) = f(x,2) = x^2 - 4x + 4 + 3 = x^2 - 4x + 7$. Using the methods of Chapter 3 ($f' = 2x - 4 = 0$ when $x = 2$, $f'' = 2 > 0$) we know $f(2,2) = 3$ is a minimum and (checking endpoints $x = 0$ and $x = 3$) that $f(0,2) = 7$ is a maximum.

On L4, $x = 0$ and $0 \le y \le 2$ so $f(0,y) = 2y + 3$ which is a linear function with minimum $f(0,0) = 3$ and maximum $f(0,2) = 7$.

On the boundary (L1, L2, L3, and L4) the minimum value is $3 = f(0,0) = f(2,2)$ and the maximum value is $12 = f(3,0)$.

Comparing the minimum and maximum values from step (2) with $f(1,1) = 4$ from step (1) we have that the absolute minimum is $3 = f(0,0) = f(2,2)$ and the absolute maximum value is $12 = f(3,0)$. See Fig. 4.

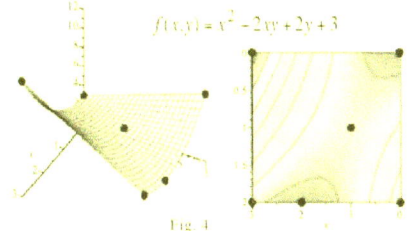

Fig. 4

To find maximums and minimums on a bounded region, we do not use the Second Derivative test, we simply evaluate the function at each critical point and select the largest and smallest values of the function

Practice 3: Find the locations and maximum and minimum values of $f(x,y) = 16 - xy$ on the elliptical domain $2x^2 + y^2 \le 36$.

Example 7: Rewrite the function $f(x,y) = x^2 - 2xy + 2y + 3$ on the boundary lines of the set $D = \{ (x,y) : 0 \le x \le 3$ and $0 \le y \le 2x \}$

Solution: The boundary of D consists of the three line segments L1 = segment from (0,0) to (3,0), L2 = segment from (3,0) to (3,6), and L3 = the segment from (0,0) to (3,6) along the line $y = 2x$.

On L1, $0 \le x \le 3$ and $y = 0$ so $f(x,y) = f(x,0) = x^2 + 3$.

On L2, $x = 3$ and $0 \le y \le 6$ so $f(x,y) = f(3,y) = 9 - 6y + 2y + 3 = 12 - 4y$.

On L3, $0 \le x \le 3$ and $y = 2x$ so $f(x,y) = f(x,2x) = x^2 - 4x^2 + 4x + 3 = -3x^2 + 4x + 3$.

To actually maximize or minimize f on D, we would now need to apply steps (1) and (2).

PROBLEMS

In problems 1 – 20, find the local maximums, minimums, and saddle points of the given function.

1. $f(x,y) = x^2 + y^2 + 4x - 6y$

2. $f(x,y) = 4x^2 + y^2 - 4x + 2y$

3. $f(x,y) = 2x^2 + y^2 + 2xy + 2x + 2y$

4. $f(x,y) = 1 + 2xy - x^2 - y^2$

5. $g(x,y) = xy^2 - x^3 + 3x - 2y^2 + 5$

6. $g(x,y) = x^3 + 2y^2 - xy^2 - 3x + 7$

7. $f(x,y) = x^2 + y^2 + x^2y + 4$

8. $f(x,y) = 2x^3 + xy^2 + 5x^2 + y^2$

9. $f(x,y) = x^3 - 3xy + y^3$

10. $f(x,y) = y\sqrt{x} - y^2 - x + 6y$

11. $f(x,y) = xy - 2x - y$

12. $f(x,y) = xy(1 - x - y)$

13. $f(x,y) = \dfrac{x^2y^2 - 8x + y}{xy}$

14. $f(x,y) = x^2 + y^2 + \dfrac{1}{x^2y^2}$

15. $f(x,y) = e^x \cdot \cos(y)$

16. $f(x,y) = (2x - x^2)(2y - y^2)$

17. $f(x,y) = 3x^2y + y^3 - 3x^2 - 3y^2 + 2$

18. $f(x,y) = xy \cdot e^{(-x^2 - y^2)}$

19. $f(x,y) = x \cdot \sin(y)$

20. $f(x,y) = 2x^3 + y^2 - 6xy + 10$

In problems 21 – 32 find the maximum and minimum values of f on the set R.

21. $f(x,y) = 5 - 3x + 4y$, R is the closed triangular region with vertices $(0,0)$, $(4,0)$, and $(4,5)$.

22. $f(x,y) = x^2 + 2xy + 3y^2$, R is the closed triangular region with vertices $(-1,1)$, $(2,1)$, and $(-1,-2)$.

23. $f(x,y) = x^2 + y^2 + x^2y + 4$, $R = \{ (x,y) \mid |x| \le 1, |y| \le 1 \}$.

24. $f(x,y) = y\sqrt{x} - y^2 - x + 6y$, $R = (x,y) \mid 0 \le x \le 9, 0 \le y \le 5 \}$.

25. $f(x,y) = 1 + xy - x - y$, R is the region bounded by the parabola $y = x^2$ and the line $y = 4$.

26. $f(x,y) = y^2 - 2x^2 + 10$. R is the region bounded by the parabola $y = x^2$ and the line $y = 4$.

27. $f(x,y) = 2x^2 - y^2 + 30$. R is the region bounded by the parabola $y = x^2$ and the line $y = 4$.

28. $f(x,y) = x^2 + 3y + 7$. R is he region shown in Fig. 5 (include the boundary).

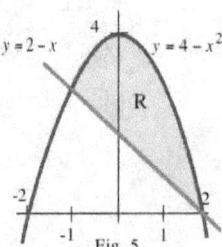

Fig. 5

29. $f(x,y) = x^2 + 3y + 7$. R is the region shown in Fig. 6 (include the boundary).

30. $f(x,y) = 2x^2 + x + y^2 - 2$, $R = \{ (x,y) \mid x^2 + y^2 \leq 4 \}$.

31. $f(x,y) = 2x^3 + y^4$, $R = \{ (x,y) \mid x^2 + y^2 \leq 1 \}$.

32. $f(x,y) = x^3 - 3x - y^3 + 12y$, R is the quadrilateral whose vertices are $(-2,3), (2,3), (2,2)$, and $(-2,2)$.

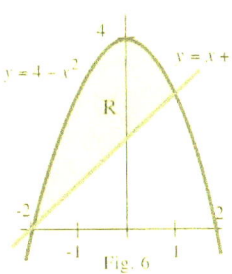
Fig. 6

33. Find the point on the plane $x + 2y + 3z = 4$ that is closest to the origin.

34. Find the point on the plane $2x - y + z = 1$ that is closest to the point $(-4, 1, 3)$.

35. Find three consecutive numbers whose sum is 100 and whose product is a maximum.

36. Find three positive numbers x, y, and z whose sum is 100 such that $x^a y^b z^c$ is a maximum.

37. Find the volume of the largest rectangular box with edges parallel to the axes that can be inscribed in the ellipsoid $9x^2 + 36y^2 + 4z^2 = 36$.

38. Solve the problem in problem 31 for a general ellipsoid $x^2/a^2 + y^2/b^2 + z^2/c^2 = 1$.

39. Find the volume of the largest rectangular box in the first octant with three faces in the coordinate planes and one vertex in the plane $x + 2y + 3z = 6$.

40. Solve the problem in problem 33 for a general plane $x/a + y/b + z/c = 1$ where a, b, and c are positive numbers.

41. Find the dimensions of a rectangular box of maximum volume such that the sum of the lengths of its 12 edges is a constant c.

Practice Solutions

Practice 1: $f(x,y) = 2x^3 + xy^2 + 5x^2 + y^2 + 3$. $f_x(x,y) = 6x^2 + y^2 + 10x$ and $f_y(x,y) = 2xy + 2y$ so we need to solve the system $\{6x^2 + y^2 + 10x = 0$ and $2xy + 2y = 0\}$. The second equation is easier: in order for $0 = 2xy + 2y = 2y(x+1)$ we know that either y=0 or x=-1.

y=0 case: Putting y=0 into f_x we have $f_x(x, 0) = 6x^2 + 10x = 0$ so $2x(3x+5)=0$ and x=0 or x= -5/3. This gives us two critical points: (0, 0) and (-5/3, 0).

x= -1 case: Putting x=-1 into f_x we have $f_x(-1, y) = 6 + y^2 - 10 = 0$ so $y^2 = 4$ and $y = \pm 2$. This

gives us two new critical points: (-1, 2) and (-1, -2).

This function has 4 critical points: (0, 0), (-5/3, 0), (-1, 2) and (-1, -2).

Practice 2: $f(x,y) = 2x^3 + xy^2 + 5x^2 + y^2 + 3$, $f_{xx}(x,y) = 12x + 10$, $f_{yy}(x,y) = 2x + 2$,

$f_{xy}(x,y) = 2y$. The information for the Second Derivative Test is organized in the table.

f has a local minimum at (0,0), saddle points at (-1,2) and (-1,-2), and a local maximum at (-5/3, 0).

point	(0, 0)	(-1, 2)	(-1,-2)	(-5/3,0)
f_{xx}	10	-2	-2	-10
f_{yy}	2	0	0	-4/3
f_{xy}	0	4	-4	0
D	20	-16	-16	40/3
result	min	saddle	saddle	max

Practice 3: $f(x,y) = 16 - xy$, $f_x = -y$, $f_y = -x$, so the only interior point with $f_x = f_y = 0$ is (0,0) and f(0,0)=16.

Boundary: $2x^2 + y^2 \leq 36$ so $y = \pm\sqrt{36 - 2x^2}$ with $-\sqrt{18} \leq x \leq \sqrt{18}$. Substituting this y into f we have $f(x,y) = 16 - x \cdot \sqrt{36 - 2x^2}$ which is a function of the single variable x. Then

$f'(x) = \frac{2x^2}{\sqrt{36 - 2x^2}} - \sqrt{36 - 2x^2}$. Setting f'(x)=0 and solving for x, we get $x = \pm 3$

so our critical points on the boundary are $(3, \sqrt{18})$, $(3, -\sqrt{18})$, $(-3, \sqrt{18})$, $(-3, -\sqrt{18})$, and the endpoints $(\sqrt{18}, 0)$ and $(-\sqrt{18}, 0)$. Evaluating f at each of these critical points we see that the maximum value of f is $16 + 9\sqrt{2} \approx 28.73$ at $(3, -\sqrt{18})$ and $(-3, \sqrt{18})$. The minimum value of f is $16 - 9\sqrt{2} \approx 3.27$ at $(3, \sqrt{18})$ and $(-3, -\sqrt{18})$.

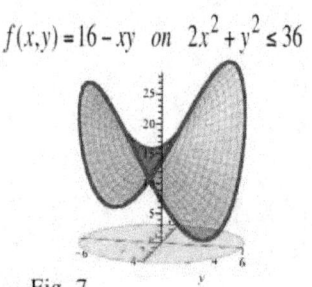

Fig. 7

Appendix: Proof of parts (i) and (ii) of the Second Derivative Test (and part (iv) too)

Suppose we have a critical point (a,b) so $\{f_x(a,b) = 0 \text{ and } f_y(a,b) = 0\}$. The proof involves calculating the second partial derivative in the direction $u = \langle h, k \rangle$ and then determining when that second partial derivative is positive and negative. The directional derivative in the direction $u = \langle h, k \rangle$ is $D_u f = \nabla f \bullet u = f_x h + f_y k$. Then the second derivative in the direction $u = \langle h, k \rangle$ is

$$\begin{aligned}
D_u^2 f = D_u(D_u f) &= \frac{\partial}{\partial x}(D_u f)h + \frac{\partial}{\partial y}(D_u f)k \\
&= \frac{\partial}{\partial x}(f_x h + f_y k)h + \frac{\partial}{\partial y}(f_x h + f_y k)k \\
&= (f_{xx}h + f_{xy}k)h + (f_{xy}h + f_{yy}k)k \\
&= f_{xx}h^2 + f_{xy}kh + f_{xy}hk + f_{yy}k^2 \\
&= f_{xx}h^2 + 2f_{xy}hk + f_{yy}k^2 \\
&= f_{xx}h^2 + 2f_{xy}hk + \frac{f_{xy}^2}{f_{xx}}k^2 + f_{yy}k^2 - \frac{f_{xy}^2}{f_{xx}}k^2 \\
&= f_{xx}\left(h^2 + 2\frac{f_{xy}}{f_{xx}}hk + \frac{f_{xy}^2}{f_{xx}^2}k^2\right) + k^2\left(f_{yy} - \frac{f_{xy}^2}{f_{xx}}\right) \\
&= f_{xx}\left(h + \frac{f_{xy}}{f_{xx}}k\right)^2 + \frac{k^2}{f_{xx}}\left(f_{xx}f_{yy} - f_{xy}^2\right) \\
&= f_{xx}\left\{\left(h + \frac{f_{xy}}{f_{xx}}k\right)^2 + \frac{k^2}{f_{xx}^2}\left(f_{xx}f_{yy} - f_{xy}^2\right)\right\}
\end{aligned}$$

If $D = f_{xx}f_{yy} - f_{xy}^2 > 0$ then the part in the curly brackets is positive, so the sign of $D_u^2 f$ is the same as the sign of f_{xx}:

(i) if $f_{xx} > 0$ then $D_u^2 f > 0$ and f is concave up in every direction u so our critical point gives a local minimum

(ii) if $f_{xx} < 0$ then $D_u^2 f < 0$ and f is concave down in every direction u so our critical point gives a local maximum.

Part (iv): $f(x,y) = x^2y^2$, $g(x,y) = -x^2y^2$, $h(x,y) = x^3y^3$ all have the critical point (0,0) and D = 0 at that critical point. But f has a local minimum at (0,0), g has a local maximum at (0,0) and h has neither a local min or max at (0,0). So D = 0 at a critical point does not tell us whether we have a local max or a local min or neither.

13.7 LAGRANGE MULTIPLIER METHOD

Suppose we go on a walk on a hillside, but we have to stay on a path. Where along this path are we at the highest elevation? That is the basic problem we consider in this section: how to find a maximum or minimum subject to a constraint (staying on a path). Our method, Lagrange Multipliers, is very algebraic, but it also has a geometric interpretation.

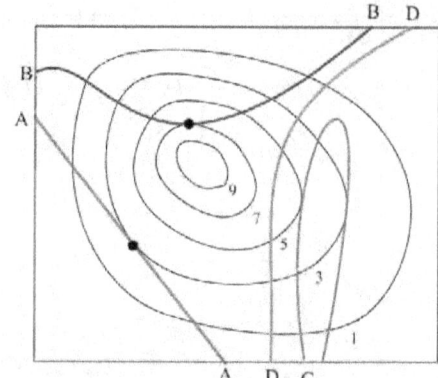

Fig. 1: A hillside and several paths

Example 1: Fig. 1 shows the level curves for a hill and several paths. The dots on path A and B are at the highest elevations along those two paths.

Practice 1: Mark the locations of maximum elevation along paths C and D.

You might have noticed that at each maximum along a path the path was tangent to a level curve. This is the basis of the Lagrange Multiplier method – find the points along a path (constraint) where the path is tangent to the level curve of the function. But finding the tangent to level curves can be difficult so instead we use the fact that if two curves have parallel tangent vectors then they have parallel gradient vectors. And it is easy to calculate the gradient vector for a function.

Lagrange Multiplier Method to Maximize/Minimize $f(x,y)$ along a path:

To find the maximum and minimum values of $f(x,y)$ subject to the constraint (condition) that $g(x,y) = k$

(a) Find all values of x, y, and λ such that $\nabla f(x,y) = \lambda \nabla g(x,y)$ and $g(x,y) = k$.

(b) Evaluate f at the points (x,y) found in step (a). The largest value of $f(x,y)$ at these points is the maximum value of f. The smallest value of $f(x,y)$ is the minimum.

Example 2: Find the maximum and minimum values of $f(x,y) = y^2 - x^2$ on the ellipse $x^2 + 4y^2 = 4$.

Solution: $f(x,y) = y^2 - x^2$ and $g(x,y) = x^2 + 4y^2 = 4$.

Then $\nabla f(x,y) = -2x\mathbf{i} + 2y\mathbf{j}$ and $\nabla g(x,y) = 2x\mathbf{i} + 8y\mathbf{j}$. The Lagrange condition that $\nabla f(x,y) = \lambda \nabla g(x,y)$ means $-2x\mathbf{i} + 2y\mathbf{j} = \lambda(2x\mathbf{i} + 8y\mathbf{j}) = 2\lambda x\mathbf{i} + 8\lambda y\mathbf{j}$ so

$\mathbf{i}$: $-2x = 2\lambda x$

$\mathbf{j}$: $2y = 8\lambda y$ (3 equations in 3 unknowns x, y, and λ)

constraint: $x^2 + 4y^2 = 4$

If $x = 0$, then $y^2 = 1$ so $y = +1$ or $y = -1$ (from constraint) and $\lambda = 1/4$ (from **j** condition).

Then we have the points $(x,y) = (0, 1)$ and $(0, -1)$

If $x \neq 0$, then $-2 = 2\lambda$ (from **i** condition) so $\lambda = -1$. Then $2y = -8y$ (from **j** condition) so $y = 0$ and $x^2 = 4$ so $x = +2$ or $x = -2$. Then we have the points $(x,y) = (2, 0)$ and $(-2, 0)$.

Finally, $f(0, 1) = 1$ $f(0, -1) = 1$ (maximum value of f is 1)

$f(2, 0) = -4$ $f(-2, 0) = -4$ (minimum value of f is –4).

Fig. 2 shows our constraint, the surface and the path along the surface.

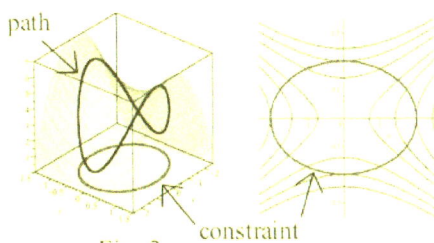
Fig. 2

The Lagrange Multiplier method allows us to trade a calculus problem for the algebra problem of solving a system of equations. But that algebra can be difficult.

Practice 2: Use the Lagrange Multiplier method to find the maximum and minimum values of $f(x,y) = 7x + 3y + 25$ on the circle $x^2 + y^2 = 9$.

Example 3: Find the maximum and minimum values of $f(x,y) = x^2 + y + 2$ on the circle $x^2 + y^2 = 1$.

Solution: $f(x,y) = x^2 + y + 2$ and $g(x,y) = x^2 + y^2 = 1$.

Then $\nabla f(x,y) = 2x\mathbf{i} + 1\mathbf{j}$ and $\nabla g(x,y) = 2x\mathbf{i} + 2y\mathbf{j}$. The Lagrange condition that $\nabla f(x,y) = \lambda \nabla g(x,y)$ means $2x\mathbf{i} + 1\mathbf{j} = \lambda(2x\mathbf{i} + 2y\mathbf{j}) = 2\lambda x\mathbf{i} + 2\lambda y\mathbf{j}$ so

i: $2x = 2\lambda x$

j: $1 = 2\lambda y$ (3 equations in 3 unknowns x, y, and λ)

constraint: $x^2 + y^2 = 1$

If $x = 0$, then $y^2 = 1$ so $y = +1$ or $y = -1$ (from the constraint $x^2 + y^2 = 1$).

If $y = +1$, then $\lambda = 1/2$ (from **j** condition) so one solution is $x = 0, y = 1$, and $\lambda = 1/2$.

If $y = -1$, then $\lambda = -1/2$ (from **j** condition) so one solution is $x = 0, y = -1$, and $\lambda = -1/2$.

Then we have the points $(x,y) = (0, 1)$ and $(0, -1)$

If $x \neq 0$, then $2x = 2\lambda x$ (from **i** condition) so $\lambda = 1$. Then $1 = 2y$ (from **j** condition) so $y = 1/2$ and $x^2 + (1/2)^2 = 1$ so $x = +\sqrt{3}/2$ or $x = -\sqrt{3}/2$. Then we have the points $(x,y) = (+\sqrt{3}/2, 1/2)$ and $(-\sqrt{3}/2, 1/2)$.

Finally, $f(0, 1) = 3$

$f(0, -1) = 1$ (minimum of f is 1)

$f(+\sqrt{3}/2, 1/2) = f(-\sqrt{3}/2, 1/2) = 13/4$

(maximum value of f is 13/4).

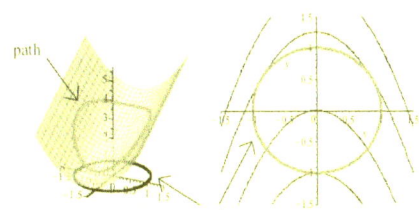
Fig. 3

Fig. 3 shows the constraint, the surface and the path.

The same idea also works for functions and constraints with three (or more) variables:

Find all values of x, y, z and λ such that $\nabla f(x,y,z) = \lambda \nabla g(x,y,z)$ and $g(x,y,z) = k$, but now we need to solve four equations in four unknowns.

Example 4: Find the volume of the largest rectangular box with a divider but no top (see Fig. 4) that can be constructed from 288 square inches of material.

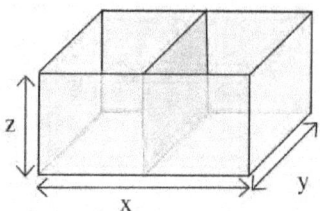

Fig. 4: Rectangular box with one divider and no top

Solution: $V(x,y,z) = xyz$ and $g(x,y,z) = xy + 2xz + 3yz = 288$.

Then $\nabla V(x,y,z) = yz\mathbf{i} + xz\mathbf{j} + xy\mathbf{k}$ and

$\nabla g(x,y,z) = (y + 2z)\mathbf{i} + (x + 3z)\mathbf{j} + (2x + 3y)\mathbf{k}$.

The Lagrange condition that $\nabla f(x,y) = \lambda \nabla g(x,y)$ means

$yz\mathbf{i} + xz\mathbf{j} + xy\mathbf{k} = \lambda(y + 2z)\mathbf{i} + \lambda(x + 3z)\mathbf{j} + \lambda(2x + 3y)\mathbf{k}$ so

i: $\quad yz = \lambda y + \lambda 2z$

j: $\quad xz = \lambda x + \lambda 3z$ (4 equations in 4 unknowns x, y, z and λ)

k: $\quad xy = \lambda 2x + \lambda 3y$

constraint: $xy + 2xz + 3yz = 288$

There are a variety of ways to solve this system, but this algebraic way is relatively easy.

Multiply the first equation by x, the second by y, and the third by z to get

i: $\quad xyz = \lambda xy + \lambda 2xz$

j: $\quad xyz = \lambda xy + \lambda 3yz$

k: $\quad xyz = \lambda 2xz + \lambda 3yz$

Then all 3 equations are equal to xyz so $\lambda xy + \lambda 2xz = \lambda xy + \lambda 3yz = \lambda 2xz + \lambda 3yz$.

Since $\lambda xy + \lambda 2xz = \lambda xy + \lambda 3yz$ then $y = \frac{2}{3}x$. Since $\lambda xy + \lambda 2xz = \lambda 2xz + \lambda 3yz$ then $z = \frac{1}{3}x$.

Putting those values for y and z into the constraint we get

$288 = x\left(\frac{2}{3}x\right) + 2x\left(\frac{1}{3}x\right) + 3\left(\frac{2}{3}x\right)\left(\frac{1}{3}x\right) = 2x^2$ so $x = 12$ inches, $y = 8$ inches and $z = 4$ inches.

The maximum volume is $V(12, 8, 4) = (12)(8)(4) = 384$ cubic inches.

Practice 3: Find the dimensions of the largest volume rectangular box with two dividers but no top (see Fig. 5) that can be constructed from 384 square centimeters of material.

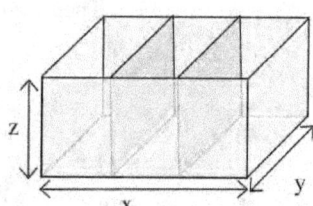

Fig. 5: Rectangular box with two dividers and no top

PROBLEMS

1. In Fig. 6 locate the maximum and minimum values of z along each path and estimate their values.

2. In Fig. 7 locate the maximum and minimum values of z along each path and estimate their values.

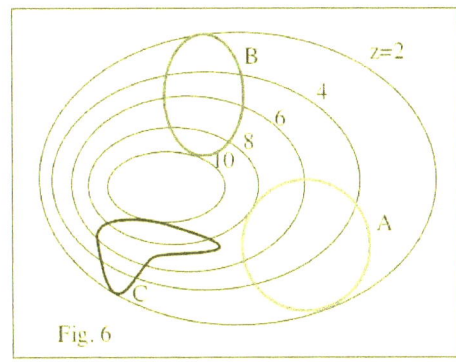
Fig. 6

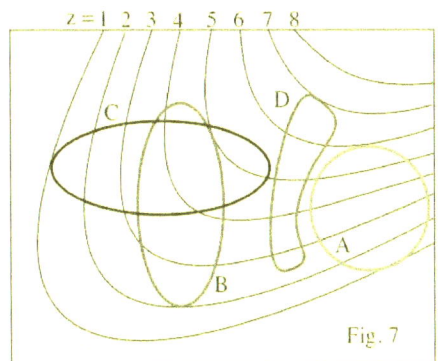
Fig. 7

In problems 3 – 11, use the Lagrange Multiplier method to find the maximum and minimum values of the given function subject to the given constraint.

3. $f(x,y) = x^2 - y^2$; $x^2 + y^2 = 1$

4. $f(x,y) = 2x + y$; $x^2 + 4y^2 = 1$

5. $f(x,y) = xy$; $9x^2 + y^2 = 4$

6. $f(x,y) = x^2 + y^2$; $x^4 + y^4 = 1$

7. $f(x,y,z) = x + 3y + 5z$; $x^2 + y^2 + z^2 = 1$

8. $f(x,y,z) = x - y + 3z$; $x^2 + y^2 + 4z^2 = 4$

9. $f(x,y,z) = xyz$; $x^2 + 2y^2 + 3z^2 = 6$

10. $f(x,y,z) = x^2 y^2 z^2$; $x^2 + y^2 + z^2 = 1$

11. $f(x,y,z) = x^2 + y^2 + z^2$; $x^4 + y^4 + z^4 = 1$

12. Find the maximum volume of a rectangular box with no top that has a surface area of 48 square inches.

13. Find the maximum volume of a rectangular box with no top that has a surface area of A square inches.

14. Using your result from problem 11*, show that the area of the bottom is A/3, the total area of the front and back sides is A/3, and the total area of the two end sides is A/3.

15. Find the maximum volume of a rectangular box with no top that be built at a cost of $15.00 if the bottom material costs $0.50/in^2 and the materials for the sides costs $0.01/in^2.

16. Find the maximum volume of a rectangular box with no top that be built at a cost of $T if the bottom material costs $B/in^2 and the materials for the 4 sides costs $S/in^2.

17. Find the maximum volume of a cylinder with no top that has a surface area of 48 square inches.

18. Find the maximum volume of a cylinder with a top that has a surface area of 48 square inches.

Practice Answers

Practice 1: The dots in Fig. P1 show the locations of the maximum elevations along paths C and D. The little square on path C is a local maximum along that path. The figure also includes part of the level curve that goes through the dot on path D.

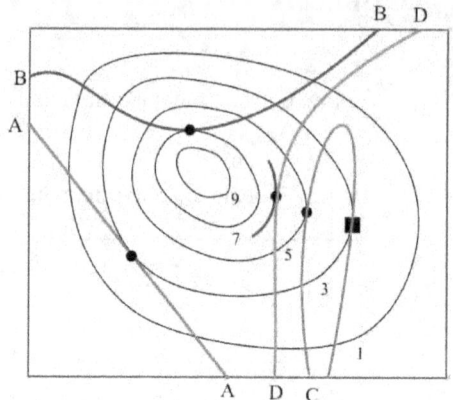

Fig. P1: A hillside and several paths

Practice 2: $f(x,y) = 7x + 3y + 25$ and $g(x,y) = x^2 + y^2$ so $\nabla g = \langle 2x, 2y \rangle$ and $\nabla f = \langle 7, 3 \rangle$. Putting these into the Lagrange equation $\nabla f = \lambda \cdot \nabla g$ we have the algebraic system

i: $\quad 7 = 2\lambda x \quad$ (so $x \neq 0$)

j: $\quad 3 = 2\lambda y \quad$ (so $y \neq 0$)

constraint: $\quad x^2 + y^2 = 9$

Then (from **i**) $x = \dfrac{7}{2\lambda}$ and (from **j**) $y = \dfrac{3}{2\lambda}$. Putting these into the constraint $\left(\dfrac{7}{2\lambda}\right)^2 + \left(\dfrac{3}{2\lambda}\right)^2 = 9$

Then $49 + 9 = 36\lambda^2$ and $\lambda = \pm \dfrac{\sqrt{58}}{6} \approx \pm 1.269$. Putting these back into x and y equations, for $\lambda = \dfrac{\sqrt{58}}{6}$ we have $x = \dfrac{21}{\sqrt{58}} \approx 2.757$, $y = \dfrac{9}{\sqrt{58}} \approx 1.182$ and $f\left(\dfrac{21}{\sqrt{58}}, \dfrac{9}{\sqrt{58}}\right) = \dfrac{169}{\sqrt{58}} + 25 \approx 57.19$ the maximum value of f on the elliptical path. For $\lambda = -\dfrac{\sqrt{58}}{6}$, we have $x = \dfrac{-21}{\sqrt{58}}$, $y = \dfrac{-9}{\sqrt{58}}$ and $f\left(\dfrac{-21}{\sqrt{58}}, \dfrac{-9}{\sqrt{58}}\right) = \dfrac{-169}{\sqrt{58}} + 25 \approx 2.81$.

Figure P2 shows the surface f, the contours for f, and the constraint g.

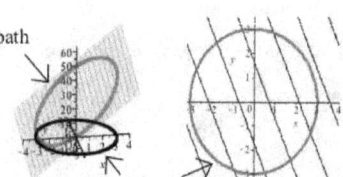

Fig. P2

Practice 3: $V(x,y,z) = xyz$ and $g(x,y,z) = xy + 2xz + 4yz = 384$ cm².

Then $\nabla V(x,y,z) = yz\mathbf{i} + xz\mathbf{j} + xy\mathbf{k}$ and

$\nabla g(x,y,z) = (y + 2z)\mathbf{i} + (x + 4z)\mathbf{j} + (2x + 4y)\mathbf{k}$ so

$yz\mathbf{i} + xz\mathbf{j} + xy\mathbf{k} = \lambda(y + 2z)\mathbf{i} + \lambda(x + 4z)\mathbf{j} + \lambda(2x + 4y)\mathbf{k}$ so

i: $\quad yz = \lambda y + \lambda 2z$

j: $\quad xz = \lambda x + \lambda 4z \quad$ (4 equations in 4 unknowns x, y, z and λ)

k: $\quad xy = \lambda 2x + \lambda 4y$

constraint: $xy + 2xz + 4yz = 384$

Multiply the first equation by x, the second by y, and the third by z,

Then all 3 equations are equal to xyz so $\lambda xy + \lambda 2xz = \lambda xy + \lambda 4yz = \lambda 2xz + \lambda 4yz$. Then $384 = x\left(\dfrac{1}{2}x\right) + 2x\left(\dfrac{1}{4}x\right) + 4\left(\dfrac{1}{2}x\right)\left(\dfrac{1}{4}x\right) = \dfrac{3}{2}x^2$ so x=16 cm, y=8 cm and z=2 cm

and he maximum volume is 256 cm³.

13.8 The Chain Rule for Functions of Several Variables

In Section 2.4 we saw the Chain Rule for a function of one variable.

Chain Rule (Leibniz notation form)

If   y is a differentiable function of u, and u is a differentiable function of x,

then   y is a differentiable function of x and $\dfrac{dy}{dx} = \dfrac{dy}{du} \cdot \dfrac{du}{dx}$.

One interpretation of this Chain Rule is that if x is a signal that is amplified by a factor of 3 by u (du/dx=3) and the signal u gets amplified by a factor of 2 by y (dy/du=3) then the total amplification of x by the combination of u followed by y is by a factor of 6:

$$\frac{dy}{dx} = \frac{dy}{du} \cdot \frac{du}{dx} = (2)(3) = 6.$$

We can also represent this pattern graphically as in Fig. 1.

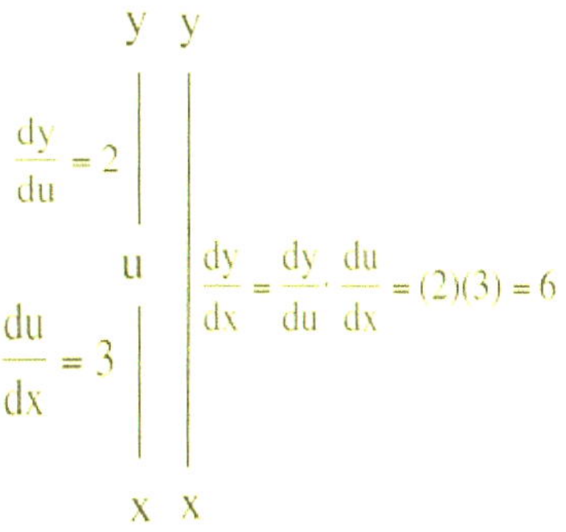

Fig. 1

f is a function of x and y, and each of x and y is a function of t

But suppose that the original signal at t is 1 db (decibel) and that there are two intermediate amplifiers x and y that feed into our final amplifier z as in Fig. 2. If x amplifies t by a factor of 3 (dx/dt=3), z amplifies x by a factor of 2 (dz/dx=2), y amplifies t by a factor of 4 (dy/dt=4), and z amplifies y by a factor of 5 (dz/dy=5), then we can ask what is the total amplification of the original signal 1 db signal at t to the final output z. The original 1 db signal at t becomes 6 db (along the txz path) and 20 db (along the tyz path) for a total output of 26 db.

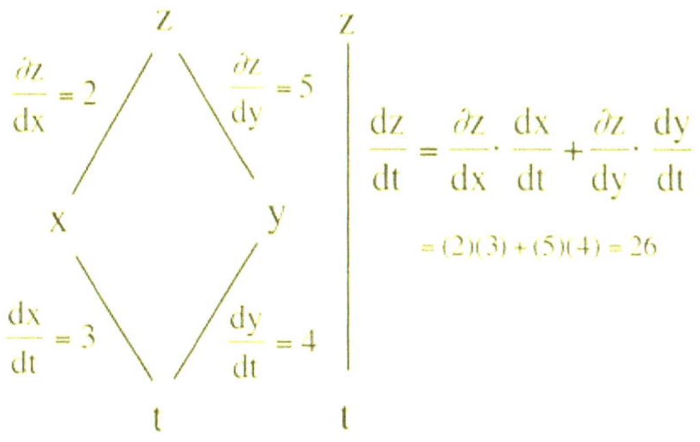

Fig. 2

This is essentially the Chain Rule for a function of two variables: we multiply the rates of change along each path and then add the results to get the total rate of change.

The Chain Rule for a Function of Two Dependent Variables

If $z = f(x, y)$ is a differentiable function of x and y, and x(t) and y(t) are differentiable functions of t, then z is a differentiable function of t, and

$$\frac{df}{dt} = \frac{dz}{dt} = \frac{\partial z}{\partial x}\frac{dx}{dt} + \frac{\partial z}{\partial y}\frac{dy}{dt} .$$

A tree diagram (Fig. 3) is a visual way to organize information that may help you remember "which derivatives go where" in the Chain Rule formula. Write z, the name of the first function at the top of the diagram. Draw branches to x and y, the names of the independent variables of z. Then draw more branches to t, the independent variable of x and y. Then add in the derivative notations as shown in the diagram below.

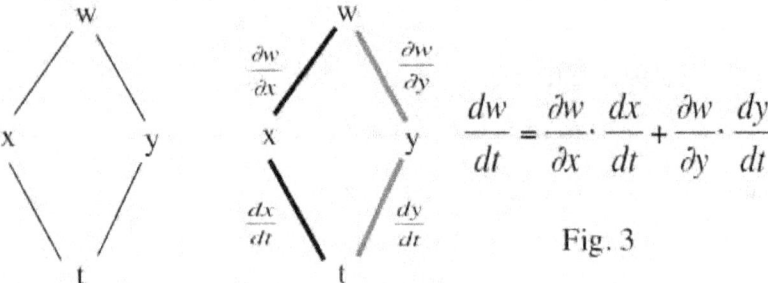

$$\frac{dw}{dt} = \frac{\partial w}{\partial x} \cdot \frac{dx}{dt} + \frac{\partial w}{\partial y} \cdot \frac{dy}{dt}$$

Fig. 3

The two paths from z to t indicate the Chain Rule formula will have two terms. The number of pieces in a path from z to t indicates the number of factors in the corresponding term. The two paths from z to t and two pieces along each path correspond to the two terms of two factors in the chain rule formula.

Example 1: (a) Use the Chain Rule to find the rate of change of $f(x,y) = xy$ with respect to t along the path $x = \cos t$, $y = \sin t$ when $t = \frac{\pi}{3}$.

(b) If the units of t are seconds, the units of x and y are meters and the units of f are °C, then what are the units of $\frac{df}{dt}$?

Solution: (a) At $t = \frac{\pi}{3}$ we have $\frac{\partial f}{\partial x} = y(t) = \sin(\pi/3) = \frac{\sqrt{3}}{2}$, $\frac{\partial f}{\partial y} = x(t) = \cos(\pi/3) = \frac{1}{2}$,

$\frac{dx}{dt} = -\sin(t) = -\sin(\pi/3) = -\frac{\sqrt{3}}{2}$ and $\frac{dy}{dt} = \cos(t) = \cos(\pi/3) = \frac{1}{2}$ so

$$\frac{df}{dt} = \frac{\partial f}{\partial x} \cdot \frac{dx}{dt} + \frac{\partial f}{\partial y} \cdot \frac{dy}{dt} = \left(\frac{\sqrt{3}}{2}\right)\left(-\frac{\sqrt{3}}{2}\right) + \left(\frac{1}{2}\right)\left(\frac{1}{2}\right) = -\frac{1}{2}.$$

(b) Clearly the units of $\frac{df}{dt} = \frac{\text{units of f}}{\text{units of t}} = \frac{°C}{\sec}$. Following the pieces of the Chain Rule

we have $\frac{df}{dt} = \frac{\partial f}{\partial x} \cdot \frac{dx}{dt} + \frac{\partial f}{\partial y} \cdot \frac{dy}{dt} = \left(\frac{°C}{m}\right)\left(\frac{m}{s}\right) + \left(\frac{°C}{m}\right)\left(\frac{m}{s}\right) = \frac{°C}{s}$ the same result.

Note: In this example we could have simply replaced x and y with the appropriate functions of t so $f(t) = \cos(t) \cdot \sin(t)$ and then differentiated, but such a replacement is not always easy.

Practice 1: Use the Chain Rule to calculate the value of $\frac{df}{dt}$ when t = 2 for the functions

$$f(x,y) = x^4 y^3 + 3x^2 y, \ x(t) = t^2 + t - 5, \ y(t) = t^3 - 2t^2 - t + 4.$$

The Chain Rule for Functions of Three Dependent Variables adds one term to our previous pattern.

> If $w = f(x, y, z)$ is a differentiable function of x, y and z, and x, y, and z are differentiable functions of t, then w is a differentiable function of t, and
>
> $$\frac{df}{dt} = \frac{dw}{dt} = \frac{\partial w}{\partial x}\frac{dx}{dt} + \frac{\partial w}{\partial y}\frac{dy}{dt} + \frac{\partial w}{\partial z}\frac{dz}{dt}.$$

Many students find that using a tree diagram (Fig/ 4) for this situation makes it easy to keep track of the pattern: multiply the derivatives along each of the three paths and then add those results together.

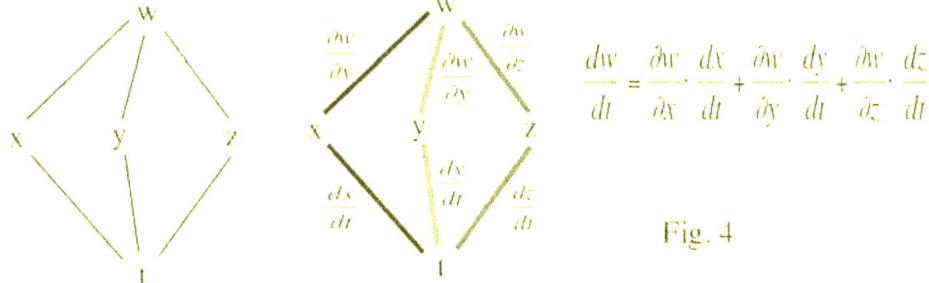

Fig. 4

Example 2: Use the Chain Rule to find the value of $\frac{df}{dt}$ for $f(x, y, z) = xy + z$ along the helix

$x(t) = \cos(t), y(t) = \sin(t), z(t) = t$. What is the derivative's value at $t = 0$?

(This is the instantaneous rate of change as our point moves along a helix.)

Solution: If $t = 0$ then $x=1, y=0, z=0$, $\frac{dx}{dt} = -\sin(t) = 0$, $\frac{dy}{dt} = \cos(t) = 1$, and $\frac{dz}{dt} = 1$. At $(1, 0, 0)$

$\frac{\partial f}{\partial x} = y = 0$, $\frac{\partial f}{\partial y} = x = 1$ and $\frac{\partial f}{\partial z} = 1$. Finally,

$\frac{df}{dt} = \frac{\partial f}{\partial x} \cdot \frac{dx}{dt} + \frac{\partial f}{\partial y} \cdot \frac{dy}{dt} + \frac{\partial f}{\partial z} \cdot \frac{dz}{dt} = (0)(0) + (1)(1) + (1)(1) = 2.$

Practice 2: Use the Chain Rule to find the value of $\frac{df}{dt}$ for the Example 2 functions when $t = \pi/3$?

In General

There are many ways in which functions of several variables can be combined. Rather than stating or memorizing a Chain Rule pattern for each new situation, just keep in mind the general pattern:

> **Build a tree dependency diagram,**
> **multiply along each path and add these path results.**

Example 3: Suppose $W = f(x,y,z)$, $x = x(r,s)$, $y = y(r,s)$, and $z = z(r)$ and that all of these functions are differentiable,

(a) Build a tree dependency diagram for these functions,

(b) Create a Chain Rule for $\dfrac{\partial w}{\partial r}$ and $\dfrac{\partial w}{\partial s}$.

Solution:

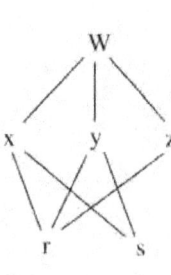

dependency tree

Fig. 5

Three paths from r to W

$$\frac{\partial W}{\partial r} = \frac{\partial W}{\partial x}\cdot\frac{\partial x}{\partial r} + \frac{\partial W}{\partial y}\cdot\frac{\partial y}{\partial r} + \frac{\partial W}{\partial z}\cdot\frac{\partial z}{\partial r}$$

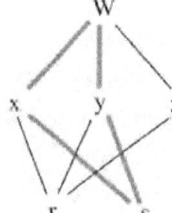

Two paths from s to W

$$\frac{\partial W}{\partial s} = \frac{\partial W}{\partial x}\cdot\frac{\partial x}{\partial s} + \frac{\partial W}{\partial y}\cdot\frac{\partial y}{\partial s}$$

Practice 3: Suppose $T = f(x, y, z)$, $x = x(p, q, r)$, $y = y(p, q)$ and $z = z(p, r)$ and that all of these functions are differentiable.

(a) Build a dependency tree diagram for these functions.

(b) Create Chain Rules for $\dfrac{\partial T}{\partial p}$ and $\dfrac{\partial T}{\partial q}$.

Example 4: (a) The voltage V in a circuit satisfies the law $V = IR$. Write the Chain Rule for $\dfrac{dV}{dt}$.

(b) If the voltage is dropping because the battery is wearing out and the resistance s increasing because the circuit is heating up, then how fast is the current I changing when $R = 500$ ohms, $I = 0.04$ amps, $dR/dt = 0.5$ ohms/sec, and $dV/dt = -0.01$ volt/sec?

Solution: (a) $\dfrac{dV}{dt} = \dfrac{\partial V}{\partial I}\cdot\dfrac{\partial I}{\partial t} + \dfrac{\partial V}{\partial R}\cdot\dfrac{\partial R}{\partial t}$

(b) $\dfrac{\partial V}{\partial I} = \dfrac{\partial(IR)}{\partial I} = R = 500$ *ohms* and $\dfrac{\partial V}{\partial R} = \dfrac{\partial(IR)}{\partial R} = I = 0.04$ *amps*.

Putting all of this information into the equation in part (a) we have

$$\left(-0.01\frac{amp\cdot ohms}{sec}\right) = (500\ ohms)\cdot\left(\frac{\partial I}{\partial t}\frac{amps}{sec}\right) + (0.04\ amps)\cdot\left(0.5\frac{ohms}{sec}\right)$$

so $\dfrac{\partial I}{\partial t} = -0.00006$ amps/sec.

PROBLEMS

In problems 1 and 2, use the information in Table 1.

1. Calculate $\dfrac{df}{dt}$ when t = 1 and t = 3

2. Calculate $\dfrac{df}{dt}$ when t = 2 and t = 4

$\dfrac{\partial f}{\partial x} = $

x\y	0	1	2	3
0	5	4	6	2
1	1	9	7	10
2	3	5	4	11
3	7	8	5	13

$\dfrac{\partial f}{\partial y} = $

x\y	0	1	2	3
0	3	7	2	4
1	6	1	8	3
2	1	4	5	6
3	5	2	9	7

t	1	2	3	4
x	2	0	1	3
y	3	1	0	2

t	1	2	3	4
$\dfrac{dx}{dt}$	-1	5	-2	6
$\dfrac{dy}{dt}$	-3	7	-1	8

Table 1

In exercises 3–7, express $\dfrac{df}{dt}$ as a function of t by using the Chain Rule. Then evaluate $\dfrac{df}{dt}$ at the given value of t.

3. $f(x,y) = x^2 + y^2$, $x = \cos t$, $y = \sin t$, $t = \pi$

4. $f(x,y) = x^2 y^2 + 3x + 4y + 1$, $x = 3 + t^2$, $y = 1 + 2t$, $t = 2$

5. $f(x,y,z) = x^2 y + yz + xz$, $x = 3 + 2t$, $y = t^2$, $z = 5t$, $t = 2$

6. $f(x,y,z) = \dfrac{x}{y} + \dfrac{y}{z} + \dfrac{z}{x}$, $x = 1 + 2t$, $y = 2 + 3t$, $z = 3 + 4t$, $t = 1$

7. $f(x,y,z) = 2y e^x - \ln z$, $x = \ln(t^2 + 1)$, $y = \tan^{-1} t$, $z = e^t$, $t = 1$

8. $f(x,y,z) = xyz$, $x = 2\cos(t)$, $y = \sin(t)$, $z = 3t$, $t = \pi$

9. $w = xy + yz + xz$, $x = u + v$, $y = u - v$, $z = uv$, $(u, v) = (-2, 0)$

   Express $\dfrac{\partial w}{\partial u}$ and $\dfrac{\partial w}{\partial v}$ as functions of u and v by using the Chain Rule.
   Then evaluate $\dfrac{\partial w}{\partial u}$ and $\dfrac{\partial w}{\partial v}$ at the point $(u, v) = (-2, 0)$.

10. Find $\dfrac{\partial w}{\partial r}$ when $r = 1, s = -1$, if $w = (x + y + z)^2$, $x = r - s$, $y = \cos(r + s)$, $z = \sin(r + s)$.

11. Find $\dfrac{\partial z}{\partial u}$ when $u = 0, v = 0$, if $z = \cos(xy) + x \cdot \sin(y)$, $x = u + v + 2$, $y = uv$.

12. The lengths a, b, and c of the edges of a rectangular box are changing with time. At the instant in question, a = 1 meter, b = 2 meters, c = 3 meters, $\dfrac{da}{dt} = \dfrac{db}{dt} = 1$ m/sec, and $\dfrac{dc}{dt} = -3$ m/sec. At what rates are the box's volume V and surface area S changing at that instant? Are the box's interior diagonals increasing or decreasing in length.?

13. In an ideal gas the pressure P (in kilopascals kPa), the volume V (in liters L), and the temperature T (in kelvin K) satisfy the equation $PV = 8.31T$. How fast is the pressure changing, dP/dt, when the temperature is 310 K and is decreasing at a rate of 0.2 K/s, and the volume is 80 L and is increasing at a rate of 0.1 L/s?

14. Given: w is a function of x, y, and z; x is function of r; y is a function of r and s; z is a function of s and t; and s is a function of t. Make a tree diagram, and then write a chain rule formula for $\dfrac{\partial w}{\partial t}$.

Practice Answers

Practice 1: $f(x,y) = x^4 y^3 + 3x^2 y$, $x(t) = t^2 + t - 5$, $y(t) = t^3 - 2t^2 - t + 4$.

If $t = 2$, then $x = 1$, $y = 2$, $\dfrac{dx}{dt} = 2t+1 = 5$ and $\dfrac{dy}{dt} = 3t^2 - 4t - 1 = 3$. At $x=1, y=2$

$\dfrac{\partial f}{\partial x} = 4x^3 y^3 + 6xy = 46$ and $\dfrac{\partial f}{\partial y} = 3x^4 y^2 + 3x^2 = 15$ so $\dfrac{df}{dt} = \dfrac{\partial f}{\partial x} \cdot \dfrac{dx}{dt} + \dfrac{\partial f}{\partial y} \cdot \dfrac{dy}{dt} = (46)(5) + (15)(3) = 275$.

Each Chain Rule problem has a lot of pieces so you need to be organized.

Note: Substituting x(t) and y(t) into f gives

$f(t) = (t^2 + t - 5)^4 (t^3 - 2t^2 - t + 4)^3 + 3(t^2 + t - 5)^2 (t^3 - 2t^2 - t + 4)$ and that would be a messy derivative.

Practice 2: $f(x,y,z) = xy + z$, $x(t) = \cos(t)$, $y(t) = \sin(t)$, $z(t) = t$.

If $t = \pi/3$ then $x=\cos(\pi/3) = \dfrac{1}{2}$, $y=\sin(\pi/3) = \dfrac{\sqrt{3}}{2}$, $z= \dfrac{\pi}{3}$, $\dfrac{dx}{dt} = -\sin(t) = -\sin(\pi/3) = -\dfrac{\sqrt{3}}{2}$,

$\dfrac{dy}{dt} = \cos(t) = \cos(\pi/3) = \dfrac{1}{2}$, and $\dfrac{dz}{dt} = 1$. At $\left(\dfrac{1}{2}, \dfrac{\sqrt{3}}{2}, \dfrac{\pi}{3}\right)$, $\dfrac{\partial f}{\partial x} = y = \dfrac{\sqrt{3}}{2}$, $\dfrac{\partial f}{\partial y} = x = \dfrac{1}{2}$, and $\dfrac{\partial f}{\partial z} = 1$.

Putting this all together $\dfrac{df}{dt} = \dfrac{\partial f}{\partial x} \cdot \dfrac{dx}{dt} + \dfrac{\partial f}{\partial y} \cdot \dfrac{dy}{dt} + \dfrac{\partial f}{\partial z} \cdot \dfrac{dz}{dt} = \left(\dfrac{\sqrt{3}}{2}\right)\left(-\dfrac{\sqrt{3}}{2}\right) + \left(\dfrac{1}{2}\right)\left(\dfrac{1}{2}\right) + (1)(1) = \dfrac{1}{2}$.

Practice 3:

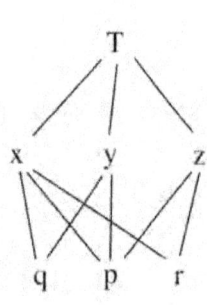

dependency tree

(there is no line from y to r or from z to q)

Fig. 6

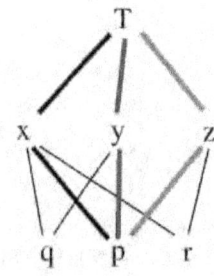

Three paths from p to T

$\dfrac{\partial T}{\partial p} = \dfrac{\partial T}{\partial x} \cdot \dfrac{\partial x}{\partial p} + \dfrac{\partial T}{\partial y} \cdot \dfrac{\partial y}{\partial p} + \dfrac{\partial T}{\partial z} \cdot \dfrac{\partial z}{\partial p}$

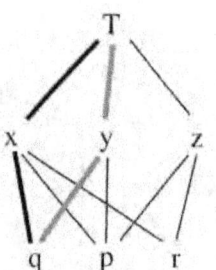

Two paths from q to T

$\dfrac{\partial T}{\partial q} = \dfrac{\partial T}{\partial x} \cdot \dfrac{\partial x}{\partial q} + \dfrac{\partial T}{\partial y} \cdot \dfrac{\partial y}{\partial q}$

Chapter 13: Selected Answers

13.1 Selected Answers

1. (a) minimum = 2, maximum = 30 (b) down 2, down 2 more, up 9 and up 5 more
   (c) up 5, up 5, up 5 up 5, then no change

3. (a) maximum = 10 m (b) depth increases by 3, decreases by 4, increase 1, decreases 1
   (c) depth decreases 2, decreases 1, decreases 1

5.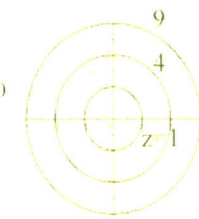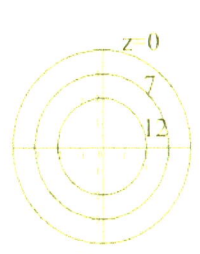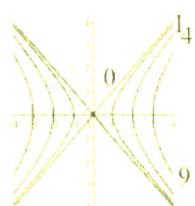

7 to 15. See figures below

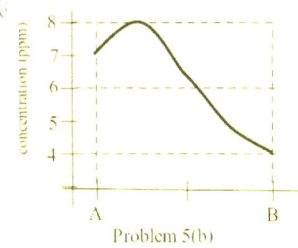 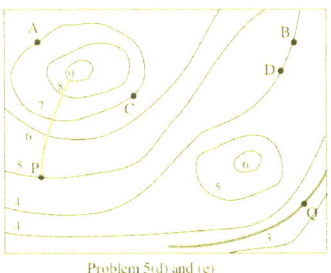

17. level curves D   18. level curves C
19. level curves A   20. level curves G
21. level curves B   22. level curves F
23. level curves E

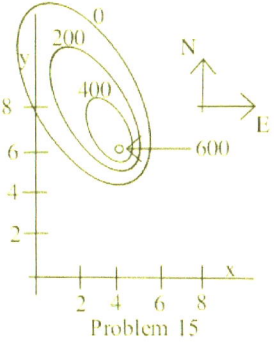

13.2 Selected Answers

1. (a) 3 (b) 2 (c) 4 (d) about 2.4   3. (a) 3 (b) 2 (c) dne (d) dne

5. -927 7. $-5/2$ 9. Π

11. 1

13. dne Paths y=0 and y=x give different limit values.

15. dne Paths y=0 and y=x give different limit values.

17. dne Paths y=0 and y=x give different limit values.

19. dne Paths y=0 and y=x give different limit values.

21. 1

23. dne Paths y=0 and y= x^2 give different limit values.

25. 2 Try rationalizing the denominator.

27. 0 The function equals $\dfrac{x(y-1)}{(x-1)^2+(y-1)^2}$ which is easier to analyze.

29. –3/5

31. dne If x=y=0 and z->0 then limit = –1. If y=z=0 and x->0 then limit = 1.

33. dne If y=z=0 and x->0 then limit = 0. If z=0 and yx=x ->0 then limit = ½..

35. Define f(2,1) = 4

37. No value of f(1,2) or f(3,2) will make f continuous at those points.
    f(1,2) = 3 makes f continuous at (1,2).

39. Not continuous (not defined) on the circle where $x^2 + y^2 = 1$.

41. Not continuous (not defined) if 2x+3y ≤ 0 or y ≤ - 2x/3 .

43. In order for T to be continuous at (x,y) we need both x+y≥0 (so y≥ -x) and x-y≥0 (so x≥y). That requires x≥0 and -x ≤ y ≤ x.

45. F is continuous at (x,y,z) if yz >0.

13.3   Selected Answers

1. $f_x(1,2) = -8$, $f_y(1,2) = -4$ 　　　　　　2. $f_x(1,0) = -1/\sqrt{3}$, $f_y(1,0) = 0$

3. $f_x(3,-1) = -27$ 　　　　　　　　　　　　4. 2

5. $\dfrac{\partial z}{\partial x} = (x^4 + 3x^2y^2 - 2xy^3)/(x^2+y^2)^2$, $\dfrac{\partial z}{\partial y} = (y^4 + 3x^2y^2 - 2x^3y)/(x^2+y^2)^2$

6. $(1/y) - (y/x^2)$ 　　　　7. $(y-z)/(x-y)$, $(x+z)/(x-y)$ 　　　8. $\dfrac{\cos(x)}{1-y\cdot e^z}$, $\dfrac{e^z}{1-y\cdot e^z}$

9. $\dfrac{\partial z}{\partial x} = \dfrac{2xz}{2yz - x^2}$ 　　$\dfrac{\partial z}{\partial y} = \dfrac{2y+z^2}{x^2 - 2yz}$ 　　11. $y\sec(xy) + xy^2 \sec(xy)\tan(xy)$

12. 0 　　　　13. y + z, x + z, x + y 　　　14. $f_x(x,y) = 3x^2y^5 - 4xy + 1$, $f_y(x,y) = 5x^3y^4 - 2x^2$

15. $f_x(x,y) = 4x^3 + 2xy^2$, $f_y(x,y) = 2x^2y + 4y^3$ 　　16. $f_x(x,y) = 2y/(x+y)^2$, $f_y(x,y) = -2x/(x+y)^2$

17. $f_x = e^x\{\tan(x-y) + \sec^2(x-y)\}$, $f_y = -e^x \sec^2(x-y)$

18. $f_s = -3s/\sqrt{2 - 3s^2 - 5t^2}$, $f_t = -5t/\sqrt{2 - 3s^2 - 5t^2}$

19. $f_u = v/(u^2 + v^2)$, $f_v = -u/(u^2 + v^2)$

20. $g_x = 2xy^4 \sec^2(x^2y^3)$, $g_y = \tan(x^2y^3) + 3x^2y^3 \sec^2(x^2y^3)$

21. $\frac{\partial z}{\partial x} = 1/\sqrt{x^2+y^2}$, $\frac{\partial z}{\partial y} = y/(x^2 + y^2 + x\sqrt{x^2+y^2})$

22. $\frac{\partial z}{\partial x} = \frac{3}{2} \cosh(\sqrt{3x+4y})/\sqrt{3x+4y}$, $\frac{\partial z}{\partial y} = 2\cosh(\sqrt{3x+4y})/\sqrt{3x+4y}$

23. $f_x = -e^{(x^2)}$, $f_y = e^{(y^2)}$

24. $f_x = 2xyz^3 + y$, $f_y = x^2z^3 + x$, $f_z = 3x^2yz^2 - 1$

25. $f_x = yz\, x^{yz-1}$, $f_y = z\, x^{yz} \ln(x)$, $f_z = y\, x^{yz} \ln(x)$

26. $u_x = -yz \cos(y/(x+z))/(x+z)^2$, $u_y = z \cos(y/(x+z))/(x+z)$,
    $u_z = \sin(y/(x+z)) - yz \cos(y/(x+z))/(x+z)^2$

27. $f_x = 1/(z-t)$, $f_y = -1/(z-t)$, $f_z = -(x-y)/(z-t)^2$, $f_t = (x-y)/(z-t)^2$

31. $\frac{\partial z}{\partial x} = f'(x)$, $\frac{\partial z}{\partial y} = g'(y)$ 32. $\frac{\partial z}{\partial x} = f'(x+y)$, $\frac{\partial z}{\partial y} = f'(x+y)$

33. $\frac{\partial z}{\partial x} = f'(x/y)(1/y)$, $\frac{\partial z}{\partial y} = f'(x/y)(-x/y^2)$

34. $f_{xx} = 2y$, $f_{xy} = 2x + 1/(2\sqrt{y}) = f_{yx}$, $f_{yy} = -x/(4y\sqrt{y})$

35. $z_{xx} = 3(2x^2 + y^2)/\sqrt{x^2+y^2}$, $z_{xy} = 3xy/\sqrt{x^2+y^2} = z_{yx}$, $z_{yy} = 3(x^2+2y^2)/\sqrt{x^2+y^2}$

13.4 Selected Answers

1. $4x + 8y - z = 8$ 3. $2x + 4y - z + 6 = 0$

5. $2x + y - z = 1$ 7. $2x - y - z + 2 = 0$

9. $dz = 2xy^3\, dx + 3x^2y^2\, dy$ 11. $dz = -2x(x^2+y^2)^{-2}\, dx - 2y(x^2+y^2)^{-2}\, dy$

13. $du = e^x \cdot (\cos(xy) - y\sin(xy))\, dx - x \cdot e^x \cdot \sin(xy)\, dy$

15. $dw = 2xy\, dx + (x^2 + 2yz)\, dy + y^2\, dz$ 17. $dw = (x^2+y^2+z^2)^{-1}(x\, dx + y\, dy + z\, dz)$

19. $\Delta z = 0.9225$ and $dz = 0.9$ 21. 2.9923.   23. -0.28

25. 5.4 cm^2 27. 16 cm^3 29. 150

13.5 Selected Answers

1. $-4 + 14\sqrt{3}$ 3. 1

5. (a) $\nabla f(x,y) = \langle 3x^2 - 8xy, -4x^2 + 2y \rangle$ (b) $\langle 0, -2 \rangle$ (c) $-8/5$

7. (a) $\nabla f(x,y,z) = \langle y^2z^3, 2xyz^3, 3xy^2z^2 \rangle$ (b) $\langle 4, -4, 12 \rangle$ (c) $20/\sqrt{3}$

9. $7/52$ 11. $29/\sqrt{13}$

13. $1/6$ 15. $\sqrt{5}$, $\langle 1/\sqrt{5}, 2/\sqrt{5} \rangle$

17. $\sqrt{17}/6$, $\langle 4/\sqrt{17}, 1/\sqrt{17} \rangle$

19. $\sqrt{(13/2)}$, $\langle -3/\sqrt{13}, -2/\sqrt{13} \rangle$

21. See Fig. 10. Note that each gradient vector is perpendicular to the level curve and points uphill.

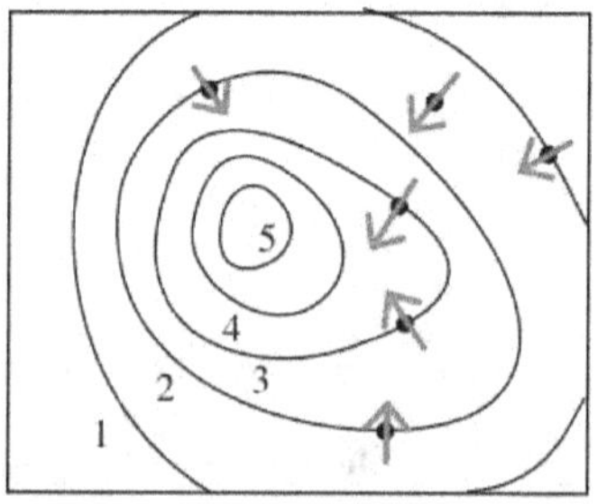

23. See Fig. 11. Note that "uphill gradient" path is always perpendicular to the level curves.

27. (a) $-40/(3\sqrt{3})$

29. (a) $32/\sqrt{3}$ (b) $u = \langle 38, 6, 12 \rangle/(2\sqrt{406})$ (c) $2\sqrt{406}$

31. $327/13$

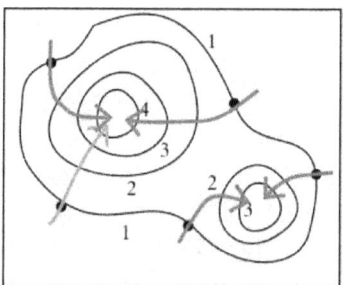

Fig. 10

Fig. 11

13.6 Selected Answers

1. Minimum $f(-2, 3) = -13$ 3. Minimum $f(0, -1) = -1$

5. Local maximum: $f(1,0) = 12$,. Saddle points: $f(-1,0) = f(2,3) = f(2,-3) = 1$

7. Local minimum: $f(0, 0) = 4$. Saddle points: $(\pm\sqrt{2}, -1)$

9. Local minimum: $f(1, 1) = -1$. Saddle point $f(0, 0) = 0$

11. Saddle point $f(1, 2) = -2$

13. Local maximum $f(-1/2, 4) = -6$ 15. None

17. Local maximum $f(0, 0) = 2$, local minimum $f(0, 2) = -2$, saddle points $(\pm 1, 1)$

19. Saddle points $(0, n\pi)$, n and integer   20. Minimum $f(4, 0) = -7$, maximum $f(4, 5) = 13$

21. Maximum $f(\pm 1, 1) = 7$, minimum $f(0, 0) = 4$

23. Maximum $f(2, 4) = 3$, minimum $f(-2, 4) = -9$

25. Critical points: f(1,1)=0, f(-2,4)= -9 minimum, f(2,4)= 3 maximum, f(-1/3, 1/9) = 32/27

27. Critical points: f(0,0) = 30, f(1,1)=f(-1,1)=31 maximum, f(-2,4)=f(2,4)=22, f(0,4) = 14 minimum

29. Critical points: f(1,3)=17, f(-2,0)=11, f(0,4) = 19 maximum, f(-3/2,1/2) = 43/4 minimum

31. Critical points: f(0,0)=0, f(1/2, 3/4)=145/256, f(1,0) = 2 maximum, f(-1,0)= -2 minimum

33. (2/7, 4/7, 6/7)

35. $\frac{100}{3}$, $\frac{100}{3}$, $\frac{100}{3}$

37. $16/\sqrt{3}$

39. 4/3

41. Cube, edge length c/12

13.7 Selected Answers

1. See Fig. 8. The solid circles marks the locations of the maximum z values along each path, and the open circles mark the locations of the minimums, A: max z=8, min z=2. B: max z=10, min z=2, C: max z=10, min z=2.

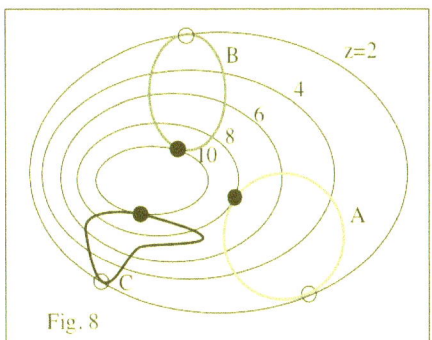
Fig. 8

3. maximum $f(\pm 1,0) = 1$, minimum $f(0,\pm 1) = -1$

4. maximum $f(\sqrt{2}/3, \sqrt{2}) = f(-\sqrt{2}/3, -\sqrt{2}) = 2/3$, minimum $f(\sqrt{2}/3, -\sqrt{2}) = f(-\sqrt{2}/3, \sqrt{2}) = -2/3$

7. maximum $f(1/\sqrt{35}, 3/\sqrt{35}, 5/\sqrt{35}) = \sqrt{35}$, minimum $f(-1/\sqrt{35}, -3/\sqrt{35}, -5/\sqrt{35}) = -\sqrt{35}$

9. $x = \pm\sqrt{2}$, $y = \pm 1$, $z = \pm\sqrt{\frac{2}{3}}$. maximum f is $2/\sqrt{3}$ (when all are positive or one is positive and two are negative), minimum f is $-2/\sqrt{3}$.

11. maximum is $\sqrt{3} = f\left(\pm\frac{1}{\sqrt{3}}, \pm\frac{1}{\sqrt{3}}, \pm\frac{1}{\sqrt{3}}\right)$, minimum is $1 = f(\pm 1,0,0) = f(0,\pm 1,0) = f(0,0,\pm 1)$

13. V=xyz with xy+2xz+2yz=A. Maximum V is $\frac{1}{2}\left(\frac{A}{3}\right)^{3/2}$ and that occurs when $x = y = \sqrt{\frac{A}{3}}$ and $z = \frac{1}{2}\sqrt{\frac{A}{3}}$.

15. V=xyz with 5xy + 1(2xz + 2yz) = 1500 (working in cents). Maximum volume is 2500 in^3 when x = y = 10 inches and z = 25 inches. Note that the cost of the bottom is $5.00, the total cost of the two ends is $5.00, and the total costs of the other two sides is $5.00 .

16. $V = xyz$ with $Bxy + S2xz + 2Syz = T$, $x = y = \sqrt{\dfrac{T}{3B}}$, $z = \dfrac{1}{2S}\sqrt{\dfrac{BT}{3}}$ and maximum volume is
$V = \dfrac{T}{3B} \cdot \dfrac{1}{2S} \cdot \sqrt{\dfrac{BT}{3}}$. The cost of the bottom is $T/3$.

17. maximum volume is $64/\sqrt{\pi}$ when $r = 4/\sqrt{\pi}$ and $h = 4/\sqrt{\pi}$.

13.8 Selected Answers

1. (a) When $t = 1$, $x=2$ and $y=3$ so $\dfrac{\partial f}{\partial x} = 11$ and $\dfrac{\partial f}{\partial y} = 6$. Also $\dfrac{dx}{dt} = -1$ and $\dfrac{dy}{dt} = -3$.

    Then $\dfrac{df}{dt} = \dfrac{\partial f}{\partial x} \cdot \dfrac{dx}{dt} + \dfrac{\partial f}{\partial y} \cdot \dfrac{dy}{dt} = (11) \cdot (-1) + (6) \cdot (-3) = -29$.

    (b) When $t = 3$, $x=1$ and $y=0$ so $\dfrac{\partial f}{\partial x} = 1$ and $\dfrac{\partial f}{\partial y} = 6$. Also $\dfrac{dx}{dt} = -2$ and $\dfrac{dy}{dt} = -1$.

    Then $\dfrac{df}{dt} = \dfrac{\partial f}{\partial x} \cdot \dfrac{dx}{dt} + \dfrac{\partial f}{\partial y} \cdot \dfrac{dy}{dt} = (1) \cdot (-2) + (6) \cdot (-1) = -8$.

3. When $t = \pi$, then $x = \cos(\pi) = -1$, $y = \sin(\pi) = 0$, $\dfrac{dx}{dt} = -\sin(t) = -\sin(\pi) = 0$,

    $\dfrac{dy}{dt} = \cos(t) = \cos(\pi) = -1$, $\dfrac{\partial f}{\partial x} = 2x = -2$, and $\dfrac{\partial f}{\partial y} = 2y = 0$ so

    $\dfrac{df}{dt} = \dfrac{\partial f}{\partial x} \cdot \dfrac{dx}{dt} + \dfrac{\partial f}{\partial y} \cdot \dfrac{dy}{dt} = (-2) \cdot (0) + (0) \cdot (-1) = 0$

5. When $t = 2$, $x = 7$, $y = 4$, $z = 10$, $\dfrac{dx}{dt} = 2$, $\dfrac{dy}{dt} = t^2 = 4$, $\dfrac{dz}{dt} = 5$,

    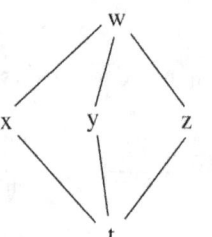

    $\dfrac{\partial f}{\partial x} = 2xy + z = 66$, $\dfrac{\partial f}{\partial y} = x^2 + z = 59$ and $\dfrac{\partial f}{\partial z} = y + x = 11$.

    $\dfrac{df}{dt} = \dfrac{\partial f}{\partial x} \cdot \dfrac{dx}{dt} + \dfrac{\partial f}{\partial y} \cdot \dfrac{dy}{dt} + \dfrac{\partial f}{\partial z} \cdot \dfrac{dz}{dt}$

    Fig. Problems 5 and 7

    $= (66) \cdot (2) + (59) \cdot (4) + (11) \cdot (5) = 423$

7. When $t = 1$, then $x = \ln(2)$, $y = \tan^{-1}(1) = \dfrac{\pi}{4}$, $z = e$, $\dfrac{dx}{dt} = \dfrac{2t}{1+t^2} = 1$, $\dfrac{dy}{dt} = \dfrac{1}{1+t^2} = \dfrac{1}{2}$,

    $\dfrac{dz}{dt} = e^t = e$, $\dfrac{\partial f}{\partial x} = 2ye^x = 2 \cdot \dfrac{\pi}{4} \cdot e^{\ln(2)} = \pi$, $\dfrac{\partial f}{\partial y} = 2e^x = 2e^{\ln(2)} = 4$ and $\dfrac{\partial f}{\partial z} = -\dfrac{1}{z} = -\dfrac{1}{e}$.

    $\dfrac{df}{dt} = \dfrac{\partial f}{\partial x} \cdot \dfrac{dx}{dt} + \dfrac{\partial f}{\partial y} \cdot \dfrac{dy}{dt} + \dfrac{\partial f}{\partial z} \cdot \dfrac{dz}{dt} = (\pi) \cdot (1) + (4) \cdot \left(\dfrac{1}{2}\right) + \left(-\dfrac{1}{e}\right)(e) = \pi + 1$

9. $w = xy + yz + xz$. When $(u, v) = (-2, 0)$ then $x = -2, y = -2, z = 0$,

$\frac{dx}{du} = 1, \frac{dx}{dv} = 1, \frac{dy}{du} = 1, \frac{dy}{dv} = -1, \frac{dz}{du} = v = 0, \frac{dz}{dv} = u = -2$

$\frac{\partial w}{\partial x} = y + z = -2, \frac{\partial w}{\partial y} = x + z = -2, \frac{\partial w}{\partial z} = y + x = -4$

$\frac{dw}{du} = \frac{\partial w}{\partial x} \cdot \frac{dx}{du} + \frac{\partial w}{\partial y} \cdot \frac{dy}{du} + \frac{\partial w}{\partial z} \cdot \frac{dz}{du} = (-2)(1) + (-2)(1) + (-4)(0) = -4$

$\frac{dw}{dv} = \frac{\partial w}{\partial x} \cdot \frac{dx}{dv} + \frac{\partial w}{\partial y} \cdot \frac{dy}{dv} + \frac{\partial w}{\partial z} \cdot \frac{dz}{dv} = (-2)(1) + (-2)(-1) + (-4)(-2) = 8$

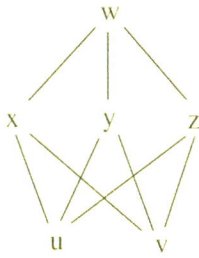

Fig. Problem 9

11. $z = \cos(xy) + x \cdot \sin(y)$, $x = u + v + 2$, $y = uv$.

When $u = 0$ and $v = 0$ then $x = 2, y = 0$ and $z = 1$.

$\frac{dx}{du} = 1, \frac{dx}{dv} = 1, \frac{dy}{du} = v = 0, \frac{dy}{dv} = u = 0$

$\frac{\partial z}{\partial x} = -y \cdot \sin(xy) + \sin(y) = 0, \frac{\partial z}{\partial y} = -x \cdot \sin(xy) + x \cdot \cos(y) = 2$

$\frac{dz}{du} = \frac{\partial z}{\partial x} \cdot \frac{dx}{du} + \frac{\partial z}{\partial y} \cdot \frac{dy}{du} = (0)(1) + (2)(0) = 0$

$\frac{dz}{dv} = \frac{\partial z}{\partial x} \cdot \frac{dx}{dv} + \frac{\partial z}{\partial y} \cdot \frac{dy}{dv} = (0)(1) + (2)(0) = 0$

Fig. Problem 11

(That was a lot of work just to get a couple 0s.)

13. We know that $T = 310$ K, $\frac{dT}{dt} = -0.2 \, \frac{K}{s}$, $V = 80$ L and $\frac{dV}{dt} = 0.1 \, \frac{L}{s}$.

$P = 8.31 \frac{T}{V}$ so $\frac{\partial P}{\partial T} = \frac{8.31}{V}$ and $\frac{\partial P}{\partial V} = -8.31 \frac{T}{V^2}$. By the Chain Rule

$\frac{dP}{dt} = \frac{\partial P}{\partial T} \cdot \frac{\partial T}{\partial t} + \frac{\partial P}{\partial V} \cdot \frac{\partial V}{\partial t} = \left(\frac{8.31}{V}\right)\left(\frac{\partial T}{\partial t}\right) + \left(-\frac{8.31 T}{V^2}\right)\left(\frac{\partial V}{\partial t}\right)$

$= \left(\frac{8.31}{80}\right)(-0.2) + \left(-\frac{8.31 \cdot 310}{80^2}\right)(0.1) = -0.061 \, \frac{kPa}{sec}$

14.0 INTRODUCTION TO DOUBLE INTEGRALS AND THEIR APPLICATIONS

Chapters 4 and 5 introduced integrals of functions of a single variable, y = f(x), as well as several of their applications. Chapter 14 introduces integrals for functions of more than one variable, shows how to calculate their values, and starts to examine some of their applications. Later chapters will use these double integrals extensively.

The applications will include calculating

* Volumes in 3D
* Total masses of solids whose densities are not constant
* Moments of solids around each axis
* Centers of mass of some solids
* Moments of inertia
* Surface areas

Moving from 2D Areas to 3D Volumes

Many of the ideas about integrals of a single variable are also important for these new integrals but things do get more complicated. In Chapter 4 we began by approximating the area under a curve by partitioning the x-axis domain, calculating the area of the rectangles over each segment of the partition, and then adding those areas together to get an approximation of the total area (Fig. 1). We begin by doing something similar to approximate the volume under a surface by portioning the 2D domain in the xy-plane into rectangles, calculating the volume of boxes over each rectangular piece, and then adding those volumes together to approximate the total volume (Fig. 2).

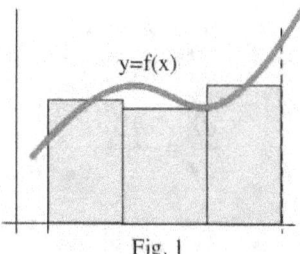
Fig. 1

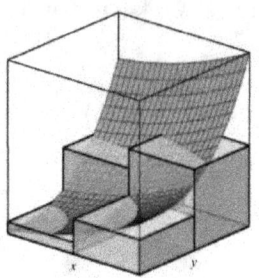
Fig. 2

The approximation of the 2D area became simply a matter of multiplying the length of the partition segment length by the height of the function and then adding those values together. The approximation of the 3D volume will be similar: multiply the area of the partition base by the height of the function, and then adding those values together.

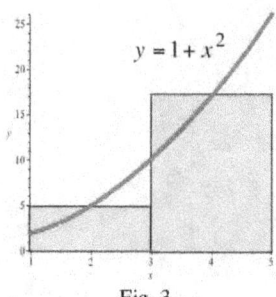
Fig. 3

Example 1: Approximate the area between the function $f(x) = 1 + x^2$ using the partition $\{1, 3, 5\}$ and evaluate the function at the midpoint of each subinterval (Fig. 3).

Solution: The partition consists of the intervals [1, 3] and [3, 5] with midpoints x=2 and x=4 so the

approximation is $\sum_{i=1}^{2} f(x_i) \cdot \Delta x_i = f(2) \cdot 2 + f(4) \cdot 2 = 10 + 34 = 44$.

(Note: We know from our earlier work that the exact value of

this area is $\int_{1}^{5} 1+x^2 \, dx = x + \frac{1}{3}x^3 \Big|_{1}^{5} = 45\frac{1}{3}$)

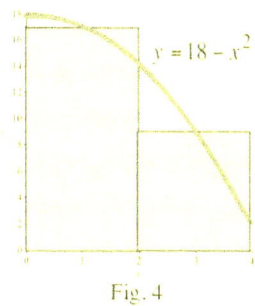

Fig. 4

Practice 1: Approximate the area between the function $g(x) = 18 - x^2$ using the partition {0, 2, 4} and evaluate the function at the midpoint of each subinterval (Fig. 4).

We can do something similar to approximate a volume.

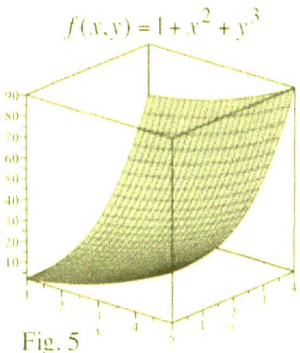

$f(x,y) = 1 + x^2 + y^3$

Example 2: Approximate the volume of between the surface

$f(x,y) = 1 + x^2 + y^3$ (Fig. 5) over the rectangle 1≤x≤5 and 0≤y≤4 by partitioning the x interval into subintervals [1,3] and [3,5] and the y interval into subintervals [0,2] and [2,4] and then evaluate the function at the midpoint of each sub-rectangle. The dots in Fig. 6 are at the midpoints of each sub-rectangle and the numbers by each dot is the value of the function at that midpoint.

Fig. 5

Solution: $\sum_{i=1}^{2} \sum_{j=1}^{2} f(x_i, y_j) \cdot \Delta x_i \cdot \Delta y_j = f(2,1) \cdot 2 \cdot 2 + f(4,1) \cdot 2 \cdot 2 + f(2,3) \cdot 2 \cdot 2 + f(4,3) \cdot 2 \cdot 2$

$= 6 \cdot 4 + 18 \cdot 4 + 32 \cdot 4 + 44 \cdot 4 = 400$

(In the next section we will be able to calculate that the exact volume

is $437\frac{1}{3}$ so our approximation was very crude.)

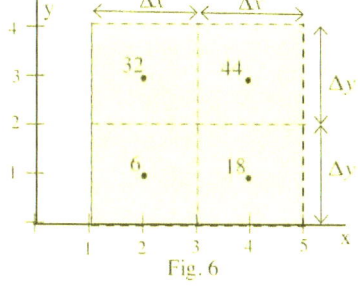

Fig. 6

Practice 2: Fig. 7 shows the depths (meters) at various locations in a backyard swimming pool. Approximate the total volume of the pool.

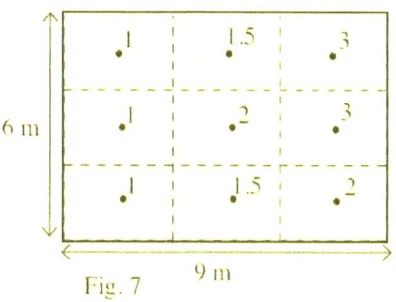

Fig. 7

If we have partitioned the domain into rectangles with dimensions Δx and Δy, then the area of each sub-rectangle is $\Delta A = \Delta x \cdot \Delta y$ and the volume of each box is $\Delta V = f(x_i, y_i) \cdot \Delta A$. But sometimes the natural partition is not rectangles.

Example 3: A gardener wants to estimate how much water a sprinkler puts out in an hour. She places several small cans at various distances from the sprinkler and measures the depth of the water in each can after one hour. Her data is given in Fig. 8. Estimate the hourly output of the sprinkler.

distance of can from sprinkler	depth of water in can
1 ft	0.3 ft
3 ft	0.4 ft
5 ft	0.3 ft
7 ft	0.2 ft

Fig. 8

Solution: In this case it is more useful to partition the circular pattern into concentric rings, and then calculate the volume of water for each ring. The area of the innermost ring (a circle) is $\Delta A_1 = \pi \cdot 2^2$, and the others are

$\Delta A_2 = \pi \cdot (4^2 - 2^2)$, $\Delta A_3 = \pi \cdot (6^2 - 4^2)$, and $\Delta A_4 = \pi \cdot (8^2 - 6^2)$.

The depths D_i are given in the table, and the total volume is

$$\sum D_i \cdot \Delta A_i = 4\pi(0.3) + 12\pi(0.4) + 20\pi(0.3) + 28\pi(0.2) \approx 55.3 \ ft^3.$$

Since each cubic foot of water is 7.5 US gallons, the sprinkler is putting out approximately 415 US gallons per hour.

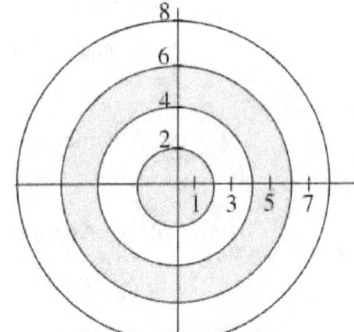

Fig. 9: Sprinkler pattern

(Note: Obviously this is a very crude estimate. She would get a much better estimate if there were more cans and the distances between the cans was smaller.)

Using Averages to Estimate Volumes

Sometimes in practice we just need an approximate value for a volume. We may need an estimate of the volume of a small pond in order to know how much of a chemical to put in the pond in order to stop an algae bloom. Or we may need an estimate of the volume of a hole on a construction site in order to know how much gravel we need to order to fill the hole. In these cases a good estimate of the volume is all that we need, and an estimate of the average depth of the pond or hole can enable us to estimate the volume.

In Section 4.7 we used integrals to calculate the average value of a function $f \geq 0$ on an interval domain as

$$\{\text{average value of } f(x) \text{ on interval } I\} = \frac{1}{\text{length of } I} \cdot \{\text{area between } f \text{ and } I\} = \frac{1}{\text{length of } I} \cdot \int_I f(x) \ dx \ .$$

A similar approach can let us estimate volume based on the average value of a function $f \geq 0$ on a domain R:

$$\{\text{average value of } f(x,y) \text{ on region } R\} = \frac{1}{\text{area of } R} \cdot \{\text{volume between } f \text{ and } R\}$$

so $\{\text{volume between } f \text{ and } R\} = \{\text{area of } R\} \cdot \{\text{average value of } f(x,y) \text{ on region } R\}$.

If we have approximate values for the area of R and for the mean value of f on R, then we simply need to multiply those values to get an approximation of the volume. It seems too easy, but it is useful.

Approximating a Volume

$$\left\{ \begin{array}{c} \text{approximate volume} \\ \text{of f over R} \end{array} \right\} = \left\{ \begin{array}{c} \text{approximate} \\ \text{area of R} \end{array} \right\} \cdot \left\{ \begin{array}{c} \text{approximate average value} \\ \text{of f(x,y) on region R} \end{array} \right\}$$

Example 4: The numbers by each dot in Fig. 6 (from Example 2) is the height of the function at that location. Use that information to approximate the volume between f and the xy-plane.

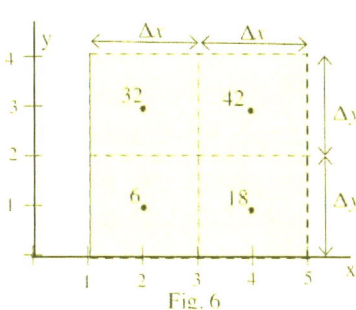
Fig. 6

Solution: The average of the four height values is (32+44+18+6)/4=25 and the area of the domain is (4)(4)=16 so our volume estimate is (25)(16)=300.

Practice 3: Use the information in Practice 2 and the average depth to estimate the volume of the pool.

Example 5: The numbers by each dot in Fig. 10 is the depth (feet) of the pond at each location. The area of the pond is approximately 12,000 square feet. Estimate the volume of the pond.

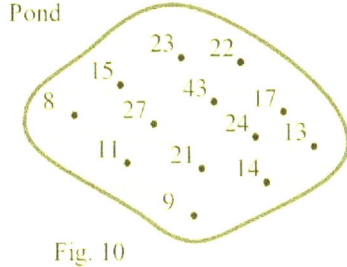
Fig. 10

Solution: The average depth is (247)/13=19 feet and the area is approximately 12,000 ft^2 so the approximate volume is (19 ft)(12,000 ft^2) = 228,000 ft^3.

In the following sections we will develop the calculus to determine volumes exactly (just as we did earlier to use integrals to calculus areas exactly), but the main ideas state as finite accumulations and lead to double integrals.

Problems

1. Fig. 11 shows the domain R of the function $z = f(x,y) = 1 + 3x + y^2$ and a partition of R into rectangles. Use the value of f at the **lower left** (x,y) point in each rectangle to approximate the volume between the graph z=f(x,y) and the xy-plane over R.

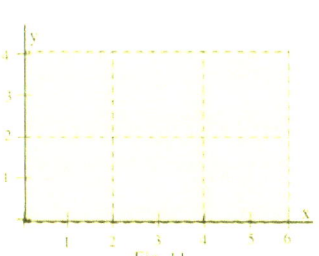
Fig. 11

2. (a) Use the same function and rectangles as in Problem 1, but evaluate the function at the **upper right** (x,y) point in each rectangle to approximate the volume between the graph z=f(x,y) and the xy-plane over R. (b) Evaluate the function at the **midpoint** (x,y) in each rectangle to approximate the volume.

3. Fig. 12 shows the domain R of the function $f(x,y) = x^2 + y^2$ and a partition of R into rectangles.

   (a) Use the value of f at the **lower left** (x,y) point in each rectangle to approximate the volume between the graph z=f(x,y) and the xy-plane over R.

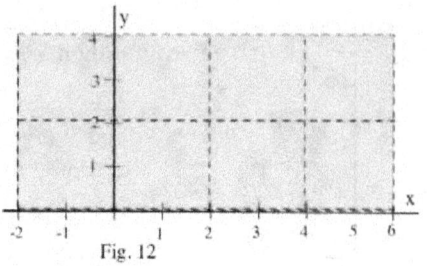
Fig. 12

   (b) Use the value of f at the **midpoint** (x,y) point in each rectangle to approximate the volume between the graph z=f(x,y) and the xy-plane over R.

4. Use the same function and rectangles as in Problem 3, but evaluate the function at the **upper right** (x,y) point in each rectangle to approximate the volume between the graph z=f(x,y) and the xy-plane over R.

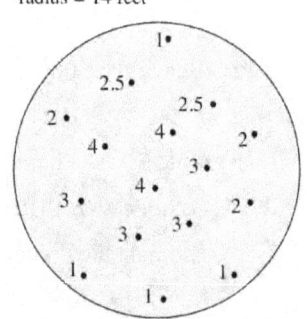
Fig. 13: water depths (inches)

5. A gardener places several small cans at a variety of locations (Fig. 13) and measures the depth of the water in each can after one hour. Estimate the hourly water output of the sprinkler.

6. A back-and-forth sprinkler has a rectangular distribution pattern. Fig. 14 shows the water depth (inches) at several locations after 15 minutes. Assume that the distribution is symmetric and estimate the hourly water output of the sprinkler.

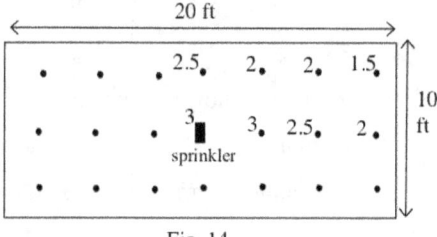
Fig. 14

7. A hole in the ground has a 4 ft. by 5 ft. rectangular opening, and the depths at the four corners are 3 ft., 4 ft. 6 ft. and 3 ft. Estimate the volume of the hole.

8. A hole in the ground has a 3 ft. by 5 ft. rectangular opening, and the depths at the four corners are 6 ft., 4 ft. 7 ft. and 6 ft. Estimate the volume of the hole.

9. A circular hole has a radius of 2 feet, and several depth measurements around the edge of the hole are 3 ft., 4 ft., 5 ft., 3 ft., and 4 ft. Estimate the volume of the hole.

10. A circular hole has a radius of 2 feet, and several depth measurements around the edge of the hole are 2 ft., 4 ft., 3 ft., 2 ft., and 3 ft. Estimate the volume of the hole.

11. Fig. 15 shows the depths (meters) at several locations in a small pond that has a surface area of 90 m^2. Approximate the volume of the pond.

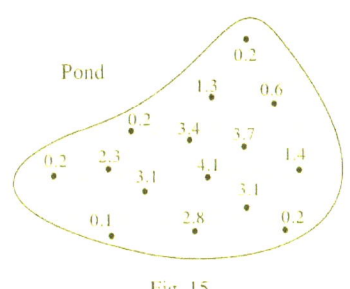

Fig. 15

12. A spacecraft has landed on a strange flat object and has measured its density ($10^3 kg/m^3$) at several locations (Fig. 16). From photographs during the approach to the object scientists have estimated the thickness of the object to be about 1.4 m thick the area to be about 3500 m^2. Approximate the mass of the object.

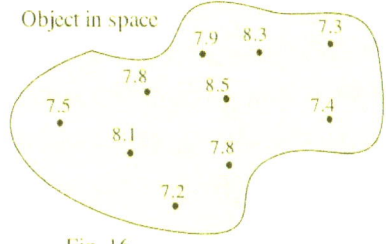

Fig. 16

13. Fig. 17 shows level elevation curves (m) for a small hill that needs to be removed for a construction project. Pick the elevations at several locations and use them to get a reasonable estimate of the volume of material to be removed. This also requires as estimate of the area of the bottom of the hill which is roughly elliptical. (ellipse area = $\frac{1}{4}\pi$(width)(length))

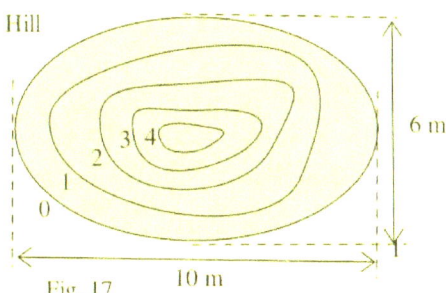

Fig. 17

14. In Example 3 (Fig. 9) we estimated the volume of the water from the sprinkler I one hour to be 55.3 ft^3. But if we use the (average value)(area) we get {average value} = (0.3+0.4+0.3+0.2)/4 =0.3 ft and the area is 64 ft^2 so our estimate for the volume is

(0.3 ft)(64 ft^2) ≈ 60.3 ft^3 .

Which estimate do you think is better and why?

15. Fig. 18 shows the level curves for z = f(x,y) on a region R. Use the midpoint sample points to estimate the value of $\iint\limits_R f(x,y)\, dA$.

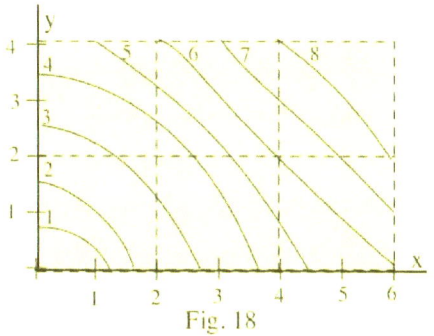

Fig. 18

16. Fig. 19 shows the level curves for z = f(x,y) on a region R. Use the midpoint sample points to estimate the value of $\iint\limits_R f(x,y)\, dA$.

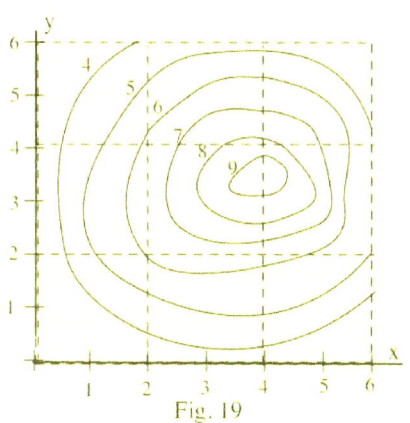

Fig. 19

Practice Answers

Practice 1: The intervals [0,2] and [2,4] have lengths 2 and 2 and midpoints x=1 and x=3. Then

$$\sum_{i=1}^{2} f(x_i) \cdot \Delta x_i = f(1) \cdot 2 + f(3) \cdot 2 = 34 + 18 = 52.$$ (The exact area is $50\frac{2}{3}$.)

Practice 2: $\sum_{i=1}^{3}\sum_{j=1}^{3} f(x_i, y_j) \cdot \Delta x_i \cdot \Delta y_j = \sum_{i=1}^{3}\sum_{j=1}^{3} f(x_i, y_j) \cdot 2 \cdot 3 = (1+1+1+1.5+2+1.5+3+3+2) \cdot 6 = 96 \ m^3.$

Practice 3: The average depth is $(1+1+1+1.5+2+1.5+3+3+2)/9 = \frac{16}{9}$ m and the area of the pool is

$(6)(9) = 54 \ m^2$ so the approximate volume is $\left(\frac{16}{9} \ m\right)\left(54 \ m^2\right) = 96 \ m^3.$

14.1 DOUBLE INTEGRALS OVER RECTANGLES

Volume of a Solid Region in 3D

To calculate the area between the curve y = f(x) and an interval on the x-axis (Fig. 1) we partitioned the interval [a,b] (Fig. 2), created an approximation of the area using a Riemann sum, and then took the limit of the Riemann sum as the widths of the subintervals approached 0 to get the area as a definite integral:

$$\text{Area} = \int_a^b f(x)\ dx$$

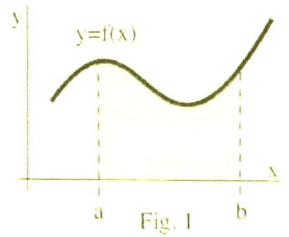

Fig. 1

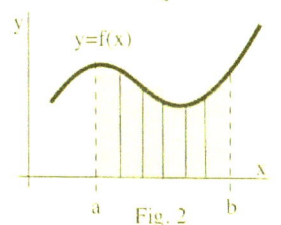

Fig. 2

In order to calculate the volume (Fig. 3) between a surface z = f(x,y) ≥ 0 and a rectangular $R = \{(x,y): a \leq x \leq b \text{ and } c \leq y \leq d\}$ of the xy-plane, we will do something similar, but now the domain of integration is a region in the xy-plane, and our result will be a double integral:

$$\text{Volume} = \iint_R f(x,y)\ dA = \int_a^b \int_c^d f(x,y)\ dy\ dx\ ,$$

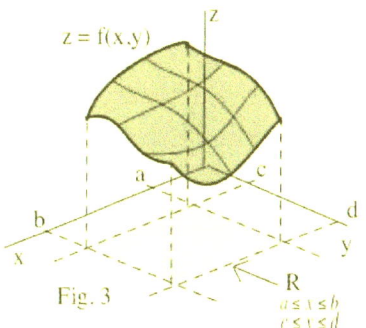

Fig. 3

For the rectangular region R we begin by partitioning along both the x-axis and along the y-axis (Fig. 4) to create small rectangles with areas $\Delta A_{ij} = \Delta x_i \cdot \Delta y_j$ and we select any point (x_i^*, y_j^*) in this rectangle. Then the volume of the box above this ij-rectangle with height $f(x_i^*, y_j^*)$ is

$$\text{Volume}_{ij} = f(x_i^*, y_j^*) \cdot \Delta x_i \cdot \Delta y_j = f(x_i^*, y_j^*) \cdot \Delta A_{ij}.$$

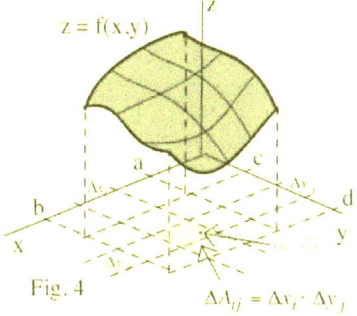

Fig. 4

$\Delta A_{ij} = \Delta x_i \cdot \Delta y_j$

Then the total approximate volume the sum of all of these little volumes:

$$\text{Approximate total volume} = \sum_j \sum_i f(x_i^*, y_j^*) \cdot \Delta x_i \cdot \Delta y_j.$$

Taking the limit as all of the Δx_i and Δy_j approach 0,

$$\text{Exact total volume} = \int_a^b \int_c^d f(x,y)\ dy\ dx = \iint_R f(x,y)\ dA\ .$$

Calculating the value of a double integral is no more difficult (nor any easier) than calculating the value of a single integral except we need to integrate twice.

Example 1: For $f(x,y) = 20 - x^2 y$ and $R = \{(x,y): 0 \leq x \leq 3, 1 \leq y \leq 2\}$ evaluate $\iint_R f(x,y)\ dA$.

Solution: In this example (Fig. 5) $\iint_R f(x,y)\, dA = \int_0^3 \int_1^2 20 - x^2 y\, dy\, dx$.

First we evaluate the inside integral $\int_1^2 20 - x^2 y\, dy$ **treating x as a constant**:

(this is just the inverse of partial differentiation)

$$\int_1^2 20 - x^2 y\, dy = 20y - \frac{1}{2} x^2 y^2 \Big|_{y=1}^{2}$$

$$= \frac{1}{2} x^2 (2)^2 - \frac{1}{2} x^2 (1)^2 = 20 - \frac{3}{2} x^2.$$

Then $\int_0^3 \left\{ \int_1^2 20 - x^2 y\, dy \right\} dx$

$$= \int_0^3 \left\{ 20 - \frac{3}{2} x^2 \right\} dx = 20x - \frac{1}{2} x^3 \Big|_{x=0}^{3} = \frac{93}{2}.$$

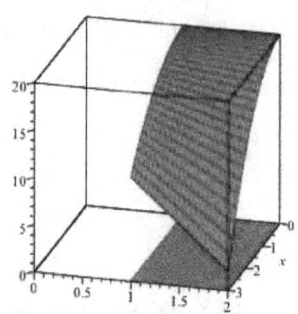

Fig. 5

Practice 1: For the same function and region R, evaluate $\int_1^2 \int_0^3 20 - x^2 y\, dx\, dy$. (This is the same solid as Example 1, but now we start by evaluating the inside integral $\int_0^3 x^2 y\, dx$, treating y as a constan.)

Theorem:

If $f(x,y) \geq 0$ and f is integrable over the rectangle R,

then   the volume V of the solid that lies above R and under the surface $z = f(x,y)$ is

$$V = \iint_R f(x,y)\, dA.$$

Note: If $f(x,y) \geq 0$ on R then the double integral gives volume. If f is sometimes negative, then the double integral gives a "signed volume" in a manner similar to how a single integral gives "signed area."

These properties of double integrals follow from the properties of summations:

(1) $\iint_R f(x,y) + g(x,y)\, dA = \iint_R f(x,y)\, dA + \iint_R g(x,y)\, dA$

(2) $\iint_R K f(x,y) \, dA = K \iint_R f(x,y) \, dA$.

(3) If $f(x,y) \geq g(x,y)$ for all (x,y) in R, then $\iint_R f(x,y) \, dA \geq \iint_R g(x,y) \, dA$.

A few important points:

* Always work from the inside out: first evaluate the inside integral.
* For $\int f(x,y) \, dx$ integrate with respect to x and **treat y as a constant**.
* For $\int f(x,y) \, dy$ integrate with respect to y and **treat x as a constant**.

It was not an accident that the answers to Example 1 and Practice 1 were the same since both versions represented the volume of the same solid.

Fubini's Theorem:

If f is integrable over the rectangle $R = \{ (x,y) : a \leq x \leq b \text{ and } c \leq y \leq d \} = [a,b] \times [c,d]$

then $\iint_R f(x,y) \, dA = \int_a^b \int_c^d f(x,y) \, dy \, dx = \int_c^d \int_a^b f(x,y) \, dx \, dy$

Fubini's Theorem says that we can integrate in either order and still get the same result — sometimes one order of integration is much easier than the other order.

Example 2: Evaluate $\iint_R 3 + y \cdot \sin(xy) \, dA$ where $R = [1,2] \times [0,\pi]$.

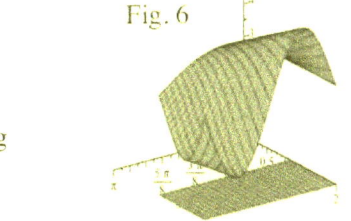
Fig. 6

Solution: The notation $R = [1,2] \times [0,\pi]$ means the rectangle $1 \leq x \leq 2$ and $0 \leq y \leq \pi$ (Fig. 6). By Fubini's Theorem we have a choice of evaluating

(a) $\int_0^\pi \int_1^2 3 + y \cdot \sin(xy) \, dx \, dy$ or (b) $\int_1^2 \int_0^\pi 3 + y \cdot \sin(xy) \, dy \, dx$.

(a) $\int_0^\pi \int_1^2 3 + y \cdot \sin(xy) \, dx \, dy = \int_0^\pi \left\{ \int_1^2 3 + y \cdot \sin(xy) \, dx \right\} dy$

$= \int_0^\pi \left\{ 3x - \cos(xy) \Big|_{x=1}^2 \right\} dy = \int_0^\pi \left\{ 6 - \cos(2y) - 3 + \cos(1y) \right\} dy$

$= 3y - \frac{1}{2} \sin(2y) + \sin(y) \Big|_{y=0}^\pi = 3\pi$

(b) $\displaystyle\int_1^2 \int_0^\pi 3 + y\cdot \sin(xy)\, dy\, dx = \int_1^2 \left\{ \int_0^\pi 3 + y\cdot \sin(xy)\, dy \right\} dx$ so first we need to

evaluate $\displaystyle\int_0^\pi 3 + y\cdot \sin(xy)\, dy$ and that requires Integration by Parts, a more difficult

situation than the method in part (a).

Example 3: Find the volume of the solid S that is bounded by the elliptic paraboloid $x^2 + 2y^2 + z = 16$, the planes $x = 2$ and $y = 2$, and the three coordinate planes (xy, xz, and yz–planes). (Fig. 7)

Solution: S lies under the surface $f(x,y) = z = 16 - x^2 - 2y^2$ and above the square $0 \le x \le 2$, $0 \le y \le 2$. Then

Fig.7

$$V = \int_0^2 \int_0^2 16 - x^2 - 2y^2\, dx\, dy = \int_0^2 \left\{ \int_0^2 16 - x^2 - 2y^2\, dx \right\} dy$$

$$= \int_0^2 \left\{ 16x - \tfrac{1}{3}x^3 - 2xy^2 \Big|_{x=0}^2 \right\} dy = \int_0^2 \left\{ \tfrac{88}{3} - 4y^2 \right\} dy = \tfrac{88}{3}y - \tfrac{4}{3}y^3 \Big|_{y=0}^2 = 48.$$

A Simple Application: Average Value of f(x,y) on R

In section 4.7 we saw that the average value of an integrable function on the interval [a,b] is $\dfrac{1}{b-a}\displaystyle\int_a^b f(x)\, dx$, that is, the average value is the integral of the function divided by the length of the domain. We have a similar result for f(x,y) on the rectangular domain D:

The average value H of an integrable function f(x,y) on domain D is $H = \dfrac{1}{\text{area of D}} \cdot \displaystyle\iint_D f(x,y)\, dA$.

The "box" (Fig. 8a) with height H={average value of f on D} has volume V ={area of D}{average height H} (Fig. 8b). But we know the volume is $v = \displaystyle\iint_D f(x,y)\, dA$ so

(area of D){average height H}$= \displaystyle\iint_D f(x,y)\, dA$ and then

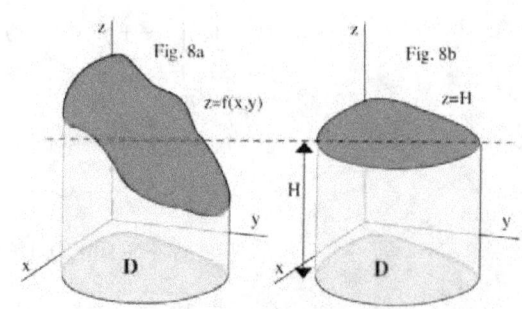

$$H = \{\text{average value of f on D}\} = \frac{1}{\text{area of D}} \cdot \iint_D f(x,y)\, dA.$$

Example 4: Determine the average value of the paraboloid $z = 16 - x^2 - 2y^2$ on the domain $D = [0,2]\times[0,2]$.

Solution: From Example 3 we know $V = \int_0^2 \int_0^2 (16 - x^2 - 2y^2)\, dx\, dy = 48$ and $\{\text{area of D}\} = 4$ so

$$H = \left(\frac{1}{4}\right)(48) = 12$$

Practice 2: Determine the average value of $f(x,y) = 3 + y \cdot \sin(xy)$ on the domain $D = [1,2]\times[0,\pi]$. (Fig. 6)

Areas, Volumes and Double Integrals

Practice 3: Evaluate $\int_a^b \int_c^d 1\, dx\, dy$ ($a<b$, $c<d$) and interpret the meaning of the result.

So far our discussion has involved double integrals of positive functions to get volumes, but sometimes it is useful to calculate the double integral of the very simple function $f(x,y) = 1$ to get an area:

$$\{\text{area of D}\} = \iint_D 1\, dA$$

This is valid even when D is not a rectangle and is especially useful in those situations.

We began the study of single integrals by trying to find the area between a positive function $y=f(x)$ and the x-axis, but then extended that idea to more general functions that were not always positive. In the more general case, $\int_a^b f(x)\, dx$ represented the "signed area" between $f(x)$ and the x-axis. If the areas above and below the x-axis were equal, then $\int_a^b f(x)\, dx = 0$. In a similar manner, if $z=f(x,y)$ is sometimes negative on D, then $\iint_D f(x,y)\, dA$ will represent the "signed volume" between $f(x,y)$ and D in the xy-plane.

Example 5: Evaluate the double integral of $f(x,y) = 2x-y$ over the rectangle $D=\{(x,y): -1\leq x\leq 1 \text{ and } 0\leq y\leq 2\}$.

Solution: $\int_0^2 \int_{-1}^1 (2x-y)\, dx\, dy = \int_0^2 (x^2 - xy)\Big|_{x=-1}^{x=1} dy = \int_0^2 -2y\, dy = -4$

so more of the volume between $f(x,y)$ and D lies below the xy-plane than lies above the xy-plane. (Fig. 9)

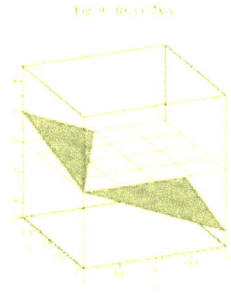

Practice 4: Suppose $f(x,y) = Ax+By$ is a plane through the origin and
$D = [-a,a] \times [-b,b]$ is a rectangle that is symmetric about the origin. Show that the volume between the plane and D that is above the xy=plane is the same of as the volume that is below the xy-plane.

This section has focused on "what is a double integral" and "how to calculate the value of a double integral over a rectangular domain." In Section 14.2 we will extend these ideas to domains that are not rectangles and in Section 14.3 we will then use these double integral calculations in several applied settings.

PROBLEMS

For problems 1 – 4, calculate $\int_0^2 f(x,y)\, dx$ and $\int_0^1 f(x,y)\, dy$

1. $f(x,y) = x^2 y^3$
2. $f(x,y) = 2xy - 3x^2$
3. $f(x,y) = x \cdot e^{x+y}$
4. $f(x,y) = \dfrac{x}{y^2 + 1}$

In problems 5 – 9, evaluate the double integrals.

5. $\displaystyle\int_{x=0}^{x=2} \int_{y=0}^{y=3} 3 + 4x + 2y \; dy \; dx$

6. $\displaystyle\int_{x=0}^{x=2} \int_{y=1}^{y=2} 7 + 3x - 6y \; dy \; dx$

7. $\displaystyle\int_{x=0}^{x=4} \int_{y=0}^{y=1} 1 + e^x + \cos(y) \; dy \; dx$

8. $\displaystyle\int_1^3 \int_2^5 x \cdot \cos(y) \; dx \; dy$

9. $\displaystyle\int_0^2 \int_0^3 x^2 + y^2 \; dx \; dy$

10. $\displaystyle\int_0^\pi \int_0^\pi \cos(x+y) \; dx \; dy$

11. $\displaystyle\int_0^4 \int_0^2 x\sqrt{y} \; dx \; dy$

12. $\displaystyle\int_{-1}^1 \int_0^1 (x^3 y^3 + 3xy^2) \; dy \; dx$

13. $\displaystyle\int_0^3 \int_0^1 \sqrt{x+y} \; dx \; dy$

14. $\displaystyle\int_0^{\pi/4} \int_0^3 \sin(x) \; dy \; dx$

15. $\displaystyle\int_0^{\ln(2)} \int_0^{\ln(5)} e^{2x - y} \; dx \; dy$

In problems 16-24 decide which order of integration is easier, $\int f(x,y) \, dx$ or $\int f(x,y) \, dy$, and then calculate the easier antiderivative.

16. $f(x,y) = x \cdot e^{xy}$
17. $f(x,y) = y \cdot \sqrt{x^2 + y^2}$
18. $f(x,y) = \dfrac{4x}{x^2 + y^2}$

19. $f(x,y) = \sin(x) \cdot e^{x+y}$
20. $f(x,y) = \dfrac{\cos(x)}{x+y}$
21. $f(x,y) = \sqrt{x+y^2}$

22. $f(x,y) = x \cdot \ln(y+3)$
23. $f(x,y) = e^{(y^2)}$
24. $f(x,y) = \cos(x^2)$

In problems 25-28, evaluate the double integrals. One order of integration may be easier than the other.

25. $\iint_R (2y^2 - 3xy^3) \, dA$ where $R = \{(x,y) : 1 \leq x \leq 2, 0 \leq y \leq 3\}$.

26. $\iint_R x \sin(y) \, dA$ where $R = \{(x,y) : 1 \leq x \leq 4, 0 \leq y \leq \pi/6\}$.

27. $\iint_R x \sin(x+y) \, dA$ where $R = [0, \pi/6] \times [0, \pi/6]$.

28. $\iint_R \frac{1}{x+y} \, dA$ where $R = [1,2] \times [0,1]$.

In problems 29-34 determine the average value of $f(x,y)$ on the given domain.

29. $f(x,y) = 3 + 4x + 2y$ on $R = [0,2] \times [0,3]$
30. $f(x,y) = 7 + 3x - 6y$ on $R = [0,2] \times [1,2]$

31. $f(x,y) = x^2 + y^2$ on $R = [0,2] \times [0,3]$
32. $f(x,y) = \cos(x+y)$ on $R = [0,\pi] \times [0,\pi]$

In problems 33 and 34 the depths (in meters) of a small rectangular pond are shown. Give a good estimate of the volume of water in the pond. You should be able to justify why your estimate is reasonable.

33. Length = 4 m. Width = 3 m. Depths at the four corners are 2.4, 3.3, 2.1 and 3.2 m.

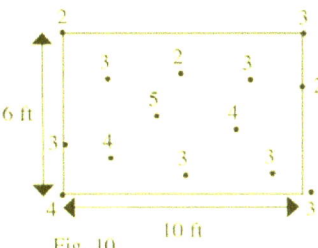
Fig. 10

34. Depths (ft) are shown at various locations (Fig. 10).

In problems 35 and 36 some height contours are shown for a small hill. Give a good estimate of the volume of the hill. You should be able to justify why your estimate is reasonable.

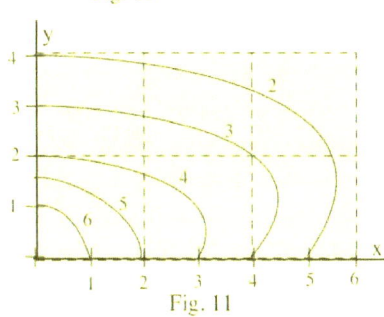
Fig. 11

35. See Fig. 11. All of the measurements are in meters.

36. See Fig. 12. All of the measurements are in meters.

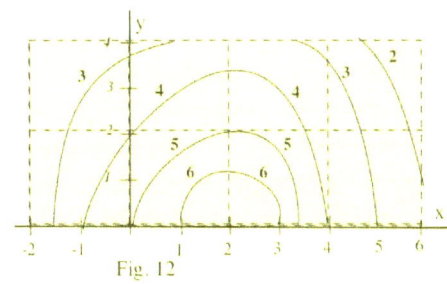
Fig. 12

Practice Answers

Practice 1: $\int_1^2 \int_0^3 20 - x^2 y \, dx \, dy$ means $\int_1^2 \left\{ \int_0^3 20 - x^2 y \, dx \right\} dy$ so first we evaluate the inside integral $\int_0^3 x^2 y \, dx$ **treating y as a constant**:

$$\int_0^3 20 - x^2 y \, dx = 20x - \frac{1}{3} x^3 y \Big|_{x=0}^{3} = 20(3) - \frac{1}{3}(3)^3 y = 60 - 9y.$$

Then $\int_1^2 \left\{ \int_0^3 x^2 y \, dx \right\} dy = \int_1^2 \left\{ 60 - 9y \right\} dy$

$$= 60y - \frac{9}{2} y^2 \Big|_{y=1}^{2} = \left(60(2) - \frac{9}{2}(2)^2 \right) - \left(60(1) - \frac{9}{2}(1)^2 \right) = (102) - \frac{111}{2} = \frac{93}{2}.$$

Practice 2: In Example 2 we evaluated $\int_0^\pi \int_1^2 3 + y \cdot \sin(xy) \, dx \, dy = 3\pi$.

The area of D is $(2-1) \cdot (\pi - 0) = \pi$ so the average value is $(3\pi)/\pi = 3$.

Practice 3: $\int_a^b \int_c^d 1 \, dx \, dy = \int_a^b x \Big|_{x=c}^{x=d} dy = \int_a^b (d-c) \, dy = (d-c) \cdot y \Big|_{y=a}^{y=b} = (d-c) \cdot (b-a)$

This is the area of the rectangular domain.

Practice 4: $\int_{-a}^{a} \int_{-b}^{b} Ax + By \, dy \, dx = \int_{-a}^{a} Axy + \frac{B}{2} y^2 \Big|_{y=-b}^{y=b} dx = \int_{-a}^{a} 2abx \, dx = abx^2 \Big|_{x=-a}^{x=a} = 0$

So the volume between the plane f(x,y)=Ax+By and D that is above the xy=plane is the same as the volume that is below the xy-plane.

14.2 DOUBLE INTEGRALS OVER GENERAL REGIONS

The double integrals over rectangular regions in the section 14.1 are relatively straightforward, but many applied situations have domains that are not rectangular, and this section considers those. In section 14.1 our focus was on finding antiderivatives (twice) and evaluating the double integrals. In this section the main difficulties involve setting up the endpoints of the integrals. This is very important and is vital for later sections.

The starting ideas here are the same as in 14.1:

* partition the general domain D into small Δx_i by Δy_j rectangles, (Fig. 1)

* create a double Riemann sum of the areas of the boxes above each rectangle,

$$\sum_j \sum_i f(x_i, y_j) \cdot \Delta x_i \cdot \Delta y_j = \sum_j \sum_i f(x_i, y_j) \cdot \Delta A_{ij}$$

* take the limits as all of the Δx_i and Δy_j values approach 0

    so they fill the domain, and

* create a double integral $\iint_D f(x,y)\, dA$.

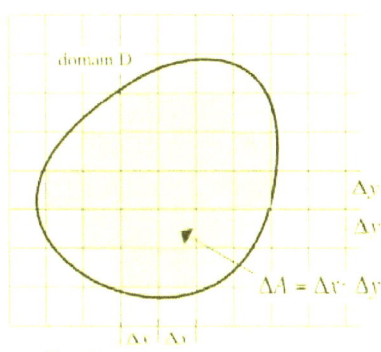

Fig. 1

The notation $\iint_D f(x,y)\, dA$ is neat and compact and is commonly used, but in order to compute the value of the double integral we need to explicitly rewrite it as $\iint_D f(x,y)\, dx\, dy$ or $\iint_D f(x,y)\, dy\, dx$ with the appropriate endpoints of integration.

Note: All of our work with double integrals assumes that the domain D is "bounded" (fits inside a finite rectangle) and that the boundary curve of D is "nice" (consists of a finite number of smooth curves with finite length). In most applications, these are not major restrictions.

Two Easiest Cases

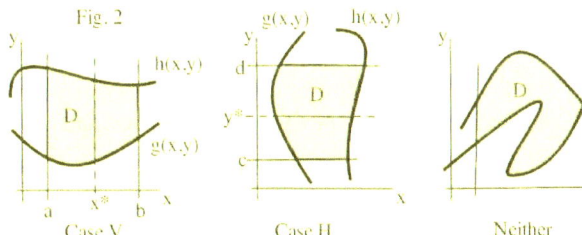

Fig. 2

Case V: For each x* value, D contains a single
    vertical segment of y values (Fig. 2).
Case H: For each y* value, D contains a single
    horizontal segment of x values .
Other: It is possible for the domain D to be both
    Case V and Case H or neither.

14.2 Double Integrals over General Regions   Contemporary Calculus   256

Case V: For each value of x between a and b, D consists of those y with $g(x) \leq y \leq h(x)$. For each value of x, we only care about f(x,y) for y values between g(x) and h(x).

Then $\iint_D f(x,y) \, dA = \int_{x=a}^{x=b} \int_{y=g(x)}^{y=h(x)} f(x,y) \, dy \, dx$ (or simply $\int_a^b \int_{g(x)}^{h(x)} f(x,y) \, dy \, dx$).

Example 1: $f(x,y) = 1 + 4x + 2y$ and D is the region in Fig. 3. Calculate the volume of the solid between D and the surface $z=f(x,y)$.

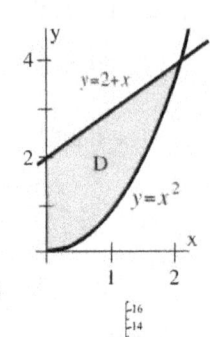

Fig. 3

Solution: For each x between 0 and 2, the y values are between $2 + x$

and x^2 so Volume = $\iint_D f(x,y) \, dA = \int_{x=0}^{x=2} \int_{y=x^2}^{y=2+x} 1+4x+2y \, dy \, dx$.

We deal with this double integral in the same way we did in section 14.1 by starting with the inside integral.

$\int_{y=x^2}^{y=2+x} 1+4x+2y \, dy = y+4xy+y^2 \Big|_{y=x^2}^{y=2+x}$

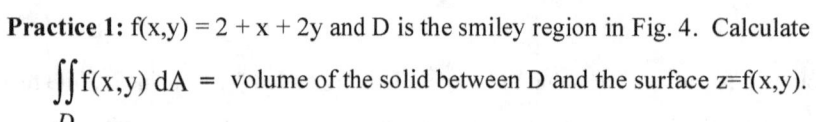

$\int_{x=0}^{x=2} \{-x^4 - 4x^3 + 4x^2 + 13x + 6\} \, dx = \frac{394}{15} \approx 26.27$

If the units of x, y and z=f(x,y) are centimeters, then the double integral units are cm^3 .

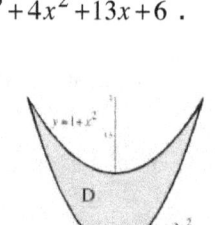

Practice 1: $f(x,y) = 2 + x + 2y$ and D is the smiley region in Fig. 4. Calculate

Fig. 4

$\iint_D f(x,y) \, dA$ = volume of the solid between D and the surface $z=f(x,y)$.

Example 2: Determine the integration endpoints for

$\iint_D f(x,y) \, dy \, dx$ on the domain D shown in Fig. 5.

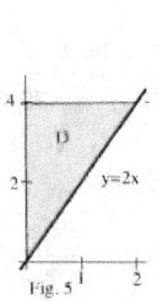

Fig. 5

Solution: Since the dx is on the outside, the endpoints of the outside integral will be from x=0 to x=2, But for each x value between 0 and 2 (Fig. 6), a vertical slice at x enters the domain D when y=2x (at the bottom) and exits when y=4 (at the top) so the endpoints of the inside integral are y=2x and y=4: 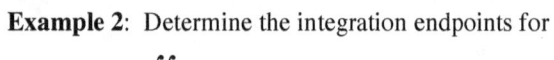 or simply $\int_0^2 \int_{2x}^4 f(x,y) \, dy \, dx$.

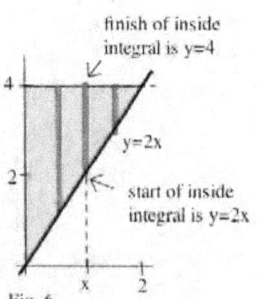

Fig. 6

Practice 2: Determine the integration endpoints for $\iint_D f(x,y)\, dy\, dx$ on the domain D shown in Fig. 7.

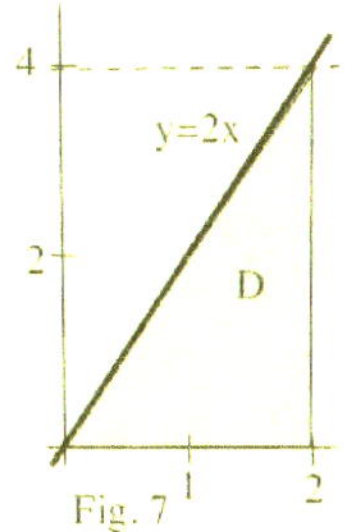

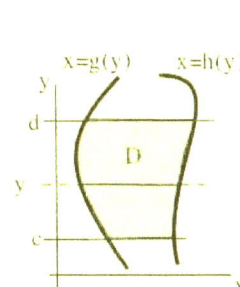
Fig. 7

Case H: For each value of y between c and d, the domain D consists of those x values with $g(y) \leq x \leq h(y)$. For each value of y, we only care about f(x,y) for x values between g(y) and h(y).

Then $\iint_D f(x,y)\, dA = \int_{y=c}^{y=d} \int_{x=g(y)}^{x=h(y)} f(x,y)\, dx\, dy$ (or simply

$\int_c^d \int_{g(y)}^{h(y)} f(x,y)\, dx\, dy$).

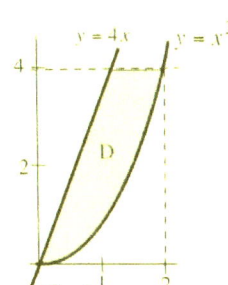
Fig. 8   Case H

Note: If the boundary function has the form y=f(x) then we need to solve for x as a function of y. For example, if the boundary is $y = f(x) = 1+2x$, then we need to solve for x to get $x = g(y) = (y-1)/2$.

Example 3: Determine the integration endpoints for $\iint_D f(x,y)\, dx\, dy$ on the domain D shown in Fig. 9.

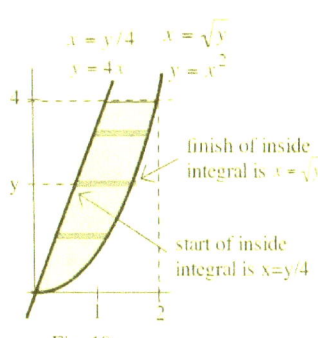
Fig. 9

Solution: Clearly the horizontal slices go from y = 0 to y = 4 so those are the endpoints of the outside integral. Then we need to convert the y=4x and $y = x^2$ from functions of x to functions of y: $x = y/4$ and $x = \sqrt{y}$. For each y value between 0 and 4 we need to see when a horizontal slice at y enters and exits the domain (Fig. 10).

The result is $\iint_D f(x,y)\, dx\, dy = \int_{y=0}^{y=4} \int_{x=y/4}^{x=\sqrt{y}} f(x,y)\, dx\, dy$.

Fig. 10

Practice 3: Determine the integration endpoints for $\iint_D f(x,y)\, dx\, dy$ on the domain D shown in Fig. 11.

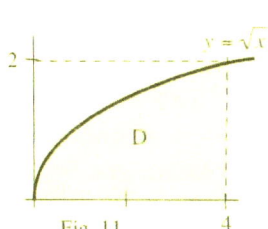
Fig. 11

Splitting the Domain & Switching the Order of Integration

Sometimes the domain requires that we do the double integral in two (or more) pieces.

Example 4: Determine the integration endpoints for $\iint_D f(x,y)\, dy\, dx$ on the shaded domain D shown in Fig. 12.

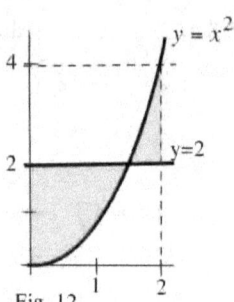

Fig. 12

Solution: For $0 \leq x \leq 1$, the bottom boundary is $y = x^2$ and the top boundary is $y=2$. For $0 < x \leq 2$ the top and bottom boundary functions are switched sp we need to use two double integrals:

$$\iint_D f(x,y)\, dy\, dx = \int_{x=0}^{x=1}\int_{y=x^2}^{y=2} f(x,y)\, dy\, dx + \int_{x=1}^{x=2}\int_{y=2}^{y=x^2} f(x,y)\, dy\, dx$$

Practice 4: Determine the integration endpoints for $\iint_D f(x,y)\, dx\, dy$ on the shaded domain D shown in Fig. 12. Note the switch to dx dy in the double integral -- this is also an example of switching the order of integration.

Example 5: Switch the order of integration of $\int_{x=-2}^{x=2}\int_{y=x^2}^{y=4} f(x,y)\, dy\, dx$.

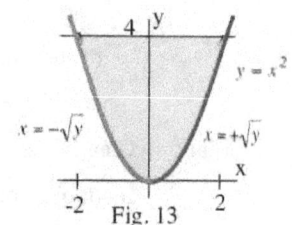

Fig. 13

Solution: From the given double integral we can create a picture of the domain (Fig. 13) and then write the double integral with the order dx dy:

$$\int_{x=-2}^{x=2}\int_{y=x^2}^{y=4} f(x,y)\, dy\, dx = \int_{y=0}^{y=4}\int_{x=-\sqrt{y}}^{x=+\sqrt{y}} f(x,y)\, dx\, dy$$

Why are we doing this?

Sometimes one order for a double integral is much easier to evaluate than the other order, and then it is worthwhile to be able to convert to the easier way.

Example 6: Evaluate $\iint_R e^{(x^2)}\, dA$ for R = {region for y between 0 and 1 and x between 3y and 3}.

14.2 Double Integrals over General Regions

Solution: It is natural to set up the integrals as $\int_{y=0}^{1}\int_{x=3y}^{3} e^{(x^2)} \, dx \, dy$ since that is the description of the region R, but as soon as we try the inside integral $\int_{x=3y}^{3} e^{(x^2)} \, dx$ we are stuck because $e^{(x^2)}$ does not have an elementary antiderivative. Fig. 14 shows the region R.

When we switch the order the integrals become $\int_{x=0}^{3}\int_{y=0}^{x/3} e^{(x^2)} \, dy \, dx$.

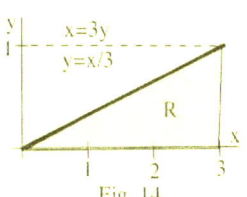

Fig. 14

Now things are easier: $e^{(x^2)}$ does not depend on y so

$\int_{y=0}^{x/3} e^{(x^2)} \, dy = y \cdot e^{(x^2)} \Big|_{y=0}^{y=x/3} = \frac{1}{3} y \cdot e^{(x^2)}$. Then the outside integral is $\int_{x=0}^{3} \frac{1}{3} x \cdot e^{(x^2)} \, dx$.

With the simple change of variable $u = x^2$ and $du = 2x \, dx$ we get

$\int \frac{1}{6} e^{(u)} \, du = \frac{1}{6} e^{(u)} = \frac{1}{6} e^{(x^2)} \Big|_{x=0}^{x=3} = \frac{1}{6} e^9 - \frac{1}{6}$.

Practice 5: Which is easier? (a) $\int_{x=0}^{1}\int_{y=x}^{1} \cos(y^2) \, dy \, dx$ or (b) $\int_{y=0}^{1}\int_{x=0}^{y} \cos(y^2) \, dx \, dy$

Properties of Double Integrals

These properties all are similar to the familiar properties of single integrals and follow from the properties of summations. We will not prove them here and have already used them in the examples. You should understand what each of them means geometrically.

(1) $\iint\limits_{D} c \cdot f(x,y) \, dA = c \cdot \iint\limits_{D} f(x,y) \, dA$

(2) $\iint\limits_{D} \{f(x,y) + g(x,y)\} \, dA = \iint\limits_{D} f(x,y) \, dA + \iint\limits_{D} g(x,y) \, dA$

(3) If $g(x,y) \leq f(x,y)$ for all (x,y) in D, then $\iint\limits_{D} g(x,y) \, dA \leq \iint\limits_{D} f(x,y) \, dA$

(4) $\iint\limits_{D} 1 \, dA = \text{area}(D)$

(5) If $m \leq f(x,y) \leq M$ for all (x,y) in D, then $m \cdot \text{area}(D) \leq \iint\limits_{D} f(x,y) \, dA \leq M \cdot \text{area}(D)$.

(6) If D1 and D2 are disjoint domains and $D = D1 \cup D2$ (= all (x,y) in $D1$ or $D2$)

then $\iint_D f(x,y) \, dA = \iint_{D1} f(x,y) \, dA + \iint_{D2} f(x,y) \, dA$.

In this section the focus has been on setting up the endpoints of integration for non-rectangular domains. Some computer programs can evaluate double integrals, but only after the user has determined the endpoints. Even with technology you need to be able to determine those endpoints. And since it is sometimes much easier to integrate in one order than the other, it is important that you be able to write the double integral in either order.

Using Technology -- MAPLE

Some of the double integrals that appear in later sections and many of them in applications are very difficult to evaluate "by hand." Fortunately technology can be very helpful. The computer program MAPLE is very powerful and relatively simple to use.

In Example 1 we were able to evaluate the double integral of $f(x,y) = 1 + 4x + 2y$ on the region $R = \{(x,y): 0 \le x \le 2, \; x^2 \le y \le 2+x\}$ by hand, but the MAPLE command

int(1+4*x+2*y, y=x^2 .. 2+x, x=0 ..1);

gives the same result, $\dfrac{379}{30}$, much quicker. And MAPLE can evaluate much more difficult integrals such as

$\displaystyle\int_{x=0}^{1} \int_{y=0}^{x+2} xy + \cos(xy) \, dy \, dx$. The MAPLE command int(x*y+cos*x*y), y=0 .. x+2, x=0 .. 1.0) : quickly

gives 3.387845686.

MAPLE syntax: int(formula for f(x,y) , y= lower endpoint .. upper endpoint, x= lower x .. upper x) ;

The program Mathematica and the online WolframAlpha can also evaluate double integrals.

PROBLEMS

For problems 1 – 6 sketch the domain of integration and evaluate the integrals.

1. $\displaystyle\int_{x=0}^{x=3} \int_{y=x}^{y=2x} x \, dy \, dx$

2. $\displaystyle\int_{x=0}^{x=2} \int_{y=0}^{y=2x} 3 + 4x + 2y \, dy \, dx$

3. $\displaystyle\int_{y=1}^{y=4} \int_{x=y}^{x=2y} 7 + 3x - 4y \, dx \, dy$

4. $\displaystyle\int_{x=0}^{x=1} \int_{y=x^2}^{y=x} x^2 + \sqrt{y} \, dy \, dx$

5. $\displaystyle\int_{x=0}^{x=1} \int_{y=0}^{y=x} \sin(x^2) \, dy \, dx$

6. $\displaystyle\int_{y=0}^{y=2} \int_{x=\sqrt{y}}^{x=3} x + y^2 \, dx \, dy$

14.2 Double Integrals over General Regions   Contemporary Calculus   261

In problems 7-12 a shaded domain D is shown. Determine the endpoints for both $\iint_D f \, dy \, dx$ and $\iint_D f \, dx \, dy$.

7. Fig. 14   8. Fig. 15   9. Fig. 16

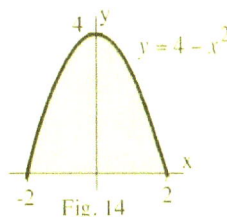
Fig. 14

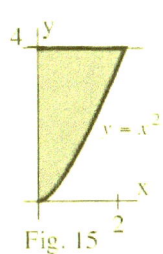
Fig. 15

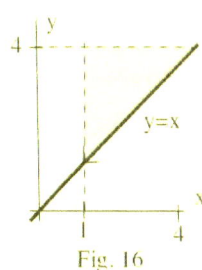
Fig. 16

10. Fig. 17   11. Fig. 18   12.

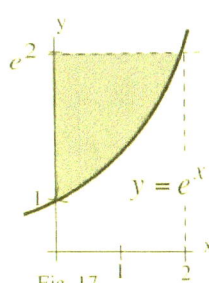
Fig. 17

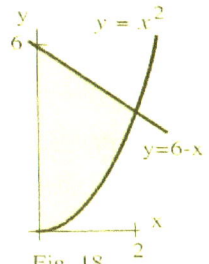
Fig. 18

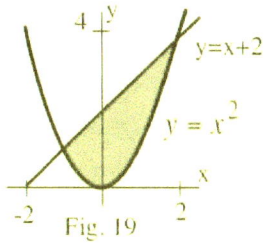
Fig. 19

In problems 13 - 20 sketch the domain of integration, set up the appropriate double integral and evaluate it.

13. $\iint_D xy \, dA$ where $D = \{ (x,y) : 0 \le x \le 1, x^2 \le y \le \sqrt{x} \}$.

14. $\iint_D (x+y) \, dA$ where $D = \{ (x,y) : \pi/6 \le x \le \pi/4, \sin(x) \le y \le \cos(x) \}$.

15. $\iint_D \left(y - xy^2 \right) dA$ where $D = \{ (x,y) : -y \le x \le 1+y, 0 \le y \le 1 \}$.

16. $\iint_D x^2 y \, dA$ where $D = \{ (x,y) : x^2 \le y \le 2+x, -1 \le x \le 2 \}$.

17. $\iint_D 1 + x \cdot \cos(y) \, dA$ where D is the bounded region between $y = 0$ and $y = 4 - x^2$.

18. $\iint_D \left(x^2 + y \right) dA$ where D is the bounded region between $y = x^2$ and $y = 8 - x^2$.

19. $\iint_D 4y^3 \, dA$ where D is the bounded region between $y = 2 - x$ and $y^2 = x$.

20. $\iint_D 6x + y \, dA$ where D is the bounded region between $y = x^2$ and $y = 4$.

In problems 21-24 determine the average value of the function f(x,y) on the given domain.

21. $f(x,y) = 3 + 4x + 2y$ where $D = \{(x,y): 0 \le x \le 2 \text{ and } 0 \le y \le 2x\}$

22. $f(x,y) = 1 + 2x + 3y$ where $D = \{(x,y): -2 \le x \le 2 \text{ and } 0 \le y \le 4 - x^2\}$

23. $f(x,y) = x + y^2$ where $D = \{(x,y): 0 \le x \le 3 \text{ and } 0 \le y \le 3 - x\}$

24. $f(x,y) = 2 + xy$ where $D = \{(x,y): 0 \le x \le 2 \text{ and } x^2 \le y \le 6 - x\}$

In problems 25 – 36, change the order of integration. (It helps to sketch the region.)

25. $\displaystyle\int_0^1 \int_0^x f(x,y)\, dy\, dx$

26. $\displaystyle\int_{-2}^2 \int_{x^2}^4 f(x,y)\, dy\, dx$

27. $\displaystyle\int_0^6 \int_0^{3-y/2} f(x,y)\, dx\, dy$

28. $\displaystyle\int_0^4 \int_0^{\sqrt{x}} f(x,y)\, dy\, dx$

29. $\displaystyle\int_1^2 \int_0^{\ln(x)} f(x,y)\, dy\, dx$

30. $\displaystyle\int_0^4 \int_{y/2}^2 f(x,y)\, dx\, dy$

31. $\displaystyle\int_0^2 \int_0^{4-2x} f\, dy\, dx$

32. $\displaystyle\int_0^3 \int_0^{9-x^2} f\, dy\, dx$

33. $\displaystyle\int_0^4 \int_{\sqrt{x}}^2 f\, dy\, dx$

34. $\displaystyle\int_1^3 \int_2^{2y} f\, dx\, dy$

35. $\displaystyle\int_0^1 \int_1^{e^x} f\, dy\, dx$

36. $\displaystyle\int_0^4 \int_{-\sqrt{y}}^{\sqrt{y}} f\, dx\, dy$

In problems 37 and 38 the depths (in meters) of a small pond are shown. Give a good estimate of the volume of water in the pond. You should be able to justify why your estimate is reasonable.

37. Fig. 20

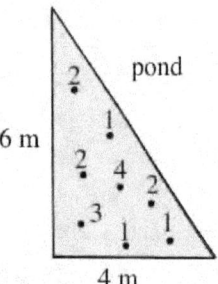

Fig. 20

38. Fig. 21

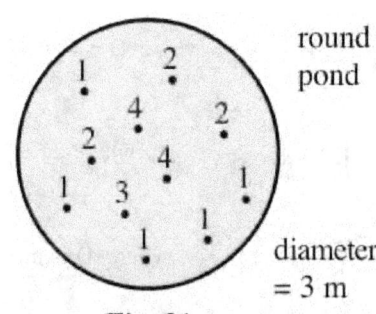

Fig. 21

PRACTICE ANSWERS

Practice 1: $\iint_D (x+2y)\, dA = \int_{-1}^{1} \int_{2x^2}^{1+x^2} (x+2y)\, dy\, dx$

Starting with the inside integral $\int_{2x^2}^{1+x^2} 2 + x + 2y \, dy = 2y + xy + y^2 \Big|_{2x^2}^{1+x^2}$

$= \{2(1+x^2) + x(1+x^2) + (1+x^2)^2\} - \{2(2x^2) + x(2x^2) + (2x^2)^2\} = -3x^4 - x^3 + x + 3$

Then $\int_{-1}^{1} -3x^4 - x^3 + x + 3 \, dx = -\frac{3}{5}x^5 - \frac{1}{4}x^4 + \frac{1}{2}x + 3x \Big|_{-1}^{1} = \left(\frac{53}{20}\right) - \left(\frac{-43}{20}\right) = \frac{24}{5}$

The Maple command int(f(x,y),y=2*x^2..1+x^2,x=-1..1); gives 24/5.

Practice 2: See Fig. 22. The outside integral (corresponding to the dx) goes

from x=0 to x=2. The inside integral begins at y=0

and ends at y=2x.

$\iint_D f(x,y)\, dy\, dx = \int_{x=0}^{x=2} \int_{y=0}^{y=2x} f(x,y)\, dy\, dx$

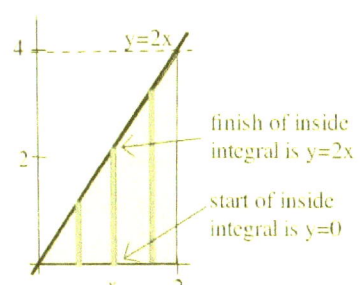

Fig. 22

Practice 3: See Fig. 23. The outside integral goes from y=0 to y=2.

$y = \sqrt{x}$ becomes $x = y^2$. Then for each y value

between 0 and 2, the horizontal slice enters the domain

at $x = y^2$ and exits at x=4.

$\iint_D f(x,y)\, dx\, dy = \int_{y=0}^{y=2} \int_{x=y^2}^{x=4} f(x,y)\, dx\, dy$

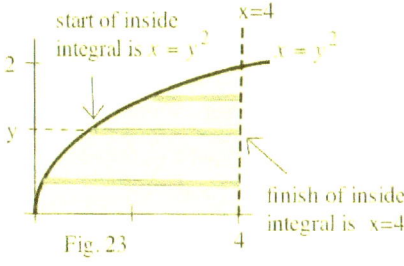

Fig. 23

Practice 4: Since the dy is on the outside, we need to partition the 0≤y≤4 interval. Then

$\iint_D f(x,y)\, dx\, dy = \int_{y=0}^{y=2} \int_{x=0}^{x=\sqrt{y}} f(x,y)\, dx\, dy + \int_{y=2}^{y=4} \int_{x=\sqrt{y}}^{x=2} f(x,y)\, dx\, dy$

Practice 5: Version (b) is easier sine we can evaluate $\int_{x=0}^{y} \cos(y^2)\, dx = x \cdot \cos(y^2) \Big|_{x=0}^{x=y} = y \cdot \cos(y^2)$

and we cannot evaluate $\int_{y=x}^{1} \cos(y^2)\, dy$.

14.3 DOUBLE INTEGRALS IN POLAR COORDINATES

As you learned in earlier classes some shapes (Fig. 1) are much simpler to describe using polar coordinates, and sometimes we need to calculate volumes of regions whose domains are those shapes. The process is straightforward, but first we need a way to partition a polar coordinate region. Since the polar variables are r and θ, it is natural to partition the domain into "polar rectangles" (Fig. 2) and then proceed as we did in Section 14.1 to build Riemann Sums and, by taking limits, get to double integrals.

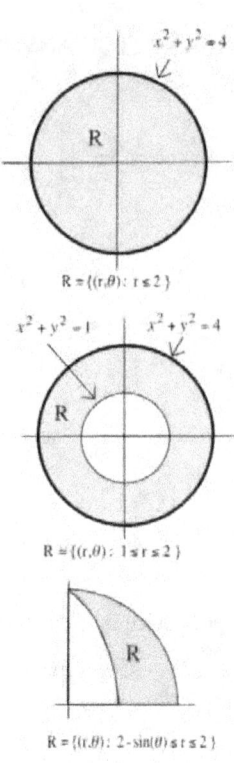

Fig. 1

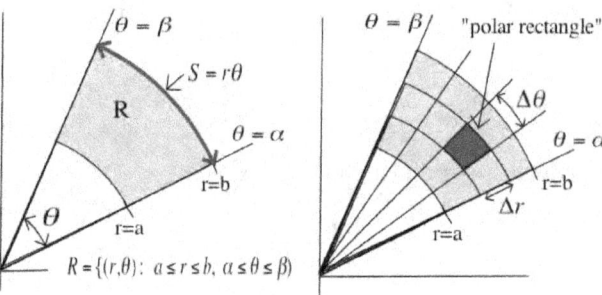

Fig. 2: Polar Rectangle

The area of each sub-rectangle is approximately $\Delta A \approx r \cdot \Delta r \cdot \Delta \theta$ (Fig. 3) so the volume above each sub-rectangle will be $\Delta V = f(x,y) \cdot \Delta A$. But we are in polar coordinates, so we rneed to eplace x and y with their polar coordinate values $x = r \cdot \cos(\theta)$ and $y = r \cdot \sin(\theta)$.

Then the total volume is approximately the value of the double Riemann sum

$$\text{volume} \approx \sum \sum f(x,y) \cdot \Delta A = \sum_r \sum_\theta f(r \cdot \cos(\theta), r \cdot \sin(\theta)) \cdot \Delta A.$$

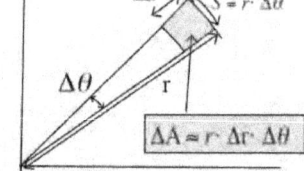

Fig. 3: Partitioned Sub-rectangle

Taking limits as Δr and $\Delta \theta$ both approach 0, the exact volume is a double integral.

Volume in Polar Coordinates

If $f(x,y) \geq 0$ is continuous on the polar rectangle $R = \{ (r,\theta): a \leq r \leq b,\ \alpha \leq \theta \leq \beta) \}$, then

The volume between the domain R and the surface $z = f(x,y)$ is

$$\text{Volume} = \iint_R f(x,y) \cdot dA = \int_{\theta=\alpha}^{\theta=\beta} \int_{r=a}^{r=b} f(r \cdot \cos(\theta),\ r \cdot \sin(\theta)) \cdot r \cdot dr \cdot d\theta$$

This may seem complicated but it is simply volume = height · (base area) with height=function (in polar coordinates) and base area = $r \cdot dr \cdot d\theta$. (Note: Don't forget the r in the base area.)

Example 1: Find the volume between the plane $f(x,y)=5+x+2y$ and the circular region $R = \{(x,y): x^2 + y^2 \leq 4\}$. (Fig. 4)

Solution: Translating the problem into polar coordinates we have
$$f(x,y) = f(r \cdot \cos(\theta), r \cdot \sin(\theta)) = 5 + r \cdot \cos(\theta) + 2 \cdot r \cdot \sin(\theta)$$
and $R = \{(r,\theta): 0 \leq r \leq 2, 0 \leq \theta \leq 2\pi\}$ so
$$\text{volume} = \iint_R f(x,y) \cdot dA$$
$$= \int_{\theta=0}^{\theta=2\pi} \int_{r=0}^{r=2} \{5 + r \cdot \cos(\theta) + 2 \cdot r \cdot \sin(\theta)\} \cdot r \cdot dr \cdot d\theta$$

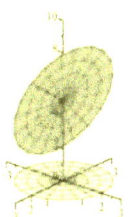

Fig. 4

which can be evaluated in the usual way starting with the inside integral.

$$\int_{r=0}^{r=2} \{5 + r \cdot \cos(\theta) + 2 \cdot r \cdot \sin(\theta)\} \cdot r \cdot dr = \int_{r=0}^{r=2} \{5r + r^2 \cdot \cos(\theta) + 2 \cdot r^2 \cdot \sin(\theta)\} \, dr$$
$$= \frac{5}{2} r^2 + \frac{1}{3} r^3 \cdot \cos(\theta) + \frac{2}{3} r^3 \cdot \sin(\theta) \Big|_{r=0}^{r=2} = \frac{5}{2}(4) + \frac{1}{3}(8) \cdot \cos(\theta) + 2 \cdot \frac{1}{3}(8) \cdot \sin(\theta)$$

Then $\int_{\theta=0}^{\theta=2\pi} 10 + \frac{8}{3} \cdot \cos(\theta) + \frac{16}{3} \cdot \sin(\theta) \, d\theta = 10\theta + \frac{8}{3} \cdot \sin(\theta) - \frac{16}{3} \cdot \cos(\theta) \Big|_{\theta=0}^{\theta=2\pi}$
$$= \{20\pi + 0 - \frac{16}{3}\} - \{0 + 0 - \frac{16}{3}\} = 20\pi \approx 62.83.$$

The volume is $20\pi \approx 62.83$.

Practice 1: If we restrict the domain to $R = \{(x,y): x^2 + y^2 \leq 4, 0 \leq y\}$ then the polar coordinate form of the domain is $R = \{(r,\theta): 0 \leq r \leq 2, 0 \leq \theta \leq \pi\}$. Verify that the volume for the same plane $f(x,y)=5+x+2y$ over this new domain is $\frac{32}{3} + 10\pi \approx 42.08$. Since the domain R in Practice 1 is half of the domain R in Example 1, why isn't the new volume half of the previous volume?

Example 2: Find the volume between the surface $f(x,y) = e^{-(x^2+y^2)}$ and the circular domain $R = \{(x,y): x^2 + y^2 \leq 4\}$. (Fig. 5)

Solution: This function looks complicated, but when we translate into polar coordinates, things get much easier.
$x = r \cdot \cos(\theta)$ and $y = r \cdot \sin(\theta)$ so $x^2 + y^2 = r^2$. And R becomes $R = \{(r,\theta): 0 \leq r \leq 2, 0 \leq \theta \leq 2\pi\}$. Then

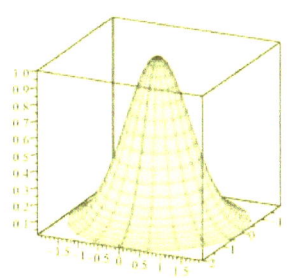

Fig. 5: $f(x,y) = e^{-(x^2+y^2)}$

$$\text{Volume} = \iint_R f \cdot dA = \int_{\theta=0}^{\theta=2\pi} \int_{r=0}^{r=2} e^{-r^2} \cdot r \cdot dr \cdot d\theta.$$

The inside integral $\int_{r=0}^{r=2} e^{-r^2} \cdot r \cdot dr = -\frac{1}{2} e^{-r^2} \Big|_{r=0}^{r=2} = \frac{1}{2}\left(1 - e^{-4}\right)$ (using the substitution $u = -r^2$)

$$\text{Volume} = \iint_R f \cdot dA = \int_{\theta=0}^{\theta=2\pi} \frac{1}{2}\left(1 - e^{-4}\right) \cdot d\theta = \frac{1}{2}\left(1 - e^{-4}\right) \cdot \theta \Big|_{\theta=0}^{\theta=2\pi}$$

$$= \frac{1}{2}\left(1 - e^{-4}\right) \cdot \theta \Big|_{\theta=0}^{\theta=2\pi} = \left(1 - e^{-4}\right)\pi \approx 0.98\pi.$$

Note: The $x^2 + y^2$ in the rectangular function is often a signal that the polar version may be easier since $x^2 + y^2 = r^2$. Then the substitution $u = r^2$ means $du = 2 \cdot r \cdot dr$ so $dA = r \cdot dr \cdot d\theta = \frac{1}{2} d\theta$.

Practice 2: (a) Find the volume between the surface $z = f(x,y) = e^{-(x^2+y^2)}$ and the circular domain

$R = \{(x,y) : x^2 + y^2 \leq A^2\}$, a circle of radius A centered at the origin.

(b) Show that as $A \to \infty$ then Volume $\to \pi$. This means the volume between the surface

$z = f(x,y) = e^{-(x^2+y^2)}$ and the entire xy-plane is π.

Areas and Average Values with Double Integrals in Polar Coordinates

If the z=f(x,y) is always equal to 1, then Volume $= \iint_R 1 \cdot dA = \{\text{base area}\}\{\text{height} = 1\} = $ base area.

Area in Polar Coordinates

If R is a closed and bounded region in the polar coordinate plane, then

Area of R $= \iint_R 1 \cdot dA = \iint_R 1 \, r \cdot dr \cdot d\theta$ In polar coordinates $dA = r \cdot dr \cdot d\theta$

The average value of a continuous function z=f(x,y) over a polar coordinate region is the same as we have used for rectangular coordinate regions: Average Value $= \frac{1}{\text{area}} \cdot$ volume.

Average Value of a Function in Polar Coordinates

If R is a closed and bounded region in the polar coordinate plane, and f(x,y) is continuous on R, then

Average Value of f on R $= \dfrac{1}{\text{area of R}} \cdot \{\text{volume between f and R}\} = \dfrac{\iint_R f \, dA}{\iint_R 1 \, dA}$

Example 3: Let $z = f(x,y) = e^{-(x^2+y^2)}$ be the height of a solid ice sculpture over the circular base $R = \{(x,y) : x^2 + y^2 \leq 4\}$. (This is Example 2.) If this sculpture is in a water-tight cylinder and then melts, how high will the resulting water be in the cylinder? (Fig. 6)

Solution: We have already done the calculus for this volume in Example 2: volume = 0.98π so we just need to divide this volume by the area of the circular base, 4π. The average height is approximately $0.98/4 = 0.245$.

Practice 3: Find the average value of $z = f(x,y) = 5 + x + 2y$ on the domain $R = \{(x,y) : x^2 + y^2 \leq 4\}$. (This is Example 1.)

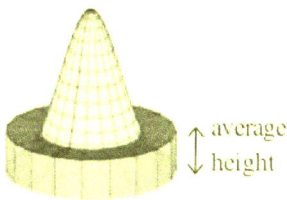

Fig. 6

Problems

For each given region R, decide whether to use polar or rectangular coordinates to evaluate a double integral with domain R.

1.
 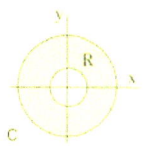 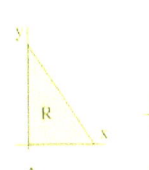

2.
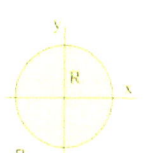 

3.
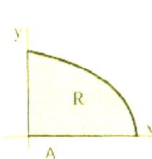 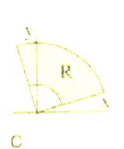 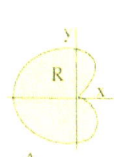

4.
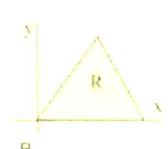 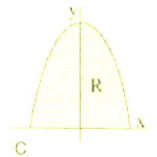

For Problems 5 to 8 find the volume between the surface $z = f(x,y)$ and the given region R.

5. $f(x,y) = 7 + 3x + 2y$ with $R = \{(x,y) : x^2 + y^2 \leq 9\}$.

6. $f(x,y) = 9 - 2x + 4y$ with $R = \{(x,y) : x^2 + y^2 \leq 5\}$.

7. $f(x,y) = A + Bx + Cy$ with $R = \{(x,y) : x^2 + y^2 \leq D^2\}$. (A, B, C, D are positive constants.)

8. $f(x,y) = A + Bx + Cy$ with $R = \{(x,y) : E^2 \leq x^2 + y^2 \leq D^2\}$. (A, B, C, D, E are positive constants.)

For Problems 9 to 12 sketch the domain of integration, use polar coordinates to evaluate each double integral.

9. Evaluate $\iint_R \sqrt{9-x^2-y^2}\, dA$ where $R = \{(x,y): x^2+y^2 \leq 9\}$.

10. Evaluate $\iint_R \sin(x^2+y^2)\, dA$ where $R = \{(x,y): 0 \leq x,\ 0 \leq y,\ x^2+y^2 \leq 9\}$.

11. Evaluate $\iint_R 2+xy\, dA$ where R is the region in inside the circle $x^2+y^2 = 4$ and above the x-axis..

12. Evaluate $\iint_R e^{x^2+y^2}\, dA$ where R is the set of (x,y) more than 1 unit and less than 3 units from the origin.

For Problems 13 to 18 use polar coordinates to find the volume of each solid.

13. Under the plane $z = 5+2x+3y$ and above the disk $x^2+y^2 \leq 16$.

14. Under the paraboloid $z = x^2+y^2$ and above the disk $x^2+y^2 \leq 9$ for (x,y) in the first quadrant.

15. Between the surfaces $z = 1+x+y$ and $z = 8+2x+3y$ for $x^2+y^2 \leq 1$ and $0 \leq y$.

16. Above the paraboloid $z = 6-x^2-y^2$ and under the plane $z = 9$ for $x^2+y^2 \leq 4$.

17. Between the surface $z = f(x,y) = \dfrac{1}{1+x^2+y^2}$ and the xy-plane for (x,y) in the first quadrant and $1 \leq x^2+y^2 \leq 4$.

18. (a) Between $z = \dfrac{1}{\left(1+x^2+y^2\right)^3}$ and the disk $x^2+y^2 \leq C^2$.

(b) Between $z = \dfrac{1}{\left(1+x^2+y^2\right)^3}$ and the entire xy-plane.

In problems 19 to 24, change the rectangular coordinate integral into an equivalent polar integral and evaluate the polar integral. It is usually helpful to sketch the domain of the integral.

19. $\displaystyle\int_{-1}^{1} \int_{0}^{\sqrt{1-x^2}} dy\, dx$

20. $\displaystyle\int_{-1}^{1} \int_{0}^{\sqrt{1-x^2}} (x^2+y^2)\, dy\, dx$

21. $\displaystyle\int_{0}^{1} \int_{0}^{\sqrt{1-x^2}} y\, dy\, dx$

22. $\displaystyle\int_{0}^{1/\sqrt{2}} \int_{y}^{\sqrt{1-y^2}} (x^2+y^2)\, dx\, dy$

23. $\displaystyle\int_{0}^{1} \int_{0}^{\sqrt{1-x^2}} e^{-(x^2+y^2)}\, dy\, dx$

24. $\displaystyle\int_{0}^{\infty} \int_{0}^{\infty} e^{-(x^2+y^2)}\, dy\, dx$

For Problems 25 to 29, find the average value of the function on the given region R.

25. $f(x,y) = 7 + 3x + 2y$ with $R = \{(x,y): x^2 + y^2 \leq 9\}$.

26. $f(x,y) = 2 + xy$ for $R = \{\text{top half of the disk } x^2 + y^2 \leq 4\}$.

27. $f(x,y) = 5 + 2x + 3y$ with $R = \{(x,y): x^2 + y^2 \leq 16\}$.

28. $f(x,y) = x^2 + y^2$ for R={part of the disk $x^2 + y^2 \leq 9$ in the first quadrant}

29. A sprinkler (located at the origin) sprays water so after one hour the depth at location (x,y) feet is
    $f(x,y) = K \cdot e^{-(x^2+y^2)}$ feet.

    (a) How much water reaches the annulus $2 \leq r \leq 4$ and the annulus $8 \leq r \leq 10$ in one hour?

    (b) What is the average amount of water (depth per square foot) of water at (x,y) in each annulus after one hour?

    (c) Why is this a poor design for a sprinkler?

30. $f(x) = K \cdot e^{-\left(\frac{x^2}{2}\right)}$ is the normal probability distribution for a population with mean 0 and standard deviation 1, and is extremely important in probability theory and applications. Unfortunately, it does not have a "nice" antiderivative in terms of elementary functions, but we can use double integrals in polar coordinates to evaluate $\int_{-\infty}^{\infty} e^{-x^2} dx$.

    (a) The rectangular coordinate double integral of $f(x,y) = e^{-(x^2+y^2)} = e^{-x^2} \cdot e^{-x^2}$ with

    R={square $-C \leq x \leq C$, $-C \leq y \leq C$} is

    $$\int_{-C}^{C} \int_{-C}^{C} e^{-(x^2+y^2)} dx \cdot dy = \int_{-C}^{C} \int_{-C}^{C} e^{-x^2} \cdot e^{-y^2} dx \cdot dy$$

    $$= \int_{-C}^{C} e^{-y^2} \left(\int_{-C}^{C} e^{-x^2} dx \right) dy = \left(\int_{-C}^{C} e^{-x^2} dx \right) \left(\int_{-C}^{C} e^{-y^2} dy \right) = \left(\int_{-C}^{C} e^{-x^2} dx \right)^2.$$

    So the

    $$\iint_{xy-plane} e^{-(x^2+y^2)} dA = \lim_{C \to \infty} \int_{-C}^{C} \int_{-C}^{C} e^{-(x^2+y^2)} dx \cdot dy = \lim_{C \to \infty} \left(\int_{-C}^{C} e^{-x^2} dx \right)^2 = \left(\int_{-\infty}^{\infty} e^{-x^2} dx \right)^2$$

    But from Practice 2 we know that $\iint_{xy-plane} e^{-(x^2+y^2)} dA = \pi$ so $\int_{-\infty}^{\infty} e^{-x^2} dx = \sqrt{\pi}$.

    Finally, changing the variable to $u = \frac{x}{\sqrt{2}}$ we get $\int_{-\infty}^{\infty} e^{-\left(\frac{x^2}{2}\right)} dx = \int_{-\infty}^{\infty} e^{-u^2} \sqrt{2} \cdot du = \sqrt{2\pi}$.

    That was a lot of work, but this is a very important integral.

Practice Answers

Practice 1: The only difference from Example 1 is that the domain angle θ goes from 0 to π instead of from 0 to 2π so the calculation is the same until we get to the final evaluation:

$$\text{volume} = \iint_R f(x,y) \cdot dA = \int_{\theta=0}^{\theta=2\pi} \int_{r=0}^{r=2} \{5 + r \cdot \cos(\theta) + 2 \cdot r \cdot \sin(\theta)\} \cdot r \cdot dr \cdot d\theta$$

$$= 10\theta + \frac{8}{3} \cdot \sin(\theta) - \frac{16}{3} \cdot \cos(\theta) \Big|_{\theta=0}^{\theta=\pi} = \left\{10\pi + 0 + \frac{16}{3}\right\} - \left\{0 + 0 - \frac{16}{3}\right\} = 10\pi + \frac{32}{3}$$

The new domain may be half the area of the original domain, but the function is larger over the new domain than over the original domain. Symmetry is very powerful, but we need to be careful and only use it when it is justified.

Practice 2: (a) The only change from Example 2 is that now the radius r goes from 0 to A instead of from 0 to 2 so most of the calculations are the same:

$$\int_{\theta=0}^{\theta=2\pi} \int_{r=0}^{r=A} e^{-r^2} \cdot r \cdot dr \cdot d\theta = \frac{1}{2}\left(1 - e^{-A^2}\right) \theta \Big|_{\theta=0}^{\theta=2\pi} = \left(1 - e^{-A^2}\right)\pi.$$

(b) $\lim\limits_{A \to \infty} \text{Volume} = \lim\limits_{A \to \infty} \left(1 - e^{-A^2}\right)\pi = \pi$

Practice 3: From Example 1 we know the volume is 20π, and the area of the circular base is 4π so the average value is 5. (Fig. 7)

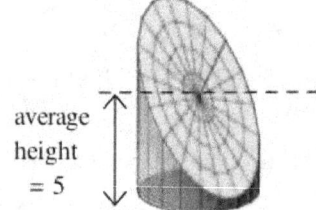

Fig. 7: f(x,y)=5+x+2y

14.4 APPLICATIONS OF DOUBLE INTEGRALS

In Section 5.4 we used single integrals to determine the mass, moments about each axis, and the center of mass of a thin plate (lamina) with uniform density δ. In this section we will use double integrals to extend those ideas and calculations to thin plates with varying densities.

Uniform density for a 2D region between f and the x-axis for $a \leq x \leq b$

Total Mass: $\quad M = \delta \cdot \int_a^b f(x)\, dx$

Moments: $\quad$ about y-axis $M_y = \delta \cdot \int_a^b x \cdot f(x)\, dx \quad$ about x-axis $M_x = \delta \cdot \int_a^b \frac{1}{2} f^2(x)\, dx$

Center of mass: $\quad \bar{x} = \dfrac{M_y}{M}\,, \quad \bar{y} = \dfrac{M_x}{M}$

If the density of the plate depends on the location (x,y) on the plate, then δ is a function of x and y: $\delta(x,y)$ with units of the form (mass)/area such as kg/m^2.

Nonuniform density for a 2D region R in the xy-plane

Area: $\quad Area = \iint_R 1\, dA$

Total Mass: $\quad M = \iint_R \delta(x,y)\, dA$

Moments: $\quad$ about y-axis $M_y = \iint_R x \cdot \delta(x,y)\, dA \quad$ about x-axis $M_x = \iint_R y \cdot \delta(x,y)\, dA$

Center of mass: $\quad \bar{x} = \dfrac{M_y}{M}\,, \quad \bar{y} = \dfrac{M_x}{M}$

Note: Be careful to use the x with M_y since the little dA piece is x units from the y-axis. Similarly, use the y with the M_x formula since the little dA piece is y units from the x-axis.

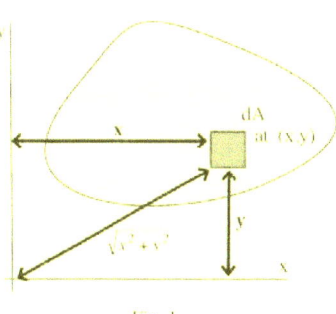

Fig. 1

Example 1: $R = \{(x,y) : 0 \le x \le 2 \text{ and } 0 \le y \le 4 - x^2\}$ (Fig. 2).

(a) Determine the center of mass of R if the region has uniform density δ.

(b) Determine the center of mass of R if $\delta(x,y) = x \cdot y^2$.

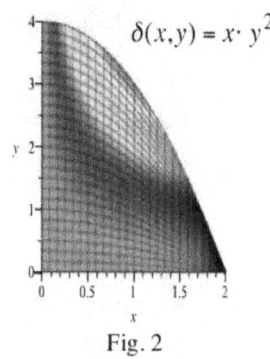

Fig. 2

Solution: (a) Using double integrals,

$$M = \iint_R \delta(x,y) \, dA = \int_{x=0}^{2} \int_{y=0}^{4-x^2} 1 \, dy \, dx = \frac{16}{3}.$$

$$M_y = \int_{x=0}^{2} \int_{y=0}^{4-x^2} x \, dy \, dx = 4 \quad \text{and} \quad M_x = \int_{x=0}^{2} \int_{y=0}^{4-x^2} y \, dy \, dx = \frac{128}{15}$$

so $\bar{x} = \frac{3}{4}$ and $\bar{y} = \frac{8}{5}$.

(b) $\delta(x,y) = x \cdot y^2$, $M = \int_{x=0}^{2} \int_{y=0}^{4-x^2} x \cdot y^2 \, dy \, dx = \frac{32}{3}$, $M_y = \int_{x=0}^{2} \int_{y=0}^{4-x^2} x \cdot x \cdot y^2 \, dy \, dx = \frac{8192}{945}$

and $M_x = \int_{x=0}^{2} \int_{y=0}^{4-x^2} y \cdot x \cdot y^2 \, dy \, dx = \frac{128}{5}$ so $\bar{x} = \frac{256}{315}$ and $\bar{y} = \frac{12}{5}$.

Practice 1: $R = \{(x,y) : 0 \le x \le 2 \text{ and } 0 \le y \le 4 - x^2\}$. Determine the center of mass of R if $\delta(x,y) = x^2 \cdot y$.

Moments of Inertia (second moments) and Radii of Gyration

Moments of inertia are needed to calculate the kinetic energy of rotating objects and also for formulas for stiffness of beams. Note that the second moments formulas are very similar to the first moment formulas but that they use the squares x^2 and y^2 of the lever arms distances of the dA piece from the axes instead of just x and y.

Moments of Inertia (second moments) of region R in the xy-plane

About the x-axis: $I_x = \iint_R y^2 \cdot \delta(x,y) \, dA$ About the y-axis: $I_y = \iint_R x^2 \cdot \delta(x,y) \, dA$

About the origin: $I_0 = \iint_R (x^2 + y^2) \cdot \delta(x,y) \, dA = I_x + I_y$

Radii of Gyration of region R in the xy-plane

About the x-axis: $R_x = \sqrt{\dfrac{I_x}{M}}$ About the y-axis: $R_y = \sqrt{\dfrac{I_y}{M}}$ About the origin: $R_0 = \sqrt{\dfrac{I_0}{M}}$

The moment of inertia about the origin, I_0, is also called the **polar moment**, and that integral uses the square of the distance of the dA piece from the origin.

The radius of gyration R_x is the distance from the x-axis that a point with mass M must be in order to give the same moment of inertia I_x: $I_x = M \cdot R_x^2$.

Example 2: If the units of x and y are meters and the units of $\delta(x,y)$ are kg/m^2, determine the units for I_x and R_x.

Solution: $dA = dx \cdot dy$ has units m^2, so $I_x = \iint_R y^2 \cdot \delta(x,y) \, dA$ has units $m^2 \cdot \frac{kg}{m^2} \cdot m^2 = kg \cdot m^2$.

$R_x = \sqrt{\frac{I_x}{M}} = \sqrt{\frac{kg \cdot m^2}{kg}} = m$.

Practice 2: If the units of x and y are feet and the units of $\delta(x,y)$ are $\frac{slug}{ft^2}$, determine the units for I_x and R_x. (One slug of mass in the British system is approximately 14.594 kg.)

Example 3: Suppose x and y are given in meters and the triangular region R={(x,y): $0 \leq x \leq 2$ and $0 \leq y \leq 4-2x$} (Fig. 3) has uniform density $\delta(x,y) = \delta \, \frac{kg}{m^2}$.

Calculate M, $\bar{x}$, I_x, and R_x.

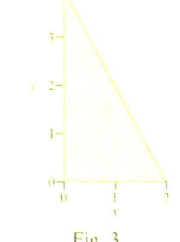

Fig. 3

Solution: $M = \iint_R \delta \, dA = \int_0^2 \int_0^{4-2x} \delta \, dy \, dx = \int_0^2 \delta \cdot (4-2x) \, dx = 4\delta$ kg.

$M_y = \iint_R x \cdot \delta \, dA = \int_0^2 \int_0^{4-2x} x \cdot \delta \, dy \, dx = \int_0^2 \delta(4x - 2x^2) \, dx = \frac{8}{3}\delta$ kg·m

$I_x = \iint_R y^2 \cdot \delta \, dA = \int_0^2 \int_0^{4-2x} y^2 \cdot \delta \, dy \, dx = \int_0^2 \frac{\delta}{3}(4-2x)^3 \, dx = \frac{32}{3}\delta$ kg·m^2

so $\bar{x} = \frac{M_y}{M} = \frac{2}{3}$ m and $R_x = \sqrt{\frac{I_x}{M}} = \sqrt{\frac{8}{3}} \, m^2 \approx 1.63$ m.

Practice 3: Calculate $\bar{y}$, I_y, and R_y for the region R in Example 3.

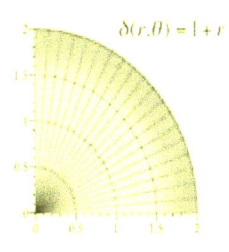

Example 4: R is the quarter circle of radius 2 in the first quadrant with density $\delta(r,\theta) = 1+r$. (Fig. 4) Calculate M, $\bar{x}$, I_x, and R_x.

Solution: $M = \iint_R \delta \, dA = \int_{\theta=0}^{\pi/2} \int_{r=0}^{2} (1+r) \cdot r \, dr \, d\theta = \int_0^{\pi/2} \frac{14}{3} \, d\theta = \frac{7}{3}\pi$

Fig. 4

$$M_y = \iint_R x \cdot \delta \, dA = \int_{\theta=0}^{\pi/2} \int_{r=0}^{2} r \cdot \cos(\theta) \cdot (1+r) \cdot r \, dr \, d\theta = \int_0^{\pi/2} \frac{20}{3} \cos(\theta) \, d\theta = \frac{20}{3}$$

$$I_x = \iint_R y^2 \cdot \delta \, dA = \int_{\theta=0}^{\pi/2} \int_{r=0}^{2} r^2 \cdot \sin^2(\theta) \cdot (1+r) \cdot r \, dr \, d\theta = \int_0^{\pi/2} \frac{52}{5} \sin^2(\theta) \, d\theta = \frac{13}{5}\pi$$

so $\bar{x} = \dfrac{M_y}{M} = \dfrac{20}{7\pi} \approx 0.909$ and $R_x = \sqrt{\dfrac{I_x}{M}} = \sqrt{\dfrac{39}{35}} \approx 1.056$.

Note: Because of the symmetry of the R and δ, then $\bar{y} = \bar{x}$, $I_y = I_x$, and $R_y = R_x$.

Problems

In problems 1 to 8 use double integrals to calculate the area, total mass, the moments about each axis, and the center of mass of the region. Plot the location of the center of mass on the region.

1. R is the rectangular region bounded by the x and y axes and the lines x=2 and y=4 with $\delta = 1$.

2. R is the rectangular region bounded by the x and y axes and the lines x=2 and y=4 with $\delta = xy$.

3. R is the shaded region in Fig. 5 and $\delta(x,y) = 1+x$.

4. R is the shaded region in Fig. 5 and $\delta(x,y) = 1+y$.

5. R={(x,y): 0≤x≤3, 0≤y≤1+x} and $\delta(x,y) = x+y$.

6. R={(x,y): 0≤x≤3, 0≤y≤1+x} and $\delta(x,y) = 1+y$.

7. R is the shaded region in Fig. 6 and $\delta(r,\theta) = 1$.

8. R is the shaded region in Fig. 6 and $\delta(r,\theta) = r$.

9. R={(r, θ): 0≤θ≤π, 0≤r≤1+cos(θ)} and $\delta(r,\theta) = r$. (Fig. 7)

10. R={(r, θ): 0≤θ≤π, 0≤r≤1+cos(θ)} and $\delta(r,\theta) = 1+r$. (Fig. 7)

11. Calculate the values of I_x and R_x for the region in Problem 1.

12. Calculate the values of I_y and R_y for the region in Problem 1.

13. Calculate the values of I_x and R_x for the region in Problem 2.

14. Calculate the values of I_y and R_y for the region in Problem 2.

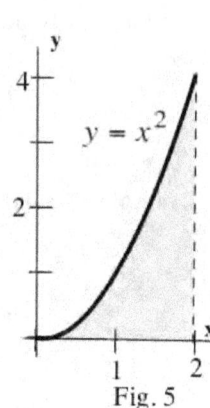

Fig. 5

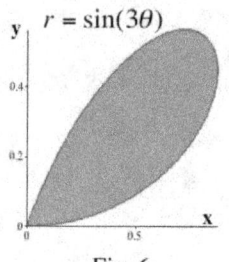

Fig. 6

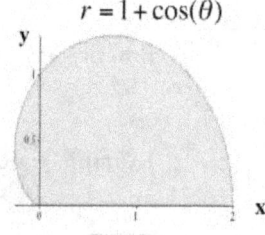

Fig. 7

14.4 Applications of Double Integrals

In Problems 15 to 20, use the fact that the attractive gravitational force between two points with masses M and m at a distance of r is $\text{force} = \dfrac{GMm}{r^2}$. (It helps to sketch the regions.)

15. Represent the total force between a point mass of 10 kg at the origin and a bar from x=2 to x=4 m on the x-axis with a mass of 8 kg. (This is a single integral.)

16. 15. Represent the total force between a point mass of M kg at the origin and a bar from a to b m (0<a<b) m on the x-axis with a mass of m kg. (This is a single integral.)

17. Represent the total force between a bar along the x-axis from 0 to 2 with mass of 10 kg and another bar on the x-axis from 4 to 7 with a mass of 9 kg. (This is a double integral.)

18. Represent the total force between a bar along the x-axis from a to b with total mass M and another bar on the x-axis from c to d with a mass of m (a<b<c<d). (This is a double integral.)

Practice Solutions

Practice 1: $R = \{(x,y) : 0 \le x \le 2 \text{ and } 0 \le y \le 4 - x^2\}$. Determine the center of mass of R if $\delta(x,y) = x^2 \cdot y$.

$(\delta(x,y) = x^2 \cdot y$, $M = \displaystyle\int_{x=0}^{2} \int_{y=0}^{4-x^2} x^2 \cdot y \, dy \, dx = \dfrac{512}{105}$, $M_y = \displaystyle\int_{x=0}^{2} \int_{y=0}^{4-x^2} x \cdot x^2 \cdot y \, dy \, dx = \dfrac{16}{3}$

and $M_x = \displaystyle\int_{x=0}^{2} \int_{y=0}^{4-x^2} y \cdot x^2 \cdot y \, dy \, dx = \dfrac{8192}{945}$ so $\bar{x} = \dfrac{35}{32}$ and $\bar{y} = \dfrac{16}{9}$.

Practice 2: $dA = dx \cdot dy$ has units ft^2, so $I_x = \displaystyle\iint_R y^2 \cdot \delta(x,y) \, dA$ has units $ft^2 \cdot \dfrac{slug}{ft^2} \cdot ft^2 = slug \cdot ft^2$.

$M = \displaystyle\iint_R \delta(x,y) \, dA = \dfrac{slug}{ft^2} \cdot ft^2 = slug$ so $R_x = \sqrt{\dfrac{I_x}{M}} = \sqrt{\dfrac{slug \cdot ft^2}{slug}} = feet$.

Practice 3: $M_x = \displaystyle\iint_R y \cdot \delta \, dA = \int_0^2 \int_0^{4-2x} y \cdot \delta \, dy \, dx = \int_0^2 \delta \cdot \dfrac{1}{2}(4-2x)^2 \, dx = \dfrac{16}{3} \delta$

so $\bar{y} = \dfrac{16/3 \, \delta}{4 \, \delta} = \dfrac{4}{3}$ m.

$I_y = \displaystyle\iint_R x^2 \cdot \delta \, dA = \int_0^2 \int_0^{4-2x} x^2 \cdot \delta \, dy \, dx = \int_0^2 \delta \cdot x \cdot (4-2x) \, dx = \dfrac{8}{3} \delta \, kg \cdot m^2$

so $R_y = \sqrt{\dfrac{I_y}{M}} = \sqrt{\dfrac{8/3 \, \delta}{4 \, \delta} \, m^2} = \sqrt{\dfrac{2}{3}} \approx 0.816$ m

14.5 SURFACE AREAS USING DOUBLE INTEGRALS

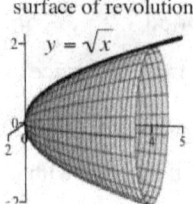

Fig. 1

In Section 5.2 we determined a method for calculating the area of a surface of revolution (Fig. 1). Here we will build a way to calculate the area of a surface of the form $z = f(x,y)$ over a region R (Fig. 2). Sometimes both methods can be used and in that case they both give the same result. There is another type of surface called a parametric surface (Fig. 3) that is even more general, and we will build a way to determine those surface areas in Section 15.7.

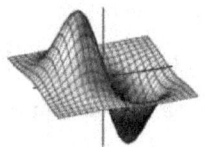

Fig. 2

To make the derivation and figures simpler, we assume that R is a rectangle in the xy-plane and that $z=f(x,y) \geq 0$ in R. As we have done before, we start by partitioning the domain into small Δx by Δy rectangles and note that the area ΔS of the tangent plane (Fig. 4) above one of these little ΔA rectangles has approximately the same area as the surface area above the ΔA rectangle:. If we can represent the sides of the tilted tangent plane above the ΔA rectangle as vectors, then we can use the cross product of those vectors to determine the area of the rectangle.

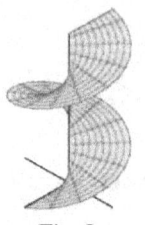

Fig. 3

Starting at one corner (x,y) of a little rectangle in the domain, and moving Δx in the x direction, the vector A along the tangent plane is $A = \langle \Delta x, 0, f_x(x,y) \cdot \Delta x \rangle$. Similarly, starting at (x,y) and moving in the y direction, the vector B along the tangent plane is $B = \langle 0, \Delta y, f_y(x,y) \cdot \Delta y \rangle$. Then the area of the tangent plane above the ΔA rectangle is the magnitude of the cross product of A and B:

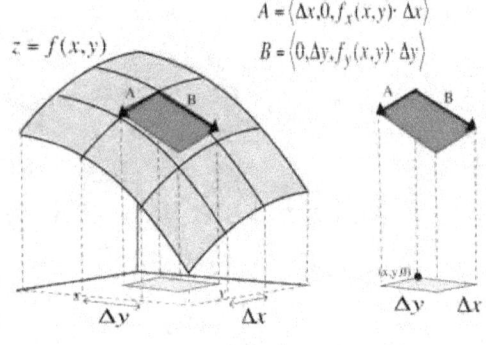

Fig. 4

$$\Delta S = \{\text{tangent plane area}\} = |A \times B|.$$

Example 1: The function in Fig. 4 is $f(x,y) = 7 - \frac{1}{2}x^2 - y^2$. Determine the area of the little rectangle at the point (2,1,4) with $\Delta x = 0.3$ and $\Delta y = 0.1$.

Solution: $f_x(2,1) = -2$ and $f_y(2,1) = -2$ so $A = \langle 0.3, 0, -0.6 \rangle$ and $B = \langle 0, 2, -0.2 \rangle$.

$$A \times B = \begin{vmatrix} i & j & k \\ 0.3 & 0 & -0.6 \\ 0 & 0.1 & -0.2 \end{vmatrix} = \langle 0.06, 0.06, 0.03 \rangle \text{ so area} = |A \times B| = 0.09,$$

Practice 1: Determine the area of the little rectangle for the Example 1 function at the point (1, 1, 5.5) with $\Delta x = 0.2$ and $\Delta y = 0.3$.

In the general case $A = \langle \Delta x, 0, f_x(x,y) \rangle$ and $B = \langle 0, \Delta y, f_y(x,y) \rangle$ so

$$A \times B = \begin{vmatrix} i & j & k \\ \Delta x & 0 & f_x(x,y) \cdot \Delta x \\ 0 & \Delta y & f_y(x,y) \cdot \Delta y \end{vmatrix} = \langle -\Delta y \cdot f_x \cdot \Delta x, -\Delta x \cdot f_y \cdot \Delta y, \Delta x \cdot \Delta y \rangle = \langle -f_x, -f_y, 1 \rangle \cdot \Delta A$$

and $|A \times B| = \left(\sqrt{(f_x)^2 + (f_y)^2 + 1} \right) \cdot \Delta A = \Delta S$.

The total surface area is the accumulation of all of these little areas:

Surface area $A \approx \sum \sum \Delta S = \sum \sum \left(\sqrt{(f_x)^2 + (f_y)^2 + 1} \right) \cdot \Delta A \rightarrow \iint\limits_R \left(\sqrt{(f_x)^2 + (f_y)^2 + 1} \right) \cdot dA$

$$\boxed{\text{For } z = f(x,y), \text{ Surface Area} = \iint\limits_R \sqrt{1 + \left(\frac{\partial z}{\partial x}\right)^2 + \left(\frac{\partial z}{\partial y}\right)^2} \cdot dA}$$

You should recognize the similarity of this formula to the formula for arc length from Section 5.2:

$$L = \int \sqrt{1 + \left(\frac{dy}{dx}\right)^2} \; dx ,$$

Example 2: Determine the surface area of the plane $f(x,y) = 1 + 2x + y$ over the rectangular region $R = \{(x,y): 0 \le x \le 2, 0 \le y \le 3\}$. (Fig. 5)

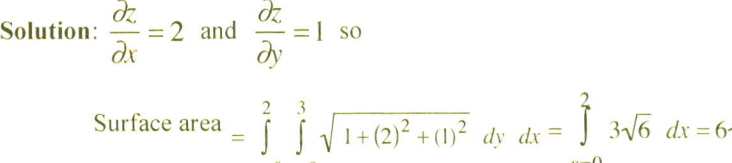

f(x,y)=1+2x+y

Fig. 5

Solution: $\dfrac{\partial z}{\partial x} = 2$ and $\dfrac{\partial z}{\partial y} = 1$ so

Surface area $= \int\limits_{x=0}^{2} \int\limits_{y=0}^{3} \sqrt{1 + (2)^2 + (1)^2} \; dy \; dx = \int\limits_{x=0}^{2} 3\sqrt{6} \; dx = 6\sqrt{6}$

Practice 2: Determine the surface area of the plane $f(x,y) = 3 + 8x + 4y$ over the rectangular region $R = \{(x,y): 0 \le x \le 4, 0 \le y \le 3\}$.

The surface area formula also works for domains that are not rectangular, and sometimes polar coordinates make the evaluation easier.

Example 3: Determine the surface area of the plane $f(x,y) = 10 - 2x - 3y$ over the circular disk $R = \{(x,y): 0 \leq x^2 + y^2 \leq 4\}$. (Fig. 6)

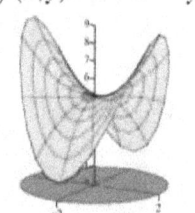

Fig. 6

Solution: $\dfrac{\partial z}{\partial x} = -2$ and $\dfrac{\partial z}{\partial y} = -3$ so Surface area = $\iint_R \sqrt{1+(-2)^2+(-3)^2} \cdot dA$

Because of the symmetry of R, this is easier to evaluate using polar coordinates:

Surface area = $\displaystyle\int_{\theta=0}^{2\pi} \int_{r=0}^{2} \sqrt{14}\; r \cdot dr \cdot d\theta = \int_{\theta=0}^{2\pi} 2\sqrt{14} \cdot d\theta = 4\pi\sqrt{14}$.

Example 4: Determine the surface area of the saddle $f(x,y) = 5 + x^2 - y^2$ over the circular disk $R = \{(x,y): 0 \leq x^2 + y^2 \leq 4\}$. (Fig. 7)

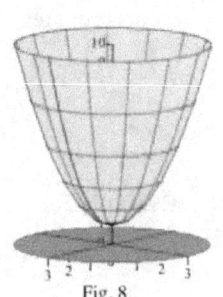

Fig. 7

Solution: $\dfrac{\partial z}{\partial x} = 2x$ and $\dfrac{\partial z}{\partial y} = -2y$ so

Surface area = $\iint_R \sqrt{1+(2x)^2+(-2y)^2} \cdot dA = \iint_R \sqrt{1+4(x^2+y^2)} \cdot dA$

Again, polar coordinates make this easier:

Surface area = $\displaystyle\int_{\theta=0}^{2\pi} \int_{r=0}^{2} \sqrt{1+4r^2} \cdot r \cdot dr \cdot d\theta$

$= \displaystyle\int_{\theta=0}^{2\pi} \dfrac{1}{12}(17)^{3/2} - \dfrac{1}{12} \; d\theta = \left(\dfrac{1}{12}(17)^{3/2} - \dfrac{1}{12} \right) \cdot 2\pi$

Practice 3: Determine the surface area of the paraboloid $f(x,y) = 1 + x^2 + y^2$ over the circular disk $R = \{(x,y): 0 \leq x^2 + y^2 \leq 9\}$. (Fig. 8)

Fig. 8

These examples and practice problems were chosen so that it was possible to evaluate the integrals "by hand." Unfortunately that is rarely the situation. Once we take partial derivatives, square them, add those squares and a 1, and then take a square root the result is usually not an integral that we can evaluate by hand, at least not easily. If we just change the Example 4 function slightly to be $f(x,y) = 5 + 2x - y^2$ then $f_x(x,y) = 2$ and $f_y(x,y) = -2y$ so the surface area integral is $\iint_R \sqrt{5+4y^2}\; dA$. This is a rather difficult antidrivative which involves the inverse hyperbolic sine function. In many cases the surface area involves integrals that do not have antiderivatives involving only elementary functions and we need to resort to software such as Maple or Mathematica.

Example 5: The formula for Fig. 2 is $f(x,y) = x \cdot e^{-x^2-y^2}$ and the graph of f is over the rectangle $-2 \leq x \leq 2$ and $-2 \leq y \leq 2$. Represent the surface area using double integrals.

Solution: $f_x(x,y) = x \cdot \left(e^{-x^2-y^2}\right)(-2x) + \left(e^{-x^2-y^2}\right) = (-2x^2+1) \cdot e^{-x^2-y^2}$ and

$$f_y(x,y) = x \cdot \left(e^{-x^2-y^2}\right)(-2y) = -2xy \cdot e^{-x^2-y^2}$$ so

Surface area $= \int_{x=-2}^{2} \int_{y=-2}^{2} \sqrt{1 + \left((-2x^2+1)^2 + 4x^2 y^2\right)\left(e^{-x^2-y^2}\right)^2} \, dy \, dx$ Yuck.

But the Maple command

int(sqrt(1+ ((1-2*x^2)^2+4*x^2*y^2)*(exp(-x^2-y^2))^2), x=-2 .. 2, y=-2 ..2);

quickly gives the result 16.72816232.

Problems

1. Find the area of the surface $f(x,y) = x^2 + y$ over the triangular domain bounded by the x-axis, the line x=2 and the line y=2x.

2. Find the area of the surface $f(x,y) = x^2 + 3y$ over the triangular domain bounded by the x-axis, the line x=2 and the line y=x.

3. Find the area of the surface $f(x,y) = 4x + y^2$ over the triangular domain with vertices (0,0), (0,4) and (2,4).

4. Find the area of the surface $f(x,y) = x + 3y^2$ over the triangular domain with vertices (0,0), (0,4) and (2,4).

5. Find the area of the surface $f(x,y) = 1 + 3x + 4y$ over the domain bounded by the x-axis and the parabola $y = 4 - x^2$.

6. Find the area of the surface $f(x,y) = 2 + 12x + y$ over the domain bounded by the x-axis and the parabola $y = 1 - x^2$.

7. Find the area of the surface $f(x,y) = 5 + 4x + 3y$ over the circular domain $R = \{(x,y): x^2 + y^2 \leq 9\}$.

8. Find the area of the surface $f(x,y) = 1 + x + y$ over the circular domain $R = \{(x,y): x^2 + y^2 \leq 4\}$.

9. Find the area of the surface $f(x,y) = 1 + x + y$ over the domain $R = \{(x,y): 1 \leq x^2 + y^2 \leq 9\}$.

10. Find the area of the cone $f(x,y) = \sqrt{x^2 + y^2}$ over the circular domain $R = \{(x,y): x^2 + y^2 \leq 4\}$.

11. Find the area of the cone $f(x,y) = \sqrt{x^2 + y^2}$ over the domain $R = \{(x,y): 1 \leq x^2 + y^2 \leq 9\}$.

12. Find the area of the paraboloid $f(x,y) = x^2 + y^2$ over the circular domain $R = \{(x,y): x^2 + y^2 \leq 4\}$.

In problems 13 to 16 set up the integrals representing the surface area of the given function on the given domain. (These may be too difficult to evaluate by hand.)

13. $f(x,y) = x \cdot y^2$ on the domain $R = \{(x,y): -2 \leq x \leq 2, x^2 \leq y \leq 4\}$.

14. $f(x,y) = x^2 + y^3$ on the domain $R = \{(x,y): -2 \leq x \leq 2, x^2 \leq y \leq 4\}$.

15. $f(x,y) = 2 + \sin(x) + \cos(y)$ on the domain $R = \{(x,y): 0 \leq x \leq 2, 0 \leq y \leq 3\}$.

16. $f(x,y) = 2 + x^3 - y^2$ on the domain $R = \{(x,y): 0 \leq x \leq 2, 0 \leq y \leq 1\}$.

Practice Answers

Practice 1: $f_x(1,1) = -1$ and $f_y(1,1) = -2$ so $A = \langle 0.2, 0, -0.2 \rangle$ and $B = \langle 0, 0.3, -0.6 \rangle$

$$A \times B = \begin{vmatrix} i & j & k \\ 0.2 & 0 & -0.2 \\ 0 & 0.3 & -0.6 \end{vmatrix} = \langle 0.06, 0.12, 0.06 \rangle \text{ so area} = |A \times B| \approx 0.146.$$

Practice 2: $f(x,y) = 3 + 8x + 4y$ so $\dfrac{\partial z}{\partial x} = 8$ and $\dfrac{\partial z}{\partial y} = 4$ then

$$\text{Surface area} = \int_{x=0}^{4} \int_{y=0}^{3} \sqrt{1 + (8)^2 + (4)^2} \, dy \, dx = \int_{x=0}^{4} 27 \, dx = 108$$

Practice 3: $\dfrac{\partial z}{\partial x} = 2x$ and $\dfrac{\partial z}{\partial y} = 2y$ so

$$\text{Surface area} = \iint_R \sqrt{1 + (2x)^2 + (2y)^2} \cdot dA = \iint_R \sqrt{1 + 4(x^2 + y^2)} \cdot dA$$

Again, polar coordinates make this easier (with a $u = 1 + 4r^2$ substitution).

$$\text{Surface area} = \int_{\theta=0}^{2\pi} \int_{r=0}^{3} \sqrt{1 + 4r^2} \cdot r \cdot dr \cdot d\theta = \int_{\theta=0}^{2\pi} \frac{1}{12}(37)^{3/2} - \frac{1}{12} \, d\theta = \left(\frac{1}{12}(37)^{3/2} - \frac{1}{12}\right) \cdot 2\pi$$

14.6 TRIPLE INTEGRALS AND APPLICATIONS

Sometimes the value of a continuous function f(x,y,z) depends on the location in 3 dimensions: perhaps we know the density (kg/m^3) at each location (x,y,z) of a 3D object and want to determine the total mass (kg) of the object. Everything in this section is in rectangular coordinates. The next section considers triple integrals in cylindrical and spherical coordinates.

Our strategy is similar to that used to create double integrals, except now our region R is a 3D solid and we partition R into small rectangular cells (boxes) by cuts parallel to the coordinate planes. (Fig. 1) Then the volume of each little box is $\Delta V = \Delta x \cdot \Delta y \cdot \Delta z$. If we want the mass of the little box, it is approximately the density δ at some point (x*, y*, z*) inside each box times the volume of the box:

$\Delta M = \delta(x^*, y^*, z^*) \cdot \Delta V = \delta(x^*, y^*, z^*) \cdot \Delta x \cdot \Delta y \cdot \Delta z$. By adding the approximate masses of all of the boxes together, a triple sum, we can approximate the total mass of the solid:

$$M \approx \sum_{\Delta z} \sum_{\Delta y} \sum_{\Delta x} \delta(x^*, y^*, z^*) \cdot \Delta x \cdot \Delta y \cdot \Delta z$$

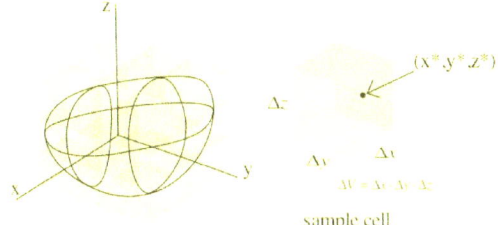

Fig. 1: Solid in 3D

sample cell

As before, letting all of the side lengths of the boxes approach 0, we get a triple integral:

$$\lim_{\Delta x, \Delta y, \Delta z \to 0} \sum_{\Delta z} \sum_{\Delta y} \sum_{\Delta x} \delta(x^*, y^*, z^*) \cdot \Delta x \cdot \Delta y \cdot \Delta z \to \iiint_R \delta(x,y,z)\, dV$$

Triple integrals have all of the properties that you might expect, and these properties follow from the properties of finite sums.

---

1) $\iiint_R k \cdot f\, dV = k \cdot \iiint_R f\, dV$

2) $\iiint_R \{f \pm g\}\, dV = \iiint_R f\, dV \pm \iiint_R g\, dV$

3) If f≥g for all points in R, then $\iiint_R f\, dV \geq \iiint_R g\, dV$

4) If the R is partitioned into two subregions R1 and R2 by a smooth surface, then $\iiint_R f\, dV = \iiint_{R1} f\, dV + \iiint_{R2} f\, dV$

5) Volume of R = $\iiint_R 1\, dV$

6) Units of $\iiint_R f\, dV$ are (units of f) . (units of x) . (units of y) . (units of z)

Evaluating Triple Integrals

Triple integrals are rarely evaluated as limits of triple sums. Instead, we evaluate single integrals, working from the inside out just as we did with double integrals.

Example 1: Evaluate $\iiint_R f\, dV$ for $f(x,y,z) = 2x + y + z^2$ on the solid
$$R = \{(x,y,z): 0 \le x \le 2,\ 1 \le y \le 4,\ 0 \le z \le 1\}$$

Solution:
$$\iiint_R f\, dV = \int_{z=0}^{1} \int_{y=1}^{4} \int_{x=0}^{2} 2x + y + z^2\ dx\, dy\, dz$$

Starting on the inside, $\int_{x=0}^{2} 2x + y + z^2\ dx = x^2 + xy + xz^2 \Big|_{x=0}^{2} = 4 + 2y + 2z^2$

Next, $\int_{y=1}^{4} \int_{x=0}^{2} 2x + y + z^2\ dx\, dy = 4y + y^2 + 2yz^2 \Big|_{y=1}^{4} = 12 + 15 + 6z^2$

Finally, $\int_{z=0}^{1} 27 + 6z^2\ dz = 27z + 2z^3 \Big|_{z=0}^{1} = 29$.

If the x, y and z units are meters, and the f units are kg/m^3, then $\iiint_R f\, dV$ = 35 kg.

3D Fubini's Theorem

If f is continuous on the domain R,

then the triple integral can be evaluated in any order that describes R.

In the case of a box [a,b]x[c,d]x[e,f], that means any order of dx, dy and dz gives the same result as long as the end points match the variable: $a \le x \le b$, $c \le y \le d$ and $e \le z \le f$.

Practice 1: Evaluate $\iiint_R x^3 + y^2 + z\ dV$ for $R = \{(x,y,z): 0 \le x \le 2,\ 0 \le y \le,\ 1 \le z \le 3\}$

using two different orders of integration.

Often the most difficult part of working with triple integrals is setting up the order and endpoints of integration. **Keep in mind that the outer integral must have constant endpoints, the middle integral can have at most one variable in the endpoints, and the inside integral can have at most two variables in the end points.**

If D is a region in the xy-plane, and $R = \{(x,y,z): (x,y) \text{ is in D and } f(x,y) \le z \le g(x,y)\}$

as in Fig. 2, then $\iiint_R F(x,y,z)\, dV = \iint_D \left\{ \int_{z=f}^{z=g} F(x,y,z)\, dz \right\} dA$

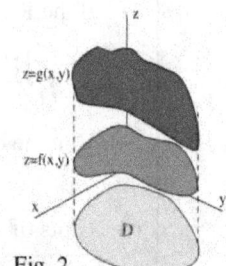

Fig 2

Example 2: Determine the volume of the region bounded by $0 \leq x \leq 1$, $0 \leq y \leq 3$ and z between plane $z = f(x,y) = 5 - 2x - y$ and the surface $z = g(x,y) = 13 - 3x^2 - y^2$ as shown in Fig. 3.

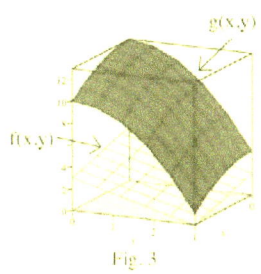

Fig. 3

Solution: The inside integral must be **dz** with z from $z = f(x,y) = 5 - 2x - y$ to $z = g(x,y) = 13 - 3x^2 - y^2$. The outer integral can be either **dx** or **dy** since each of them is bounded by constants.

$$\text{volume} = \iiint_R 1 \, dV = \int_{x=0}^{1} \int_{y=0}^{3} \int_{z=5-2x-y}^{z=13-3x^2-y^2} 1 \, dz \, dy \, dx = \int_{x=0}^{1} \int_{y=0}^{3} (8 - 3x^2 - y^2 + 2x + y) \, dy \, dx$$

$$= \int_{x=0}^{1} (-9x^2 + \frac{39}{2} + 6x) \, dx = \frac{39}{2}$$

Practice 2: Write the triple integral that represents the volume of the region bounded by $0 \leq x \leq 1$, $0 \leq y \leq 3-3x$ and z between plane $z = f(x,y) = 5 - 2x - y$ and the surface $z = g(x,y) = 13 - 3x^2 - y^2$. How does this region differ from the one in Example 2?

Example 3: Write an iterated triple integral for f(x,y,z) in the solid bounded by the paraboloid $z = x^2 + y^2$ and the plane $z=4$ (Fig. 4).

Fig. 4: Solid paraboloid

Solution: The domain in the xy-plane is the circle $x^2 + y^2 \leq 4$ which can be described as $-2 \leq x \leq 2$ and $-\sqrt{4-x^2} \leq y \leq \sqrt{4-x^2}$. And z goes from $x^2 + y^2$ to 4. Putting this information together, the triple integral is

$$\iiint_R f(x,y,z) \, dV = \int_{x=-2}^{2} \int_{y=-\sqrt{4-x^2}}^{\sqrt{4-x^2}} \int_{z=x^2+y^2}^{4} f(x,y,z) \, dz \, dy \, dx$$

Note that the endpoints of the outside integral had no variables, the middle endpoints have one variable, and the inside endpoints have two variables.

Practice 3: (a) Write an iterated triple integral for f(x,y,z) in the solid bounded below by the cone $z = \sqrt{x^2 + y^2}$ and above by the sphere $x^2 + y^2 + z^2 = 8$ (Fig. 5).

Fig. 5: Cone topped by sphere

(b) Write an iterated triple integral for this function and solid that is only in the first octant.

Applications of Triple Integrals

These are similar to the applications of single and double integrals and are very useful in some applications.

Volume of a solid region R = $\iiint\limits_R 1\, dV$

Average value of f on a solid region R = $\dfrac{1}{\text{volume of R}} \cdot \iiint\limits_R f\, dV$

If $\delta = \delta(x,y,z)$ is the density of a solid region R at the location (x,y,z) then Mass = M = $\iiint\limits_R \delta\, dV$

First moments about the coordinate planes:

$M_{yz} = \iiint\limits_R x \cdot \delta\, dV$ $\qquad M_{xz} = \iiint\limits_R y \cdot \delta\, dV$ $\qquad M_{xy} = \iiint\limits_R z \cdot \delta\, dV$

Center of Mass: $\bar{x} = \dfrac{M_{yz}}{M}$ $\qquad \bar{y} = \dfrac{M_{xz}}{M}$ $\qquad \bar{z} = \dfrac{M_{xy}}{M}$

Second moments (moments of inertia):

$I_x = \iiint\limits_R (y^2 + z^2) \cdot \delta\, dV$ $\qquad I_y = \iiint\limits_R (x^2 + z^2) \cdot \delta\, dV$ $\qquad I_z = \iiint\limits_R (x^2 + y^2) \cdot \delta\, dV$

about line L: $I_L = \iiint\limits_R r^2 \cdot \delta\, dV$ where r(x,y,z) = distance of (x,y,z) from line L

Radius of gyration about a line L: $R_L = \sqrt{I_L / M}$

Example 4: A 1 cm by 1 cm ($0 \le y \le 1$, $0 \le z \le 1$) bar along the x-axis has a length of 6 cm ($0 \le x \le 6$). The density of the bar is $\delta(x,y,z) = 1 + x$ g/ cm^3. Determine (a) the mass M, (b) M_{yz}, (c) $\bar{x}$, (d) I_z and (e) R_z.

Solution: (a) mass = $\int\limits_{z=0}^{1} \int\limits_{y=0}^{1} \int\limits_{x=0}^{6} (1+x)\, dx\, dy\, dz = 24$ g

(b) $M_{yz} = \iiint\limits_R x \cdot \delta\, dV = \int\limits_{z=0}^{1} \int\limits_{y=0}^{1} \int\limits_{x=0}^{6} x \cdot (1+x)\, dx\, dy\, dz = 90$ g·cm

Fig. 6: Bar along the x-axis

(c) $\bar{x} = \dfrac{M_{yz}}{M} = \dfrac{90 \text{ g·cm}}{24} = \dfrac{15}{4} = 3.75$ cm The bar balances on a fulcrum at location x=3.75 cm.

(d) $I_z = \iiint\limits_R (x^2 + y^2) \cdot \delta\, dV = \int\limits_{z=0}^{1} \int\limits_{y=0}^{1} \int\limits_{x=0}^{6} (x^2 + y^2) \cdot (1+x)\, dx\, dy\, dz = 404$ g·cm^2

(e) $R_z = \sqrt{I_z / M} = \sqrt{404/24} \approx 4.10$ cm The kinetic energy of this bar rotating around the z-axis is the same as a point mass of 24 g located at x=4.1 cm rotating around the z-axis with the same angular speed.

Practice 4: A cube ($0 \leq x \leq 2$, $0 \leq y \leq 2$, $0 \leq z \leq 2$ cm) has density $\delta(x,y,z) = 1 + x + y + z^2$ g/cm^3.

Determine (a) the mass M, (b) the moment about each coordinate plane, and (c) the center of mass of the cube.

Example 5: $\iiint_R \sqrt{x^2 + y^2} \, dV$ where R is the solid bounded by the paraboloid $z = x^2 + y^2$ and the plane $z = 4$ (Fig. 4).

Solution: This is the solid from Example 2 so

$$\iiint_R f(x,y,z) \, dV = \int_{x=-2}^{2} \int_{y=-\sqrt{4-x^2}}^{\sqrt{4-x^2}} \int_{z=x^2+y^2}^{4} \sqrt{x^2 + y^2} \, dz \, dy \, dx$$

$$= \int_{x=-2}^{2} \int_{y=-\sqrt{4-x^2}}^{\sqrt{4-x^2}} (4 - (x^2 + y^2)) \cdot \sqrt{x^2 + y^2} \, dy \, dx$$

But the domain of this remaining double integral is the disk $0 \leq x^2 + y^2 \leq 2$ so it is useful to switch to polar coordinates. Then we have

$$= \int_{\theta=0}^{2\pi} \int_{r=0}^{2} \left((4 - r^2) \cdot \sqrt{r^2} \right) r \, dr \, d\theta = \int_{\theta=0}^{2\pi} \int_{r=0}^{2} \left(4r^2 - r^4 \right) dr \, d\theta = \int_{\theta=0}^{2\pi} \frac{64}{15} \, d\theta = \frac{128}{15}\pi .$$

Problems

In Problems 1 to 6, set up the appropriate iterated integrals for $\iiint_R f \, dV$ on the indicated domains.

1. R is the solid prism in Fig. 7.

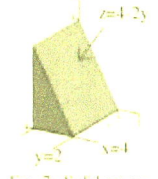
Fig. 7: Solid prism

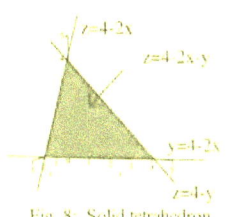
Fig. 8: Solid tetrahedron

2. R is the solid tetrahedron in Fig. 8.

3. R is the solid cone in Fig. 9.

Fig. 9: Solid cone

Fig. 10: Truncated cylinder

4. R is the solid sliced cylinder in Fig. 10.

5. R is the solid in Fig. 11.

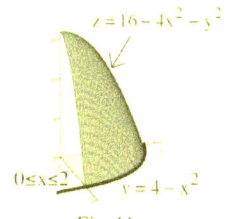
Fig. 11

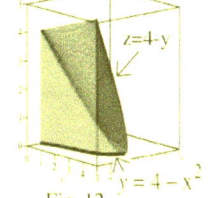
Fig. 12

6. R is the solid in Fig. 12.

In problems 7 to 18, evaluate the integrals.

7. $\displaystyle\int_{x=0}^{1}\int_{y=0}^{1}\int_{z=0}^{1} (x^2+y^2+z)\, dz\, dy\, dx$

8. $\displaystyle\int_{z=1}^{e}\int_{y=1}^{e}\int_{x=1}^{e} \frac{1}{xyz}\, dx\, dy\, dz$

9. $\displaystyle\int_{0}^{1}\int_{0}^{x-2}\int_{0}^{3-x-y} dz\, dy\, dx$

10. $\displaystyle\int_{0}^{1}\int_{0}^{\pi}\int_{0}^{\pi} y\cdot\cos(z)\, dx\, dy\, dz$

11. $\displaystyle\int_{0}^{1}\int_{0}^{2}\int_{0}^{x+z} (12xz)\, dy\, dx\, dz$

12. $\displaystyle\int_{0}^{2}\int_{x}^{2x}\int_{0}^{y} (6xyz)\, dz\, dy\, dx$

13. $\displaystyle\int_{0}^{3}\int_{0}^{1}\int_{0}^{1-z} z\cdot e^{y}\, dx\, dz\, dy$

14. $\displaystyle\int_{0}^{3}\int_{0}^{1}\int_{0}^{1-z} x\cdot e^{y}\, dx\, dz\, dy$

15. $\displaystyle\int_{0}^{2}\int_{0}^{\sqrt{4-x^2}}\int_{0}^{\sqrt{4-x^2}} 1\, dy\, dz\, dx$

16. $\displaystyle\int_{0}^{4}\int_{0}^{\ln(y)}\int_{\ln(y)}^{\ln(3y)} e^{x+y-z}\, dx\, dz\, dy$

17. $\displaystyle\int_{0}^{2}\int_{0}^{3}\int_{0}^{y} (2x+4y+6z)\, dz\, dy\, dx$

18. $\displaystyle\int_{0}^{3}\int_{0}^{1}\int_{0}^{y} (4xy)\, dx\, dz\, dy$

For problems 19 to 22 write the Maple command to evaluate the triple integral.

19. f(x,y)=2x+3 on the domain of Problem 19.
20. f(x,y)=sin(xy)+z on the domain of Problem 20.
21. f(x,y)=xyz on the domain of Problem 21.
22. f(x,y)=1 on the domain of Problem 22.

Practice Answers

Practice 1: Evaluate $\iiint_R x^3 + y^2 + z \, dV$ for $R = \{(x,y,z): 0 \le x \le 2, 0 \le y \le, 1 \le z \le 3\}$.

$$\iiint_R x^3 + y^2 + z \, dV = \int_{z=1}^{3} \int_{y=0}^{1} \int_{x=0}^{2} x^3 + y^2 + z \, dx \, dy \, dz$$

$$\int_{x=0}^{2} x^3 + y^2 + z \, dx = 4 + 2y^2 + 2z, \quad \int_{y=0}^{1} 4 + 2y^2 + 2z \, dy = 4 + \frac{2}{3} + 2z,$$

and finally $\int_{z=1}^{3} \frac{14}{3} + 2z \, dz = \frac{28}{3} + 8 \cdot$

Check that integrating in some other order gives the same result.

Try $\int_{x=0}^{2} \int_{z=1}^{3} \int_{y=0}^{1} (x^3 + y^2 + z) \, dy \, dz \, dx$ and $\int_{y=0}^{1} \int_{x=0}^{2} \int_{z=1}^{3} (2x + y + z^2) \, dz \, dx \, dy$

Practice 2: volume = $\int_{x=0}^{1} \int_{y=0}^{3-3x} \int_{z=5-2x-y}^{z=13-3x^2-y^2} 1 \, dz \, dy \, dx$ (= 23/2)

The domain is now a triangle in the xy-plane (Fig. P2).

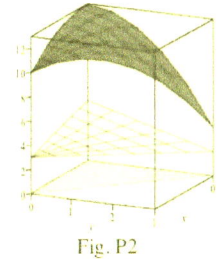

Fig. P2

Practice 3: (a) $\iiint_R f \, dV = \int_{x=-2}^{2} \int_{y=-\sqrt{4-x^2}}^{\sqrt{4-x^2}} \int_{z=\sqrt{x^2+y^2}}^{\sqrt{8-x^2-y^2}} f(x,y,z) \, dz \, dy \, dx$

(b) $\int_{x=0}^{2} \int_{y=0}^{\sqrt{4-x^2}} \int_{z=\sqrt{x^2+y^2}}^{\sqrt{8-x^2-y^2}} f(x,y,z) \, dz \, dy \, dx$

Practice 4: Since x, y and z have constant endpoints, the integration can be done in any order.

(a) Mass = $\int_{x=0}^{2} \int_{y=0}^{2} \int_{z=0}^{2} \left(1 + x + y + z^2\right) dz \, dy \, dx = \frac{104}{3}$ g

(b) $M_{yz} = \int_{x=0}^{2} \int_{y=0}^{2} \int_{z=0}^{2} x \cdot \left(1 + x + y + z^2\right) dz \, dy \, dx = \frac{112}{3}$ g·cm, $M_{xz} = M_{yz}$

$M_{xy} = \int_{x=0}^{2} \int_{y=0}^{2} \int_{z=0}^{2} z \cdot \left(1 + x + y + z^2\right) dz \, dy \, dx = 40$ g·cm

(c) $\bar{x} = \frac{M_{yz}}{M} = \frac{14}{13}$ cm, $\bar{y} = \bar{x} = \frac{14}{13}$ cm, $\bar{z} = \frac{M_{xy}}{M} = \frac{15}{13}$ cm

14.7 TRIPLE INTEGRALS IN CYLINDRICAL AND SPHERICAL COORDINATES

"In physics everything is straight, flat or round." Statement by a physics teacher

Maybe not, but lots of applications have pieces of round or spherical domains, and using cylindrical or spherical coordinates can make many triple integrals much easier to evaluate.

$\iiint f \, dV$ **in Cylindrical Coordinates**

If our domain of integration is round or is easily described using polar coordinates (r, θ), then a triple integral in cylindrical coordinates (r, θ, z) is often the best method, and it begins with the form of the ΔV. Fig. 1 illustrates that if we partition each of r, θ and z, then the volume of each little cell is $\Delta V = (r \cdot \Delta\theta) \cdot \Delta r \cdot \Delta z$. Then the triple Riemann sum is

$\sum \sum \sum f(r^*, \theta^*, z^*) \cdot r \cdot \Delta r \cdot \Delta\theta \cdot \Delta z$. If f is continuous, then the limit of this Riemann sum as $\Delta r, \Delta\theta, \Delta z \to 0$ is the triple integral in cylindrical coordinates

$$\iiint_R f \, dV = \int_z \int_\theta \int_r f(r,\theta,z) \, r \, dr \, d\theta \, dz \, .$$

As before, we can alter the order of the integrals as long as we accurately describe the domain of integration. Also, the outer integral endpoints must be constants, the middle integral endpoints can have only one variable, and the inner integral endpoints can have two variables.

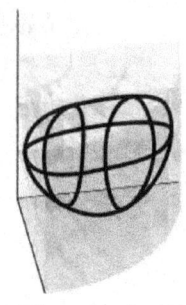

Partition of r, θ and z

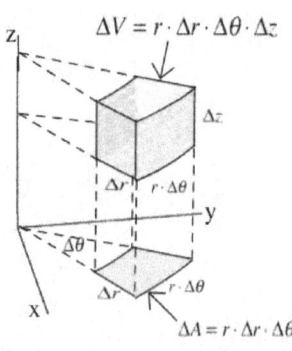

Fig. 1

(Note: Recall that $x = r \cdot \cos(\theta)$, $y = r \cdot \sin(\theta)$ so $x^2 + y^2 = r^2$ and $z = z$.)

Example 1: Evaluate (a) $\iiint_R f \, dV$ for $f(r,\theta,z) = r \cdot z$ with $R = \{1 \le r \le z, 0 \le \theta \le \pi, 0 \le z \le 2\}$

and (b) $\int_{r=0}^{3} \int_{\theta=0}^{2\pi} \int_{z=0}^{e^{-r^2}} 1 \, r \, dz \, d\theta \, dr$

Solution: (a) $\iiint_R f \, dV = \int_{z=0}^{2} \int_{\theta=0}^{\pi} \int_{r=1}^{z} (r \cdot z) \, r \, dr \, d\theta \, dz = \int_{z=0}^{2} \int_{\theta=0}^{\pi} \left(\frac{1}{3}r^3\right)\Big|_{r=1}^{z} \, d\theta \, dz = \int_{z=0}^{2} \frac{1}{3}\left(z^4 - z\right)\pi \, dz = \frac{22}{15}\pi$

(b) $\int_{r=0}^{3} \int_{\theta=0}^{2\pi} \int_{z=0}^{e^{-r^2}} 1 \, r \, dz \, d\theta \, dr = \int_{r=0}^{3} \int_{\theta=0}^{2\pi} \left(e^{-r^2}\right) r \, d\theta \, dr = \int_{r=0}^{3} 2\pi \left(e^{-r^2}\right) r \, dr = -\pi e^{-r^2}\Big|_{r=0}^{3} = \pi\left(1 - e^{-9}\right)$

Practice 1: Evaluate (a) $\iiint_R dV$ for $R = \{0 \le r \le 2, 0 \le \theta \le \pi/2, 0 \le z \le 8-r^3\}$ and

(b) the volume of the solid cylinder above the disk $x^2 + y^2 \le 4$ and below the plane $z=4-y$.

Example 2: R is the solid bounded by the paraboloid $z = x^2 + y^2$ and the plane $z=4$ (Fig. 2).

(a) Write and evaluate an iterated triple integral for the volume of R.

(b) Write and evaluate an iterated triple integral for the mass of R if the density is $\delta(x,y,x) = 1+z$.

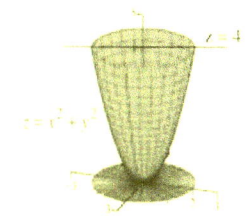

Fig. 2: Solid paraboloid

Solution: (a) The domain of this integral is the circle $x^2 + y^2 \le 4$ so $r^2 \le 4$ and $0 \le \theta \le 2\pi$:

$$\iiint_R f\, dV = \int_{\theta=0}^{2\pi} \int_{r=0}^{2} \int_{z=r^2}^{4} (1)\, r\, dz\, dr\, d\theta = \int_{\theta=0}^{2\pi} \int_{r=0}^{2} (z \cdot r)\Big|_{z=r^2}^{4}\, dr\, d\theta = \int_{\theta=0}^{2\pi} \int_{r=0}^{2} \left(4r - r^3\right)\, dr\, d\theta = 8\pi$$

(b) mass $= \int_{\theta=0}^{2\pi} \int_{r=0}^{2} \int_{z=r^2}^{4} (1+z) \cdot r\, dz\, dr\, d\theta = \dfrac{88}{3}\pi$

Practice 2: R is the solid hemisphere $x^2 + y^2 + z^2 \le 4$ with $z \ge 0$ (Fig. 3).

(a) Write and evaluate an iterated triple integral for the volume of R.

(b) Write and evaluate an iterated triple integral for the mass of R if the density is $\delta(x,y,x) = 1+z$.

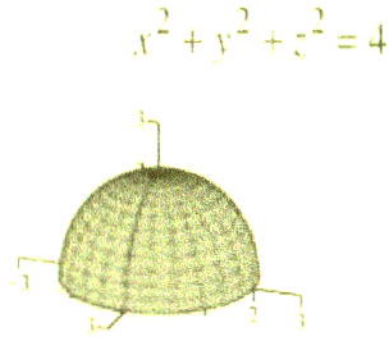

Fig. 3: Solid hemisphere

Example 3: Find the centroid of the solid that is bounded below by the disk $x^2 + y^2 \le 9$ and above by the paraboloid $z = x^2 + y^2$.

Solution: $x^2 + y^2 \le 9$ means $0 \le r \le 3$, and $z = x^2 + y^2$ means $z = r^2$, the domain is

$$R = \{(r,\theta,z): 0 \le r \le 3, 0 \le \theta \le 2\pi, 0 \le z \le r^2\}$$

$$\text{mass} = \int_{\theta=0}^{2\pi} \int_{r=0}^{3} \int_{z=0}^{r^2} r \cdot dz \cdot dr \cdot d\theta = \int_{\theta=0}^{2\pi} \int_{r=0}^{3} r \cdot z\Big|_{z=0}^{r^2}\, dr \cdot d\theta = \int_{\theta=0}^{2\pi} \int_{r=0}^{3} r^3\, dr \cdot d\theta = \dfrac{81}{2}\pi$$

$$M_{xy} = \int_{\theta=0}^{2\pi} \int_{r=0}^{3} \int_{z=0}^{r^2} z \cdot r\, dz \cdot dr \cdot d\theta = \int_{\theta=0}^{2\pi} \int_{r=0}^{3} r \cdot \dfrac{z^2}{2}\Big|_{z=0}^{r^2}\, dr \cdot d\theta = \int_{\theta=0}^{2\pi} \int_{r=0}^{3} \dfrac{r^5}{2}\, dr \cdot d\theta$$

$$= \int_{\theta=0}^{2\pi} \int_{r=0}^{3} \dfrac{r^6}{12}\Big|_{r=0}^{3}\, d\theta = 2\pi\left(\dfrac{729}{12}\right) = \dfrac{243}{2}\pi .$$

Then $\bar{z} = \dfrac{(243/2)\pi}{(81/2)\pi} = 3$. Because of the symmetry about the z-axis, $\bar{x}$ and $\bar{y}$ are both 0 so the centroid is (0, 0, 3).

$\iiint f\,dV$ in Spherical Coordinates

This development is very similar to what was done for cylindrical coordinates. First we partition our domain R into $(\Delta\rho, \Delta\theta, \Delta\varphi)$ cells, pick a representative point $(\rho^*, \theta^*, \varphi^*)$ in each cell, form the triple Riemann sum $\sum\sum\sum f(\rho^*, \theta^*, \varphi^*)\,\Delta V$, and, finally, take the limit as all of the cell dimensions approach 0 in order to form a triple integral: $\lim_{\Delta \to 0} \sum\sum\sum f(\rho^*, \theta^*, \varphi^*)\,\Delta V \to \iiint_R f(\rho, \theta, \varphi)\cdot dV$

But before we can actually use this idea, we first need to determine dV in terms of the variables ρ, θ and φ. That is a bit complicated and is derived in the Appendix of this section as well as in the next section when Jacobeans are introduced. In either case, $dV = \rho^2 \cdot \sin(\varphi)\cdot d\rho\cdot d\theta\cdot d\varphi$, and then

$$\iiint_R f(\rho,\theta,\varphi)\cdot dV = \iiint_R f(\rho,\theta,\varphi)\cdot \rho^2 \cdot \sin(\varphi)\cdot d\rho\cdot d\theta\cdot d\varphi$$

As before, the we can use any order of integration that describes the domain as long as the outside integral has constant endpoints, and the middle integral has at most one variable endpoint The inside integral can have two variable endpoints.

Example 4: Represent each domain R using iterated triple integrals.
 (a) R is shown in Fig. 4a, (b) R is shown in Fig. 4b.

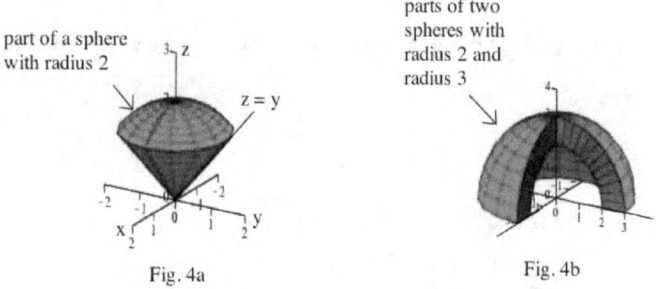

Fig. 4a — part of a sphere with radius 2, $z = y$

Fig. 4b — parts of two spheres with radius 2 and radius 3

Solution: (a) $R = \{(\rho,\theta,\varphi): 0 \le \rho \le 2,\ 0 \le \theta \le 2\pi,\ 0 \le \varphi \le \pi/4\}$

$$\iiint_R f\,dV = \int_{\varphi=0}^{\pi/4} \int_{\theta=0}^{2\pi} \int_{\rho=0}^{2} f(\rho,\theta,\varphi)\,\rho^2\cdot\sin(\varphi)\cdot d\rho\cdot d\theta\cdot d\varphi$$

(b) $R = \{(\rho,\theta,\varphi): 2 \le \rho \le 3,\ \pi/2 \le \theta \le 2\pi,\ 0 \le \varphi \le \pi/2\}$

$$\iiint_R f\,dV = \int_{\varphi=0}^{\pi/2} \int_{\theta=\pi/2}^{2\pi} \int_{\rho=2}^{3} f(\rho,\theta,\varphi)\,\rho^2\cdot\sin(\varphi)\cdot d\rho\cdot d\theta\cdot d\varphi$$

All of these integral endpoints are constants so the integrals can done be in any order.

Practice 3: Represent each domain R using iterated triple integrals.

(a) R is shown in Fig. 5a, (b) R is shown in Fig. 5b.

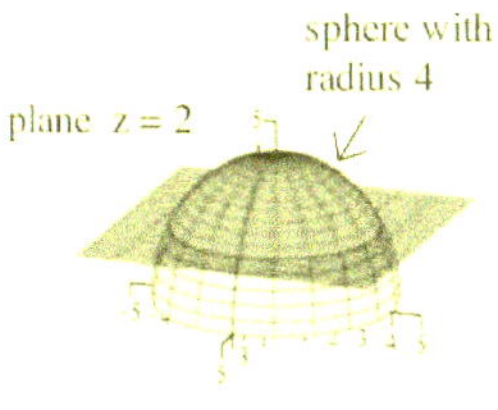

Example 5: Determine the mass and $\bar{z}$ for the solid hemisphere with radius 2 that is above the xy-plane and has density $\delta(x,y,x) = 1+z$.

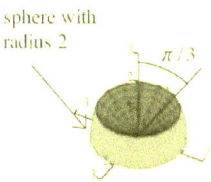

Fig. 5a

Fig. 5b

Solution: $R = \{(\rho,\theta,\varphi): 0 \le \rho \le 2, 0 \le \theta \le 2\pi, 0 \le \varphi \le \pi/2\}$ so

$$\text{mass} = \int_{\varphi=0}^{\pi/2} \int_{\theta=0}^{2\pi} \int_{\rho=0}^{2} (1+z) \; \rho^2 \cdot \sin(\varphi) \cdot d\rho \cdot d\theta \cdot d\varphi$$

$$= \int_{\varphi=0}^{\pi/2} \int_{\theta=0}^{2\pi} \int_{\rho=0}^{2} (1+\rho\cdot\cos(\varphi)) \; \rho^2 \cdot \sin(\varphi) \cdot d\rho \cdot d\theta \cdot d\varphi$$

$$= \int_{\varphi=0}^{\pi/2} \int_{\theta=0}^{2\pi} \int_{\rho=0}^{2} \left(\rho^2 \cdot \sin(\varphi) + \rho^3 \cdot \cos(\varphi) \cdot \sin(\varphi)\right) \; d\rho \cdot d\theta \cdot d\varphi$$

$$= \int_{\varphi=0}^{\pi/2} \int_{\theta=0}^{2\pi} \left(\frac{8}{3}\cdot\sin(\varphi) + 4\cdot\cos(\varphi)\cdot\sin(\varphi)\right) \cdot d\theta \cdot d\varphi$$

$$= \int_{\varphi=0}^{\pi/2} 2\pi\left(\frac{8}{3}\cdot\sin(\varphi) + 4\cdot\cos(\varphi)\cdot\sin(\varphi)\right) \cdot d\varphi = 2\pi\left(2\cdot\sin^2(\varphi) - \frac{8}{3}\cdot\cos(\varphi)\right)\bigg|_{\varphi=0}^{\pi/2}$$

$$= 2\pi(2) + 2\pi\left(\frac{8}{3}\right) = \frac{28}{3}\pi$$

$$M_{xy} = \int_{\varphi=0}^{\pi/2} \int_{\theta=0}^{2\pi} \int_{\rho=0}^{2} z\cdot(1+z) \; \rho^2 \cdot \sin(\varphi) \cdot d\rho \cdot d\theta \cdot d\varphi = \frac{124}{15}\pi \quad \text{(using Maple)}$$

so $\bar{z} = \dfrac{(124/15)\pi}{(28/3)\pi} = \dfrac{32}{35} \approx 0.914$

Conclusion: Even a simple looking problem can take a long time.

These conversion formulas for cylindrical and spherical coordinates are useful.

Coordinate conversion formulas

Cylindrical to Rectangular	Spherical to Cylindrical	Spherical to Rectangular
$x = r\cdot\cos(\theta)$	$r = \rho\cdot\sin(\varphi)$	$x = \rho\cdot\sin(\varphi)\cdot\cos(\theta)$
$y = r\cdot\sin(\theta)$	$z = \rho\cdot\cos(\varphi)$	$y = \rho\cdot\sin(\varphi)\cdot\sin(\theta)$
$z = z$	$\theta = \theta$	$z = \rho\cdot\cos(\varphi)$

$$dV = dx\cdot dy\cdot dz = r\cdot dr\cdot d\theta\cdot dz = \rho^2\cdot\sin(\varphi)\cdot d\rho\cdot d\theta\cdot d\varphi$$

Problems

In problems 1-8, evaluate the triple integrals in cylindrical coordinates.

1. $\displaystyle\int_0^\pi \int_0^1 \int_0^{\sqrt{2-r^2}} r \, dz \cdot dr \cdot d\theta$

2. $\displaystyle\int_0^4 \int_0^\pi \int_r^6 r \, dz \cdot d\theta \cdot dr$

3. $\displaystyle\int_0^\pi \int_0^{\theta/\pi} \int_0^{\sqrt{4-r^2}} z \, dz \cdot dr \cdot d\theta$

4. $\displaystyle\int_0^\pi \int_0^1 \int_{-1/2}^{1/2} \left(r^2 \cdot \sin(\theta) + z^2\right) dz \cdot dr \cdot d\theta$

5. $\displaystyle\int_0^{2\pi} \int_0^3 \int_0^{\pi/3} r^3 \, dr \cdot dz \cdot d\theta$

6. $\displaystyle\int_{-1}^1 \int_0^{2\pi} \int_0^{1+\sin(\theta)} 2r \, dr \cdot d\theta \cdot dz$

7. $\displaystyle\int_0^2 \int_{r-2}^{\sqrt{4-r^2}} \int_0^{2\pi} (r \cdot \sin(\theta) + 1) \cdot r \, d\theta \cdot dz \cdot dr$

8. $\displaystyle\int_0^\pi \int_r^{2r} \int_0^\pi r \cdot \cos(\theta) \, d\theta \cdot dz \cdot dr$

In problems 9 to 12, evaluate the integrals in cylindrical coordinates.

9. $\displaystyle\int_0^4 \int_0^{\sqrt{2}/2} \int_x^{\sqrt{1-x^2}} e^{-(x^2+y^2)} \, dy \cdot dx \cdot dz$

10. $\displaystyle\int_0^4 \int_{-1}^1 \int_{-\sqrt{1-x^2}}^{\sqrt{1-x^2}} 1 \, dy \cdot dx \cdot dz$

11. $\displaystyle\int_0^1 \int_0^{\sqrt{1-x^2}} \int_0^{4-y} 1 \, dz \cdot dy \cdot dx$

12. $\displaystyle\int_{-2}^2 \int_0^{\sqrt{4-x^2}} \int_0^1 \cos(x^2 + y^2) \, dz \cdot dy \cdot dx$

In problems 13 to 18, set up and evaluate the triple integrals in cylindrical coordinates.

13. $f(x,y) = \sqrt{x^2 + y^2}$. R is the region inside the cylinder $x^2 + y^2 = 9$ and between the planes z=3 and z=5.

14. $f(x,y) = (x^3 + xy^2)$. R is the region in the first octant and under the paraboloid $z = 4 - x^2 - y^2$.

15. $f = e^z$. R is the region enclosed by paraboloid $z = 1 + x^2 + y^2$, the cylinder $x^2 + y^2 = 7$, and the xy-plane.

16. $f = x^2$. R is the region inside the cylinder $x^2 + y^2 = 4$, below the cone $z^2 = 9x^2 + 9y^2$ and above the xy-plane.

17. Find the volume of the region R in first octant below the $z = x^2 + y^2$ and above $z = 36 - 3x^2 - 3y^2$.

18. $f(x,y) = 6 + 4x^2 + 4y^2$. R is the region in the 2nd, 3rd and 4th octants, inside the cylinder $x^2 + y^2 = 1$, and between the planes z=2 and z=3.

Now spherical

In problems 19 to 26 evaluate the integrals in spherical coordinates.

19. $\displaystyle\int_0^\pi \int_0^\pi \int_0^{2\cdot\cos(\varphi)} \rho^2 \cdot \sin(\varphi) \, d\rho \cdot d\varphi \cdot d\theta$

20. $\displaystyle\int_0^{2\pi} \int_0^{\pi/4} \int_0^2 \rho^3 \cdot \sin^2(\varphi) \, d\rho \cdot d\varphi \cdot d\theta$

21. $\displaystyle\int_0^\pi \int_0^\pi \int_0^1 5\rho^3 \cdot \sin^3(\varphi) \, d\rho \cdot d\varphi \cdot d\theta$

22. $\displaystyle\int_0^\pi \int_0^{\pi/3} \int_{\sec(\varphi)}^1 3\rho^2 \cdot \sin(\varphi) \, d\rho \cdot d\varphi \cdot d\theta$

23. $\int_{0}^{\pi/2}\int_{0}^{\pi}\int_{1}^{2} \rho^2 \cdot \sin(\varphi)\ d\rho\cdot d\theta\cdot d\varphi$

24. $\int_{0}^{\pi}\int_{0}^{\pi}\int_{0}^{1} e^{(\rho^2)}\cdot\rho\cdot\sin(\varphi)\ d\rho\cdot d\theta\cdot d\varphi$

25. $\int_{0}^{\pi}\int_{0}^{\pi/3}\int_{0}^{\cos(\varphi)} 4\rho^3\cdot\sin(\varphi)\ d\rho\cdot d\varphi\cdot d\theta$

26. $\int_{0}^{2\pi}\int_{0}^{\pi/2}\int_{0}^{\csc(\varphi)} \sin(\varphi)\ d\rho\cdot d\varphi\cdot d\theta$

Practice Answers

Practice 1: (a) $\int_{r=0}^{2}\int_{\theta=0}^{\pi/2}\int_{z=0}^{8-r^3} r\ dz\ d\theta\ dr = \int_{z=0}^{2}\int_{\theta=0}^{\pi/2} (z\cdot r)\Big|_{z=0}^{8-r^3} d\theta\ dr = \int_{z=0}^{2}\int_{\theta=0}^{\pi/2} 8r - r^4\ d\theta\ dr$

$= \int_{z=0}^{2} \frac{\pi}{2}(8r - r^4)\ dr = \frac{\pi}{2}\left(\frac{48}{5}\right)$

(b) $y = r\cdot\sin(\theta)$ so

volume $= \int_{r=0}^{2}\int_{\theta=0}^{2\pi}\int_{z=0}^{4-r\cdot\sin(\theta)} 1\ r\ dz\ d\theta\ dr = \int_{r=0}^{2}\int_{\theta=0}^{2\pi} (4 - r\cdot\sin(\theta))\cdot r\ d\theta\ dr = \int_{r=0}^{2} 2\pi(4r)\ dr = 16\pi$

Practice 2: $z^2 \le 4 - (x^2 + y^2) = 4 - r^2$, and, as in Example 1, $0 \le r \le 2$ and $0 \le \theta \le 2\pi$.

(a) volume $= \int_{\theta=0}^{2\pi}\int_{r=0}^{2}\int_{z=0}^{\sqrt{4-r^2}} (1)\cdot r\ dz\ dr\ d\theta = \int_{\theta=0}^{2\pi}\int_{r=0}^{2} z\cdot r\Big|_{z=0}^{\sqrt{4-r^2}} dr\ d\theta = \int_{\theta=0}^{2\pi}\int_{r=0}^{2} \sqrt{4-r^2}\cdot r\ dr\ d\theta$

$= \int_{\theta=0}^{2\pi}\int_{r=0}^{2} \sqrt{4-r^2}\cdot r\ dr\ d\theta = \int_{\theta=0}^{2\pi} \left(-\frac{1}{3}(4-r^2)^{3/2}\right)\Big|_{r=0}^{2} d\theta = \int_{\theta=0}^{2\pi} \frac{8}{3}\ d\theta = \frac{16}{3}\pi$

(b) mass $= \int_{\theta=0}^{2\pi}\int_{r=0}^{2}\int_{z=0}^{\sqrt{4-r^2}} (1+z)\cdot r\ dz\ dr\ d\theta = \frac{28}{3}\pi$

Practice 3: (a) $R = \{(\rho,\theta,\varphi):\ 0\le\rho\le 2,\ 0\le\theta\le 2\pi,\ \pi/3 \le \varphi \le \pi/2\}$

$\int_{\varphi=\pi/3}^{\pi/2}\int_{\theta=0}^{2\pi}\int_{\rho=0}^{2} f(\rho,\theta,\varphi)\ \rho^2\cdot\sin(\varphi)\cdot d\rho\cdot d\theta\cdot d\varphi$

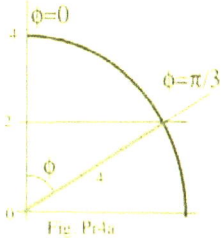

Fig. Pr4a

(b) Clearly $0 \le \theta \le 2\pi$, but φ and ρ require a bit of work (Fig. Pr4):

$R = \{(\rho,\theta,\varphi):\ 2\sec(\varphi)\le\rho\le 4,\ 0\le\theta\le 2\pi,\ 0\le\varphi\le\pi/3\}$

$\int_{\varphi=0}^{\pi/3}\int_{\theta=0}^{2\pi}\int_{\rho=2\sec(\varphi)}^{4} f(\rho,\theta,\varphi)\ \rho^2\cdot\sin(\varphi)\cdot d\rho\cdot d\theta\cdot d\varphi$

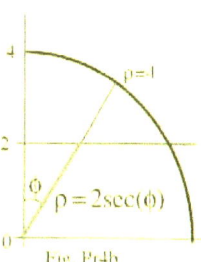

Fig. Pr4b

Appendix: Why $dV = \rho^2 \cdot \sin(\varphi) \cdot d\rho \cdot d\theta \cdot d\varphi$

The figures A1-A3 are an attempt to explain where this strange equation comes from. We need to partition the space by partitioning each of the three variables ρ, θ and φ. This results in cells as shown greatly magnified in Fig. A1. Using different views of a typical cell in Fig. A2, it is possible to determine the lengths of the sides of this cell. Putting all of this together in Fig. A3, the volume of the cell, ΔV ia the product of the lengths of the sides:

$$dV = \rho^2 \cdot \sin(\varphi) \cdot d\rho \cdot d\theta \cdot d\varphi$$

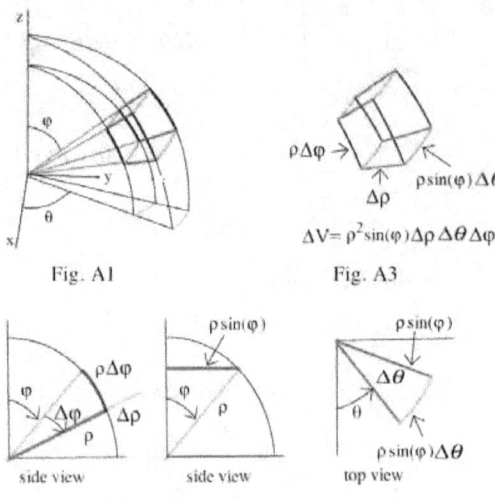

14.8 Changing Variables in Double and Triple Integrals

In section 4.5 we saw how a change of variables could make some integrals easier to evaluate: the integral $\int x\sqrt{3-x^2} \, dx$ is made easier by using the substitution $u = 3-x^2$. Similarly, in section 14.3 we saw that converting some integrals from rectangular to polar coordinates can make them easier: $\iint \sqrt{x^2+y^2} \, dx \, dy$ is made easier by replacing x, y and dx dy with $r \cdot \cos(\theta)$, $r \cdot \sin(\theta)$ and $r \cdot dr \cdot d\theta$ respectively. Then $\iint_R \sqrt{x^2+y^2} \, dA = \iint_G r \cdot r \cdot dr \, d\theta$. There are other substitutions that can make double and triple integrals easier, and this section shows how to make some of those transformations.

Changing Variables in Double Integrals

Since a double integral depends on x and y, we will typically need two substitution variables, u and v, and formulas that replace x=x(u,v) and y=y(u,v). Then the domain S in the uv-plane will be mapped into a region R in the xy-plane. Reversing the transformation, we can map R from the xy-plane to S in the uv-plane. The goal is to map a complicated xy domain to an easier uv domain.

Example 1: Suppose $S = \{u,v\}: 0 \le u \le 2$ and $1 \le v \le 2\}$ in the uv-plane, and that the transformation T is given by $x = x(u,v) = 2u + v$ and $y = y(u,v) = u - v$. (a) What is the R= T(S) region in the xy-plane?
(b) What is the inverse transformation u=u(x,y), v=v(x,y) that maps R back onto S?

Solution: (a) S is shown in Fig. 1a. The corners of S, moving counterclockwise, are (0,1), (2,1), (2,2) and (0,2) and these are mapped by T to the (x,y) points (1,–1), (5,1), (6,0) and (2,–2) respectively, and these become the corners of R = T(S) in the xy-plane as shown in Fig. 1b. Since the transformation T is linear, the straight line boundaries of S are mapped to straight lines of R.
(b) Solving x=2u+v and y=u-v for u and v, we get u=(x+y)/3 and v=(x–2y)/3. It is difficult to describe the domain of integration for R in terms of x and y, but quite easy for S in terms of u and v.

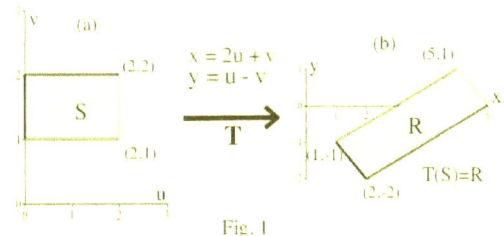

Fig. 1

Practice 1: Suppose $S = \{u,v\}: 1 \le u \le 3$ and $0 \le v \le 2\}$ in the uv-plane, and T is given by $x = x(u,v) = u + 2v$ and $y = y(u,v) = 2u - v$. (a) What is the R= T(S) region in the xy-plane?
(b) What is the inverse transformation u=u(x,y), v=v(x,y) that maps R back onto S?

Example 2: Suppose $S = \{(u,v): 0 \leq u \leq 1 \text{ and } 1 \leq v \leq 2\}$ in the uv-plane, and T is given by $x=u+v$ and $y=u/v$. (a) What is the $R = T(S)$ region in the xy-plane?

(b) What is the inverse transformation $u=u(x,y)$, $v=v(x,y)$ that maps R back onto S?

Solution: (a) S is shown in Fig. 2. The corners of S, moving counterclockwise, are (0,1), (1,1), (1,2) and (0,2) and these are mapped by T to the (x,y) points (1,0), (2,1), (3, 1/2) and (2,0) respectively, and these become the corners of $R = T(S)$ in the xy-plane as shown in Fig. 2.

The boundary: If v=1 and $0 \leq u \leq 1$, then y=x-1 for $1 \leq x \leq 2$. If v=2 and $0 \leq u \leq 1$, then y=x/2-1 for $2 \leq x \leq 3$. If u=0 and $1 \leq v \leq 2$, then y=0 and $1 \leq x \leq 2$. Finally, the interesting boundary: if u=1 and $1 \leq v \leq 2$, then y=1/(x-1) for $2 \leq x \leq 3$.

(b) The inverse transformation (solving x=u+v and y=u/v for u and v) is $u=xy/(1+y)$ and $v=x/(1+y)$.

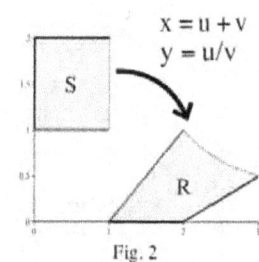

Fig. 2

Practice 2: Suppose $S = \{(u,v): 0 \leq u \leq \pi \text{ and } 1 \leq v \leq 2\}$ in the uv-plane, and T is given by and $x = 1 + v \cdot \sin(u)$ and $y=u$. (a) What is the $R = T(S)$ region in the xy-plane? (b) What is the inverse transformation $u=u(x,y)$, $v=v(x,y)$ that maps R back onto S?

Our goal in this section is not to simply map regions into other regions, but it is to do substitutions that make double integrals easier, either by making the integral domain easier or by making the integrand function easier or both. However, we need one more piece, the Jacobian.

Definition

The **Jacobian** of the transformation T:(u,v)->(x,y) by the substitution $x=x(u,v)$ and $y=y(u,v)$ is

$$J(u,v) = \frac{\partial(x,y)}{\partial(u,v)} = \begin{vmatrix} \frac{\partial x}{\partial u} & \frac{\partial x}{\partial v} \\ \frac{\partial y}{\partial u} & \frac{\partial y}{\partial v} \end{vmatrix} = \frac{\partial x}{\partial u} \cdot \frac{\partial y}{\partial v} - \frac{\partial y}{\partial u} \cdot \frac{\partial x}{\partial v}$$

Note: partials are with respect to u and v

The Jacobian is the determinant of the 2x2 matrix of partial derivatives. It is required that $x(u,v)$ and $y(u,v)$ have continuous first partial derivatives on the region S in the uv-plane.

The Jacobian of the inverse transformation with $u=u(x,y)$ and $v=v(x,y)$ is

$$J(x,y) = \frac{\partial(u,v)}{\partial(x,y)} = \begin{vmatrix} \frac{\partial u}{\partial x} & \frac{\partial u}{\partial y} \\ \frac{\partial v}{\partial x} & \frac{\partial v}{\partial y} \end{vmatrix} = \frac{\partial u}{\partial x} \cdot \frac{\partial v}{\partial y} - \frac{\partial v}{\partial x} \cdot \frac{\partial u}{\partial y}$$

The derivation of this Jacobian formula is intricate and is given in the Appendix.

Example 3: (a) Calculate the Jacobian $J(u,v)$ of the transformation $x = x(u,v) = 2u + v$ and $y = y(u,v) = u - v$ from Example 1.

(b) Also calculate the Jacobian $J(x,y)$ of the inverse transformation.

Solution: (a) $\frac{\partial x}{\partial u} = 2, \frac{\partial x}{\partial v} = 1, \frac{\partial y}{\partial u} = 1$ and $\frac{\partial y}{\partial v} = -1$ so $J(u,v) = \frac{\partial(x,y)}{\partial(u,v)} = \begin{vmatrix} 2 & 1 \\ 1 & -1 \end{vmatrix} = -3$.

(b) The inverse transformation (using algebra) is $u=(x+y)/3$ and $v=(x-2y)/3$ so

$\frac{\partial u}{\partial x} = \frac{1}{3}, \frac{\partial u}{\partial y} = \frac{1}{3}, \frac{\partial v}{\partial x} = \frac{1}{3}$ and $\frac{\partial v}{\partial y} = -\frac{2}{3}$ and $J(x,y) = \frac{\partial(u,v)}{\partial(x,y)} = \begin{vmatrix} 1/3 & 1/3 \\ 1/3 & -2/3 \end{vmatrix} = -\frac{1}{3}$.

Practice 3: (a) Calculate the Jacobian $J(u,v)$ of the transformation $x=u+2v$ and $y=2u-v$ from Practice 1.

(b) Also calculate the Jacobian $J(x,y)$ of the inverse transformation.

Fact: You might have noticed that $J(x,y) = \frac{1}{J(u,v)}$ in the previous Examples and Practices. That is true in general and can make some computations much easier. A proof of this for the 2D Jacobian is given in the Appendix.

Now we finally get to change variables in double integrals.

Change of Variables Theorem

If the region G in the (u,v) plane is transformed into the region R in the xy-plane by $x=x(u,v)$ and $y=y(u,v)$, and if $x(u,v)$ and $y(u,v)$ have continuous first partial derivatives,

then $\iint\limits_R F(x,y) \, dx \, dy = \iint\limits_S F(x(u,v),y(u,v)) \, |J(u,v)| \, du \, dv$

for any continuous function F.

Example 4: (a) Calculate the area of the region R in Example 1.

(b) Calculate the mass of the region in Example 1 when the density is $\delta(x,y) = 5+y$.

Solution: $x=2u+v$ and $y=u-v$ so $J(u,v) = \frac{\partial(x,y)}{\partial(u,v)} = \begin{vmatrix} \frac{\partial x}{\partial u} & \frac{\partial x}{\partial v} \\ \frac{\partial y}{\partial u} & \frac{\partial y}{\partial v} \end{vmatrix} = \begin{vmatrix} 2 & 1 \\ 1 & -1 \end{vmatrix} = -3$.

(a) $\{\text{area of R}\} = \iint\limits_R 1 \, dA = \iint\limits_S 1 \cdot |J(u,v)| \, du \, dv = \int_{v=1}^{2} \int_{u=0}^{2} 1 \cdot (3) \, du \, dv = \int_{v=1}^{2} 6 \, dv = 6$

(b) $\text{Mass} = \iint\limits_R \delta \, dA = \iint\limits_S (5+y) \, |J(u,v)| \, du \, dv = \int_{v=1}^{2} \int_{u=0}^{2} (5+u-v) \cdot (3) \, du \, dy = \int_{v=1}^{2} 36 - 6v \, dv = 27$

Practice 4: (a) Calculate the area of the region R in Practice 1.

(b) Calculate the mass of the region in Practice 1 when the density is $\delta(x,y) = x + y$.

Example 5: Let $\iint\limits_{R} \dfrac{2x-y}{2}\, dA = \int_{y=0}^{4} \int_{x=y/2}^{(y/2)+1} \dfrac{2x-y}{2}\, dx\, dy$

(a) Sketch the region R.

(b) Make the substitutions $u = \dfrac{2x-y}{2}$, $v = \dfrac{y}{2}$ and solve for $x = x(u,v)$ and $y = y(u,v)$.

(c) Calculate the Jacobian $J(u,v)$ and (d) Rewrite the xy-integral in terms of u and v and evaluate this integral.

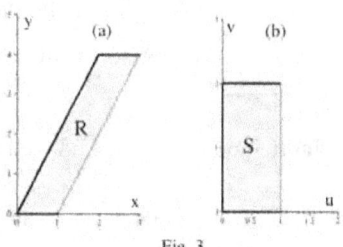

Fig. 3

Solution: (a) Since this is a linear transformation, corners get mapped to corners and straight line boundaries get mapped to straight line boundaries. The corners in the xy-plane of R are (0,0), (2,4), (1,0) and (3,4) as shown in Fig. 3a. Under the change of variables, (0,0)-->(0,0), (2,4)-->(0,2), (1,0)-->(1,0), and (3,4)-->(1,2) so the integration region in the uv-plane is the rectangle S in Fig. 3b.

(b) Simple algebra gives $x = u + v$ and $y = 2v$. (c) $J(u,v) = \dfrac{\partial(x,y)}{\partial(u,v)} = \begin{vmatrix} \dfrac{\partial x}{\partial u} & \dfrac{\partial x}{\partial v} \\ \dfrac{\partial y}{\partial u} & \dfrac{\partial y}{\partial v} \end{vmatrix} = \begin{vmatrix} 1 & 1 \\ 0 & 2 \end{vmatrix} = 2$

(d) $\int_{y=0}^{4} \int_{x=y/2}^{(y/2)+1} \dfrac{2x-y}{2}\, dx\, dy = \int_{v=0}^{2} \int_{u=0}^{1} u\, |J(u,v)|\, du\, dv = \int_{v=0}^{2} \int_{u=0}^{1} u \cdot 2\, du\, dv = \int_{v=0}^{2} 1\, dv = 2$.

Practice 5: Suppose the area of S is $\iint\limits_{S} 1\, dV = 20$, and $J(u,v) = 2$ for the transformation $T(S) = R$. Determine the area of region R.

In Section 14.3 we discussed double integrals in polar coordinates and used geometry to find the conversion formula between the two types of double integrals. The conversion from polar to rectangular coordinates is simply the transformation $x = r \cdot \cos(\theta)$ and $y = r \cdot \sin(\theta)$. Then

$J(r,\theta) = \dfrac{\partial(x,y)}{\partial(r,\theta)} = \begin{vmatrix} \dfrac{\partial x}{\partial r} & \dfrac{\partial x}{\partial \theta} \\ \dfrac{\partial y}{\partial r} & \dfrac{\partial y}{\partial \theta} \end{vmatrix} = \begin{vmatrix} \cos(\theta) & -r \cdot \sin(\theta) \\ \sin(\theta) & r \cdot \cos(\theta) \end{vmatrix} = r \cdot \cos^2(\theta) + r \cdot \sin^2(\theta) = r$ and

$\iint\limits_{R} F(x,y)\, dx\, dy = \iint\limits_{S} F(r \cdot \cos(\theta), r \cdot \sin(\theta))\, r\, dr\, d\theta$, the same result we got in this special case in 14.3.

Creating the Desired Transformation for Double Integrals

So far in this section all of the transformations have been given. But often we have a "strange" xy-domain and need to pick a transformation that maps it to something nice such as a rectangle in the uv-plane. Sometimes that is very difficult, but there are a few situations and ideas that are much easier. Technology and software can help us evaluate double integrals once they are set up, but it is usually up to us to set up those integrals first.

(1) Region R is bounded by parallel lines

In Example 1 the region R (Fig. 4) is bounded by the pair of parallel lines y=6–x and y=–x which can be rewritten as x+y=6 and x+y=0. This suggests that we might set x+y=**u** and let u vary from 0 to 6. Similarly with the parallel pair 2y=x–3 and 2y=x–6, rewritten as x-2y=3 and x–2y=6, suggesting that we put x–2y=**v** with v going from 3 to 6. It is straight forward to verify that the transformation u=x+y, v=x-2y transforms R into the uv-plane rectangle shown in Fig. 5. (The transformation in Example 1 was u=(x+y)/3 and v=(x-2y)/3 which leads to a smaller rectangle in the uv plane.)

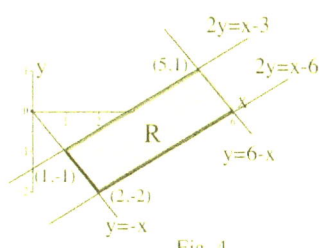

Fig. 4

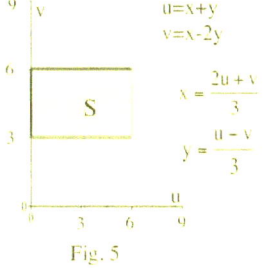

Fig. 5

Practice 6: Find a transformation of the region R bounded by the lines y=x, y=x+2, y=6–2x and y=9–2x (Fig. 6) into a rectangle S in the uv-plane.

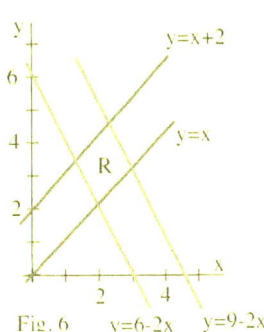

Fig. 6

(2) Region R is bounded by shifted curves

Example 6: R is the region in the xy plane bounded by the lines y=x and y=x+3 and the parabolas $y = 9 - x^2$ and $y = 16 - x^2$ (Fig. 7). Find a transformation of R into a rectangle S in the uv-plane.

Solution: y-x=0 and y-x=3 suggests putting u=y-x (so u goes from 0 to 3). The parabolas $x^2 + y = 9$ and $x^2 + y = 16$ suggests putting v= $x^2 + y$ (so v goes from 9 to 16). You can verify that this works.

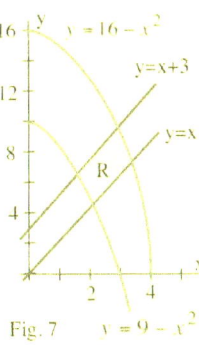

Fig. 7

Practice 7: R is the region in the xy-plane bounded by the four parabolas $y = 9 - x^2$, $y = 16 - x^2$, $y = x^2$ and $y = x^2 + 4$. Sketch the region R and find a transformation of R into a rectangle in the uv-plane.

(3) Two parameter family of curves

If the bounding curves of the region R can be written using the two parameters u and v, then we can usually transform R into a rectangle in the uv-plane.

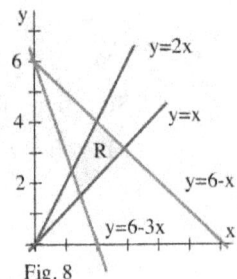

Example 7: R is the region in the xy-plane bounded by the four lines y=x, y=2x, y=6–x and y=6–3x in Fig. 8. Find a transformation of R into a rectangle S in the uv-plane.

Fig. 8

Solution: y=1x and y=2x can be rewritten with a single parameter u as y=ux for u from 1 to 2. y=6–1x and y=6–3x can be rewritten with the single parameter y=6–vx with v going from 1 to 3. Solving for u and v, we get the transformation u=y/x and v=(6–y)/x which takes R to a rectangle in the uv-plane. The inverse transformation is x=6/(u+v) and y=6u/(u+v).

Changing Variables in Triple Integrals

Changing variables for triple integrals is very similar to the situation for double integrals. If T is a transformation from an uvw-space region S to an xyz-space region R (so x, y and z are each differentiable functions of u, v and w: x=g(u,v,w), y=h(u,v,w) and z=k(u,v,w)) then the

$$\text{3D Jacobian is } J(u,v,w) = \begin{vmatrix} \dfrac{\partial x}{\partial u} & \dfrac{\partial x}{\partial v} & \dfrac{\partial x}{\partial w} \\ \dfrac{\partial y}{\partial u} & \dfrac{\partial y}{\partial v} & \dfrac{\partial y}{\partial w} \\ \dfrac{\partial z}{\partial u} & \dfrac{\partial z}{\partial v} & \dfrac{\partial z}{\partial w} \end{vmatrix}$$

The 3D change of variables formula is

$$\iiint_R f(x,y,z)\, dx \cdot dy \cdot dz = \iiint_S f(g(u,v,w), h(u,v,w), k(u,v,w)) \cdot |J(u,v,w)|\, du \cdot dv \cdot dw$$

Example 8: Suppose f(x,y,z)=2x+4z on the box-like region (Fig. 9)

$$R = \{(x,y,z): x \le y \le x+3,\ x \le z \le x+2 \text{ and } 1-x \le z \le 3-x\}.$$ Evaluate

$\iiint_R f(x,y,z)\, dx \cdot dy \cdot dz$ by using the transformation T: u=y–x (0≤u≤3),

v=z–x (0≤v≤2) w=z+x (1≤w≤3) and then evaluating the new integral.

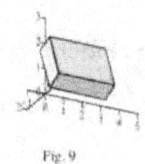

Fig. 9

Solution: We need several pieces. The inverse transformation (after a bit of algebra) is x= –v/2+w/2, y= u–v/2+w/2, z= v/2+w/2, f(x,y,z)=2x+4z=v+3w, S={(u,v,w): 0≤u≤3, 0≤v≤2, 1≤w≤3},

and the Jacobean is
$$J(u,v,w) = \begin{vmatrix} 0 & -\frac{1}{2} & \frac{1}{2} \\ 1 & -\frac{1}{2} & \frac{1}{2} \\ 0 & \frac{1}{2} & \frac{1}{2} \end{vmatrix} = \frac{1}{2}$$

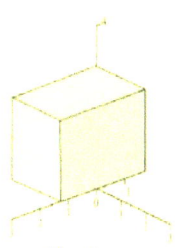

Fig. 10

The new domain in uvw-space is shown in Fug. 10. Finally,
$$\iiint_R f(x,y,z)\,dx\cdot dy\cdot dz \quad \int_1^3\int_0^2\int_0^3 (v+3w)\left(\frac{1}{2}\right)du\cdot dv\cdot dw$$
$$\int_1^3\int_0^2 3(v+3w)\left(\frac{1}{2}\right)dv\cdot dw = \int_1^3 3+9w\,dw = 42.$$

Practice 8: Evaluate $\iiint_R (x+z)\,dx\cdot dy\cdot dz$ on the region

R = {(x,y,z): x+1 ≤ y ≤ x+3, x ≤ z ≤ 2x and 0 ≤ z ≤ 3} by using the transformation

T: u=y–x (1≤u≤3), v=x/z (1≤v≤2) w=z (0≤w≤3).

Rectangular to Spherical: In section 14.7 we transformed some integrals in rectangular coordinates into ones in spherical coordinates, and we derived the $dV = \rho^2\cdot\sin(\varphi)\,d\rho\cdot d\theta\cdot d\varphi$ formula for the transformation geometrically. Instead, we can use the Jacobean. For spherical coordinates $x = \rho\cdot\sin(\varphi)\cdot\cos(\theta)$, $y = \rho\cdot\sin(\varphi)\cdot\sin(\theta)$, and $z = \rho\cdot\cos(\varphi)$. Then the Jacobean J(u,v,w) is

$$J(\rho,\theta,\varphi) = \begin{vmatrix} \frac{\partial x}{\partial \rho} & \frac{\partial x}{\partial \theta} & \frac{\partial x}{\partial \varphi} \\ \frac{\partial y}{\partial \rho} & \frac{\partial y}{\partial \theta} & \frac{\partial y}{\partial \varphi} \\ \frac{\partial z}{\partial \rho} & \frac{\partial z}{\partial \theta} & \frac{\partial z}{\partial \varphi} \end{vmatrix} = \begin{vmatrix} \sin(\varphi)\cdot\cos(\theta) & -\rho\cdot\sin(\varphi)\cdot\sin(\theta) & \rho\cdot\cos(\varphi)\cdot\cos(\theta) \\ \sin(\varphi)\cdot\sin(\theta) & \rho\cdot\sin(\varphi)\cdot\cos(\theta) & \sin(\varphi)\cdot\sin(\theta) \\ \cos(\varphi) & 0 & -\rho\cdot\sin(\varphi) \end{vmatrix} = \ldots = \rho^2\cdot\sin(\varphi)\,d\rho\cdot d\theta\cdot d\varphi.$$

The "..." is simply a matter of carefully calculating the determinant and then using some fundamental trigonometric identities to simplify the result.

Practice 9: Calculate $\sin(\varphi)\cdot\cos(\theta)\cdot\begin{vmatrix} \rho\cdot\sin(\varphi)\cdot\cos(\theta) & \sin(\varphi)\cdot\sin(\theta) \\ 0 & -\rho\cdot\sin(\varphi) \end{vmatrix}$

Other triple integral transformations are possible, but rectangular to cylindrical or spherical are the most common.

Problems

In problems 1 to 4, (a) sketch the given set S and the image of S under the given transformation, (b) calculate the Jacobians, J(x,y) and J(u,v), and rewrite $\iint_R f(x,y)\, dx\, dy$ as $\iint_S \underline{\quad} du\, dv$.

1. $S = \{(u,v): 0 \le u \le 2, 1 \le v \le 4\}$ under $x = u+v$ and $y = 2u-v$.

2. $S = \{(u,v): 1 \le u \le 2, 0 \le v \le 2\}$ under $x = 2u-3v$ and $y = u+2v$.

3. $S = \{(u,v): 0 \le u \le 1, 0 \le v \le 1\}$ under $x = au+bv$ and $y = cu+dv$.

4. $S = \{(u,v): 2 \le u \le 4, \frac{\pi}{6} \le v \le \frac{\pi}{2}\}$ under $x = u \cdot \cos(v)$ and $y = u \cdot \sin(v)$.

In problems 5 to 10, (a) sketch the given set S and the image of S under the given transformation, and (b) calculate the Jacobians, J(x,y) and J(u,v).

5. $S = \{(x,y): 0 \le x \le 2, 1 \le y \le 3\}$ under $u = \frac{3x-3y}{4}$ and $v = \frac{y}{3}$.

6. $S = \{(x,y): 0 \le x \le 2, 1 \le y \le 3\}$ under $u = x+y$ and $v = \frac{x}{y}$.

7. $S = \{(x,y): 0 \le x \le 1, 0 \le y \le 1\}$ under $u = x^2 - y^2$ and $v = 2xy$.

8. $S = \{$region bounded by the hyperbolas $y = 1/x$ and $y = 4/x$ and the lines $y = x$ and $y = 4x\}$ with $x = u/v$ and $y = uv$.

9. $S = \{$trapezoid with vertices $(x,y) = (1,0), (2,0), (0,-2)$ and $(0,-1)\}$ under $u = x-y$ and $v = x+y$.

10. $S = \{$triangle with vertices $(x,y) = (1,0), (3,1)$ and $(0,4)\}$ under $u = x-y$ and $v = x+y$.

11. Suppose $\iint_S 1\, du\, dv = 14$, and $J(u,v) = 7$ for the transformation $T(S) = R$. Evaluate $\iint_R 1\, dV$.

12. Suppose $\iint_S 1\, du\, dv = 30$, and $J(u,v) = 5$ for the transformation $T(S) = R$. Evaluate $\iint_R 1\, dA$.

13. Suppose $\iint_S 1\, du\, dv = 15$, and $J(x,y) = 3$ for the transformation $T(S) = R$. Evaluate $\iint_R 1\, dA$.

14. Suppose $\iint_S 1\, du\, dv = 24$, and $J(x,y) = 4$ for the transformation $T(S) = R$. Evaluate $\iint_R 1\, dA$.

In problems 15-24, use the given transformation to evaluate the integral.

15. $\iint_R (x+3y)\, dA$. R is the triangular region with vertices (0,0), (2,1) and (1,2). Use $x = 2u+v$ and $y = u+2v$ (so $u = (2x-y)/3$ and $v = (2y-x)/3$).

16. $\iint_R x\, dA$. R is the ellipse $\frac{x^2}{9} + \frac{y^2}{4} \le 1$. Use $x = 3u$ and $y = 2u$.

17. $\iint_R \frac{x+y}{3}\, dA$ where R is the region shown in Fig. 4.

18. $\iint_R (x-2y)\, dA$ where R is the region shown in Fig. 4.

19. $\iint_R (3x+6y)\, dA$ where R is the region shown in Fig. 6.

20. $\iint_R (6x-3y)\, dA$ where R is the region bounded by the lines y=x, y=2x, y=6-2x and y=9-2x (Fig. 6).

21. $\iint_R \sqrt{\frac{y}{x}} + \sqrt{xy}\, dA$ where R={region bounded by the hyperbolas y=1/x and y=4/x and the lines y=x and

    y=4x} Use the substitution x=u/v and y=uv.

22. $\iint_R (6x-3y)\, dA$ where R is the region bounded by the lines y=x, y=2x, y=6-x and y=6-3x (Fig. 8).

    Use u=y/x and v=(6-y)/x.

23. The area of the ellipse $\frac{x^2}{a^2} + \frac{y^2}{b^2} \leq 1$ can be found using earlier methods, but the integral requires a

    trigonometric substitution. Instead, transform the ellipse into a circle using the substitution x=au and
    y=bv, write the new uv integral and evaluate it to show that the area of the ellipse is πab .

24. Evaluate $\iint_R xy\, dA$ where R is the square with vertices at (0,0), (1,1), (2,0) and (1,-1).

Practice Solutions

Practice 1: See Fig. P1 for the S and R regions. u=(x+2y)/5
and v=(2x–y)/5 .

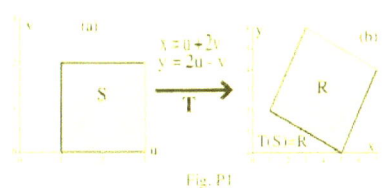

Fig. P1

Practice 2: See Fig. P2 for the S and R regions. u=y and
v=x/(1+sin(y)).

Practice 3: $\frac{\partial x}{\partial u}=1$, $\frac{\partial x}{\partial v}=2$, $\frac{\partial y}{\partial u}=2$ and $\frac{\partial y}{\partial v}=-1$ so $J(u,v)=\frac{\partial(x,y)}{\partial(u,v)}=\begin{vmatrix}1 & 2\\ 2 & -1\end{vmatrix}=-5$

The inverse transformation is u=(x+2y)/5 and v=(2x-y)/5 so

$\frac{\partial u}{\partial x}=\frac{1}{5}$, $\frac{\partial u}{\partial y}=\frac{2}{5}$, $\frac{\partial v}{\partial x}=\frac{2}{5}$ and $\frac{\partial v}{\partial y}=-\frac{1}{5}$ and $J(x,y)=\frac{\partial(u,v)}{\partial(x,y)}=\begin{vmatrix}1/5 & 2/5\\ 2/5 & -1/5\end{vmatrix}=-\frac{1}{5}$

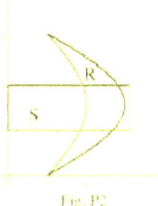

Fig. P2

Practice 4: $\delta(x,y)=x+y$. $x=x(u,v)=u+2v$ and $y=y(u,v)=u-v$ so J(u,v)=5 (see Practice 3)

(a) $\{\text{area of R}\} = \iint_R 1\, dA = \iint_S 1\cdot J(u,v)|\, du\, dv = \int_{v=0}^{2}\int_{u=1}^{3}(5)\, du\, dv = \int_{v=1}^{2} 10\, dv = 10$

(b) Mass $= \iint_R \delta\, dA = \iint_S (x+y)\cdot J(u,v)|\, du\, dv = \int_{v=0}^{2}\int_{u=1}^{3}(2u+v)(5)\, du\, dv = \int_{v=1}^{2} 40+10v\, dv = 65$

Practice 5: The area of R is $\iint_R 1\, dV = \iint_S 1 \cdot |J(u,v)|\, dV \iint_S 2\, dV = 2(20) = 40$.

Practice 6: The first two line equations can be written as y-x=0 and y-x=2 so put u=y-x (then u goes from 0 to 2). The other two parallel lines can be written as 2x+y=6 and 2x+y=9 so put v=2x+y (then v goes from 6 to 9). This transformation takes R in the xy-plane to the $0 \leq u \leq 2$ and $6 \leq v \leq 9$ rectangle in the uv-plane.

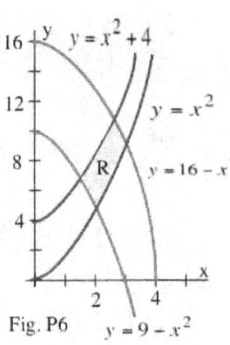

Fig. P6

Practice 7: R is shown in Fig. P6. $x^2 + y = 9$ and $x^2 + y = 16$ so put $u = x^2 + y$ (then u goes from 9 to 16). Similarly, $x^2 - y = 0$ and $x^2 - y = 4$ so put $u = x^2 - y$ (then v goes from 0 to 4).

Practice 8: The inverse transformation is x=uv, y=vw+u and z=w. x+z=uv+w and

$$J(u,v,w) = \begin{vmatrix} v & u & 0 \\ 1 & w & 0 \\ 0 & 0 & 1 \end{vmatrix} = vw$$ so the new integral is

$$\iiint_R (x+z)\, dx \cdot dy \cdot dz = \int_0^3 \int_1^2 \int_1^3 (uv+w)(vw)\, du \cdot dv \cdot dw = \int_0^3 \int_1^2 \int_1^3 (uv^2w + vw^2)\, du \cdot dv \cdot dw$$

$$= \int_0^3 \int_1^2 \left(4v^2w + 2vw^2\right) dv \cdot dw = \int_0^3 \left(\frac{28}{3}w + 3w^2\right) dw = 69$$

Practice 9: $\sin(\varphi) \cdot \cos(\theta) \cdot \begin{vmatrix} \rho \cdot \sin(\varphi) \cdot \cos(\theta) & \sin(\varphi) \cdot \sin(\theta) \\ 0 & -\rho \cdot \sin(\varphi) \end{vmatrix} = \sin(\varphi) \cdot \cos(\theta)\left[-\rho^2 \cdot \sin^2(\varphi) \cdot \cos(\theta)\right] = -\rho^2 \cdot \sin^3(\varphi) \cdot \cos^2(\theta)$

Two Transformations, Same Result

Integral of $f(x,y) = \sqrt{\dfrac{y}{x}} + \sqrt{xy}$ on

$R = \{(x,y)$ bounded by $y = 1/x$, $y = 4/x$, $y = x$ and $y = 4x\}$ (see figure)

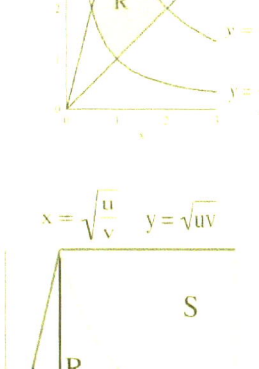

(A) The "natural" transformation (pages 5 and 6):

$y = vx$ for $v = 1..4$ so $v = y/x$. $y = u/x$ for $u = 1..4$ so $u = xy$.

Then $x = \sqrt{\dfrac{u}{v}}$ and $y = \sqrt{uv}$.

The Jacobian is
$$J(x,y) = \begin{vmatrix} \dfrac{\partial u}{\partial x} & \dfrac{\partial u}{\partial y} \\ \dfrac{\partial u}{\partial x} & \dfrac{\partial u}{\partial x} \end{vmatrix} = \begin{vmatrix} y & x \\ -\dfrac{y}{x^2} & \dfrac{1}{x} \end{vmatrix} = \dfrac{2y}{x} = 2v.$$

$$\iint_R \sqrt{\dfrac{y}{x}} + \sqrt{xy}\, dx\, dy = \int_{u=1}^{4} \int_{v=1}^{4} [\sqrt{v} + \sqrt{u}]\left(\dfrac{1}{2v}\right) dv\, du$$

$$= \int_{u=1}^{4} \int_{v=1}^{4} \left[\dfrac{1}{2\sqrt{v}} + \dfrac{\sqrt{u}}{2v}\right] dv\, du = \int_{u=1}^{4} \left[\sqrt{v} + \dfrac{\sqrt{u}}{2}\ln(v)\right]_{v=1}^{v=4} du$$

$$= \int_{u=1}^{4} \left[1 + \dfrac{\sqrt{u}}{2}\ln(4)\right]_{v=1}^{v=4} du = u + \dfrac{1}{2}\cdot\dfrac{2}{3}(u)^{3/2}\ln(4) \Big|_{u=1}^{u=4} = 3 + \dfrac{7}{3}\ln(4)$$

(B) The "suggested" transformation (Problem 8):

$x = u/v$ and $y = uv$. Then $u = xv$ so

$y = (xv)v = x\cdot v^2$ with $v^2 = 1..4$ and $v = 1..2$ and $v = \sqrt{\dfrac{y}{x}}$.

$xy = \left(\dfrac{u}{v}\right)(uv) = u^2$ with $u^2 = 1..4$ and $u = 1..2$ and $u = \sqrt{xy}$.

The Jacobian is
$$J(u,v) = \begin{vmatrix} \dfrac{\partial x}{\partial u} & \dfrac{\partial x}{\partial v} \\ \dfrac{\partial y}{\partial u} & \dfrac{\partial y}{\partial v} \end{vmatrix} = \begin{vmatrix} \dfrac{1}{v} & -\dfrac{u}{v^2} \\ v & u \end{vmatrix} = \dfrac{2u}{v}.$$

$$\iint_R \sqrt{\dfrac{y}{x}} + \sqrt{xy}\, dx\, dy = \int_{u=1}^{2} \int_{v=1}^{2} [v + u]\left(\dfrac{2u}{v}\right) dv\, du = \int_{u=1}^{2} \int_{v=1}^{2} \left[2u + \dfrac{2u^2}{v}\right] dv\, du$$

$$= \int_{u=1}^{2} \left[2uv + 2u^2\ln(v)\right]_{v=1}^{v=2} du = \int_{u=1}^{2} \left[2u + 2u^2\ln(2)\right] du = u^2 + \dfrac{2}{3}u^3\ln(2)\Big|_{u=1}^{u=2}$$

$$= u^2 + \dfrac{2}{3}u^3\ln(2)\Big|_{u=1}^{u=2} = 3 + \dfrac{14}{3}\ln(2) = 3 + \dfrac{7}{3}\ln(4).$$

Conclusion: These two transformations have different Jacobians (magnifications) and lead to different S regions ($[1,4]\times[1,4]$ verses $[1,2]\times[1,2]$) but the resulting integral values are the same.

Appendix: Derivation of the 2D Jacobian Formula

Suppose T is a transformation from a rectangular region S in the uv-plane to a region R in the xy-plane given by $x=x(u,v)$ and $y=y(u,v)$ as in Fig A1.

The corners of the rectangular region S are (u,v), $(u+\Delta u, v)$, $(u, v+\Delta v)$ and $(u+\Delta u, v+\Delta v)$.

Then $T:(u,v) \rightarrow (x(u,v), y(u,v)) = A$, a point in the xy-plane (Fig. A2)
$T: (u+\Delta u, v) \rightarrow (x(u+\Delta u, v), y(u+\Delta u, v)) = B$
$T: (u, v+\Delta v) \rightarrow (x(u, v+\Delta v), y(u, v+\Delta v)) = C$

Next we want to approximate the area of the region R using the cross product.
Put $P = $ vector $AB = \langle x(u+\Delta u, v) - x(u,v), y(u+\Delta u, v) - y(u,v) \rangle$
$= \left\langle \dfrac{x(u+\Delta u,v) - x(u,v)}{\Delta u}, \dfrac{y(u+\Delta u,v) - y(u,v)}{\Delta u} \right\rangle \Delta u = \left\langle \dfrac{\Delta x}{\Delta u}, \dfrac{\Delta y}{\Delta u} \right\rangle \Delta u$

and $Q = $ vector $AC = \langle x(u, v+\Delta v) - x(u,v), y(u, v+\Delta v) - y(u,v) \rangle$
$= \left\langle \dfrac{x(u, v+\Delta v) - x(u,v)}{\Delta v}, \dfrac{y(u, v+\Delta v) - y(u,v)}{\Delta v} \right\rangle \Delta v = \left\langle \dfrac{\Delta x}{\Delta v}, \dfrac{\Delta y}{\Delta v} \right\rangle \Delta v$

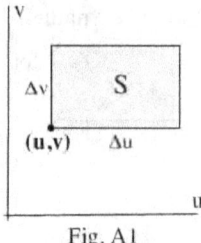

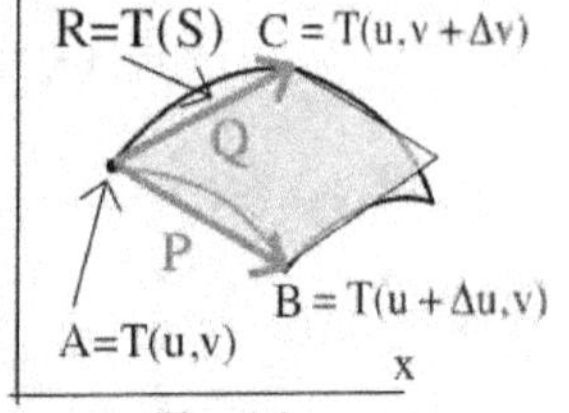

Fig. A1

Fig. A2

Then the area of region R is approximately $|P \times Q| = \begin{vmatrix} \dfrac{\Delta x}{\Delta u} & \dfrac{\Delta y}{\Delta u} \\ \dfrac{\Delta x}{\Delta v} & \dfrac{\Delta y}{\Delta v} \end{vmatrix} \Delta u \Delta v$

As $\Delta u, \Delta v \rightarrow 0$, the fractions in the determinant approach the partial derivatives.

So $dA = dx \cdot dy = \begin{vmatrix} \dfrac{\partial x}{\partial u} & \dfrac{\partial x}{\partial v} \\ \dfrac{\partial y}{\partial u} & \dfrac{\partial y}{\partial v} \end{vmatrix} du \cdot dv = |J(u,v)| du \cdot dv$

Appendix: Simple proof that $J(x,y) = \dfrac{1}{J(u,v)}$ for a linear transformation in 2D (could be a student assignment)

Theorem; If $T:(u,v) \rightarrow (x,y)$ is a linear transformation ($x=au+bv$ and $y=cu+dv$) then $J(u,v)=1/J(x,y)$.

Proof: $J(x,y) = \begin{vmatrix} \dfrac{\partial x}{\partial u} & \dfrac{\partial x}{\partial v} \\ \dfrac{\partial y}{\partial u} & \dfrac{\partial y}{\partial v} \end{vmatrix} = \begin{vmatrix} a & b \\ c & d \end{vmatrix} = ad - bc$.

Using elementary algebra and solving for u and v we get $u = \dfrac{dx - by}{ad - bc}$ and $v = \dfrac{-cx + ay}{ad - bc}$ so

$J(u,v) = \begin{vmatrix} \dfrac{\partial u}{\partial x} & \dfrac{\partial u}{\partial y} \\ \dfrac{\partial v}{\partial x} & \dfrac{\partial v}{\partial y} \end{vmatrix} = \begin{vmatrix} \dfrac{d}{ad-bc} & \dfrac{-b}{ad-bc} \\ \dfrac{-c}{ad-bc} & \dfrac{a}{ad-bc} \end{vmatrix} = \dfrac{ad-bc}{(ad-bc)^2} = \dfrac{1}{ad-bc} = \dfrac{1}{J(x,y)}$.

Appendix: Proof that $J(x,y) = \dfrac{1}{J(u,v)}$ for a general linear transformation in 2D.

The following results are true for larger matrices and the proofs are similar.

14.8 Changing Variables in Double & Triple Integrals

Lemma: For 2x2 matrices S and T, $|S \cdot T| = |S| \cdot |T|$

Proof: If $S = \begin{pmatrix} a & b \\ c & d \end{pmatrix}$ and $T = \begin{pmatrix} A & B \\ C & D \end{pmatrix}$ then $|S| = ad - bc$ and $|T| = AD - BC$.

By matrix multiplication, $S \cdot T = \begin{pmatrix} aA + bC & aB + bD \\ cA + dC & cB + dD \end{pmatrix}$ so

$|S \cdot T| = (aA + bC)(cB + dD) - (cA + dC)(aB + bD)$

$= (aAcB + aAdD + bCcB + bCdD) - (cAaB + cAbD + dCaB + dCbD)$

$= acAB + adAD + bcBC + bdCD - acAB - bcAD - adBC - bdCD$

$= adAD + bcBC - bcAD - adBC = (ad - bc)(AD - BC) = |S| \cdot |T|$

Lemma: If $S:(x,y) \to (u,v)$ and $T:(u,v) \to (w,z)$ then $J(T \circ S) = J(S) \cdot J(T)$.

Proof. $J(S) = \begin{vmatrix} \frac{\partial x}{\partial u} & \frac{\partial x}{\partial v} \\ \frac{\partial y}{\partial u} & \frac{\partial y}{\partial v} \end{vmatrix}$ and $J(S) = \begin{vmatrix} \frac{\partial x}{\partial u} & \frac{\partial x}{\partial v} \\ \frac{\partial y}{\partial u} & \frac{\partial y}{\partial v} \end{vmatrix}$.

Then $T \circ S : (x,y) \to (u,v) \to (w,z)$ so $J(T \circ S) = \begin{vmatrix} \frac{\partial x}{\partial w} & \frac{\partial x}{\partial z} \\ \frac{\partial y}{\partial w} & \frac{\partial y}{\partial z} \end{vmatrix}$.

But by the Chain Rule for functions of several variables,

$$\frac{\partial x}{\partial w} = \frac{\partial x}{\partial u} \cdot \frac{\partial u}{\partial w} + \frac{\partial x}{\partial v} \cdot \frac{\partial v}{\partial w} \qquad \frac{\partial y}{\partial w} = \frac{\partial y}{\partial u} \cdot \frac{\partial u}{\partial w} + \frac{\partial y}{\partial v} \cdot \frac{\partial v}{\partial w}$$

$$\frac{\partial x}{\partial z} = \frac{\partial x}{\partial u} \cdot \frac{\partial u}{\partial z} + \frac{\partial x}{\partial v} \cdot \frac{\partial v}{\partial z} \qquad \frac{\partial y}{\partial z} = \frac{\partial y}{\partial u} \cdot \frac{\partial u}{\partial z} + \frac{\partial y}{\partial v} \cdot \frac{\partial v}{\partial z}$$

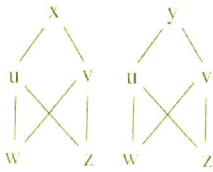

so $J(T \circ S) = \begin{vmatrix} \frac{\partial x}{\partial w} & \frac{\partial x}{\partial z} \\ \frac{\partial y}{\partial w} & \frac{\partial y}{\partial z} \end{vmatrix} = \begin{vmatrix} \frac{\partial x}{\partial u} \cdot \frac{\partial u}{\partial w} + \frac{\partial x}{\partial v} \cdot \frac{\partial v}{\partial w} & \frac{\partial x}{\partial u} \cdot \frac{\partial u}{\partial z} + \frac{\partial x}{\partial v} \cdot \frac{\partial v}{\partial z} \\ \frac{\partial y}{\partial u} \cdot \frac{\partial u}{\partial w} + \frac{\partial y}{\partial v} \cdot \frac{\partial v}{\partial w} & \frac{\partial y}{\partial u} \cdot \frac{\partial u}{\partial z} + \frac{\partial y}{\partial v} \cdot \frac{\partial v}{\partial z} \end{vmatrix}$

$= \begin{pmatrix} \frac{\partial x}{\partial u} & \frac{\partial x}{\partial v} \\ \frac{\partial y}{\partial u} & \frac{\partial y}{\partial v} \end{pmatrix} \begin{pmatrix} \frac{\partial u}{\partial w} & \frac{\partial u}{\partial z} \\ \frac{\partial v}{\partial w} & \frac{\partial v}{\partial z} \end{pmatrix} = \begin{vmatrix} \frac{\partial x}{\partial u} & \frac{\partial x}{\partial v} \\ \frac{\partial y}{\partial u} & \frac{\partial y}{\partial v} \end{vmatrix} \cdot \begin{vmatrix} \frac{\partial u}{\partial w} & \frac{\partial u}{\partial z} \\ \frac{\partial v}{\partial w} & \frac{\partial v}{\partial z} \end{vmatrix} = J(S) \cdot J(T)$

Theorem: If S and T are inverse transformations $\left(T \circ S = I : (x,y) \to (x,y) = \begin{pmatrix} 1 & 0 \\ 0 & 1 \end{pmatrix} \right)$, then $J(S) \cdot J(T) = 1$.

Proof: Since $T \circ S = I$ then $J(T \circ S) = J(I) = \begin{vmatrix} 1 & 0 \\ 0 & 1 \end{vmatrix} = 1$ and $J(S) \cdot J(T) = J(T \circ S) = 1$.

14.0 Odd Answers

1. v=(24)(54/6)=216 (actual value is 368 – find out how in section 14.1)
3. (a) v=(32)(64/8)=256   (b) v=(32)(112/8)=448 (actual = 1184/3 = 393.667)
5. sum=39 inches, avg=39/16=2.4375 inches= 0.203125 ft, total = (avg)(area)=125.075 ft^3
7. v=(16/4)(4)(5)=80 ft^3
9. v=(19/5)(PI*2^2)=(76/5)*Pi=47.752 f^3
11. v=(26.7/15)*90=160.2 m^3
13. crude avg ht= 2, so vol=(2)(6)(10)(Pi)/4=30Pi = 94.25 m^3
15. avg =30/6=5, v=5*6*4= 120

14.1 Odd Answers

1. $\frac{8}{3} y^3, \frac{1}{4} x^2$ 3. $e^{2+y} + e^y = e^y(e^2 + 1)$, $xe^{x+1} - xe^x = xe^x(e - 1)$ 5. 60

7. $3+e^4 +4\sin(1) \approx 60.96$ 9. 26   11. 32/3   13. $\frac{4}{15}(31 - 9\sqrt{3})$ 15. 6

17. Easier $\int y \cdot \sqrt{x^2 + y^2} \, dy = \frac{1}{3}(x^2 + y^2)^{3/2} + C(x)$

19. $\int \sin(x) \cdot e^{x+y} \, dy = \int \sin(x) \cdot e^x \cdot e^y \, dy = \sin(x) \cdot e^x \cdot e^y + C(x)$

21. $\int \sqrt{x + y^2} \, dx = \frac{2}{3}(x + y^2)^{3/2} + C(y)$ 23. $\int e^{(y^2)} \, dx = x \cdot e^{(y^2)} + C(y)$

25. $-\frac{585}{8}$ 27. $\frac{\pi}{12} - \frac{3}{2} + \sqrt{3} - \frac{\pi}{12}\sqrt{3} \approx 0.0404$ 29. $\frac{1}{6}(60) = 10$ 31. $\frac{1}{6}(26) = \frac{13}{3}$

33. One reasonable approximation is 33 m^3. Your approximation should be close to this number, and you should be able to justify why your method is reasonable.

35. Using midpoints, one reasonable estimate is 88 m^3.

14.2 Odd Answers

1. 9   3. 63   5. $\frac{1}{2} - \frac{1}{2}\cos(1)$

7. $\int_{x=-2}^{x=2} \int_{y=0}^{y=4-x^2} f \, dy \, dx$, $\int_{y=0}^{y=4} \int_{x=-\sqrt{4-y}}^{x=\sqrt{4-y}} f \, dx \, dy$ 9. $\int_1^4 \int_x^4 f \, dy \, dx$, $\int_1^4 \int_1^y f \, dx \, dy$

11. $\int_0^2 \int_{x^2}^{6-x} f \, dy \, dx$, $\int_0^4 \int_0^{\sqrt{y}} f \, dx \, dy + \int_4^6 \int_0^{6-y} f \, dx \, dy$

13. $\frac{1}{12}$ 15. $\frac{3}{4}$ 17. $\frac{32}{3}$ 19. $-\frac{72}{5}$

21. avg. value = volume/area = $\frac{44}{4} = 11$ 23. volume= $\frac{45}{4}$, area = $\frac{9}{2}$, avg. value = $\frac{5}{2}$

25. $\displaystyle\int_0^1 \int_y^1 f(x,y)\ dx\ dy$ 27. $\displaystyle\int_0^3 \int_0^{6-2x} f(x,y)\ dy\ dx$ 29. $\displaystyle\int_0^{\ln(2)} \int_{e^y}^{2} f(x,y)\ dx\ dy$

31. $\displaystyle\int_0^4 \int_0^{2-y/2} f\ dx\ dy$ 33. $\displaystyle\int_0^2 \int_0^{y^2} f\ dx\ dy$ 35. $\displaystyle\int_1^e \int_{\ln(y)}^{1} f\ dx\ dy$

37. One reasonable approximation of the volume is $24\ m^3$. Your approximation should be close to this.

14.3 Odd Answers

1. A: rectangular B: polar C: polar

3. A: rectangular B: rectangular (since the circle is not centered at the origin) C: polar

5. 63π 7. $A \cdot D^2 \cdot \pi$

9. $V = \displaystyle\iint_R \sqrt{9 - x^2 - y^2}\ dA$ where $R = \{(x,y):\ x^2 + y^2 \le 9\}$.

$V = \displaystyle\int_{\theta=0}^{2\pi} \int_{r=0}^{3} \sqrt{9 - r^2}\ r \cdot dr \cdot d\theta = \int_{\theta=0}^{2\pi} 9\ d\theta = 18\pi$.

11. $V = \displaystyle\iint_R 2 + xy\ dA$ where R is the region in inside the circle $x^2 + y^2 = 4$ and above the x-axis..

$V = \displaystyle\int_{\theta=0}^{\pi} \int_{r=0}^{2} \{2 + r\cdot\cos(\theta)\cdot r\cdot\sin(\theta)\}\cdot r\cdot dr\cdot d\theta = \int_{\theta=0}^{\pi} \{4 + 4\cos(\theta)\sin(\theta)\}\ d\theta = 4\pi$.

13. Under the plane $z = 5 + 2x + 3y$ and above the disk $x^2 + y^2 \le 16$.

$V = \displaystyle\int_{\theta=0}^{2\pi} \int_{r=0}^{4} \{5 + 2\cos(\theta) + 3\sin(\theta)\}\ r\cdot dr\cdot d\theta = \int_{\theta=0}^{2\pi} \{40 + 16\cos(\theta) + 24\sin(\theta)\}\ d\theta = 80\pi$

15. Between the surfaces $z = 1 + x + y$ and $z2 = 8 + 2x + 3y$ for $x^2 + y^2 \le 1$ and $0 \le y$.

$V = \displaystyle\iint_R \{z2 - z1\}\ dA = \int_{\theta=0}^{\pi} \int_{r=0}^{1} \{z2 - z1\}\ dA = \int_{\theta=0}^{\pi} \int_{r=0}^{1} \{7 + x + 2y\}\ r\cdot dr\cdot d\theta$

$= \displaystyle\int_{\theta=0}^{\pi} \int_{r=0}^{1} \{7 + r\cdot\cos(\theta) + 2r\cdot\sin(\theta)\}\ r\cdot dr\cdot d\theta = \int_{\theta=0}^{\pi} \left\{\frac{7}{2} + \frac{1}{3}\cos(\theta) + \frac{2}{3}\sin(\theta)\right\}\ d\theta = \frac{4}{3} + \frac{7}{2}\pi$

17. Between the surface $z = f(x,y) = \dfrac{1}{1 + x^2 + y^2}$ and the xy-plane for (x,y) in the first quadrant and

$1 \le x^2 + y^2 \le 4$. The domain is $R = \{(x,y): 1 \le x^2 + y^2 \le 4\} = \left\{(r,\theta): 1 \le r \le 2,\ 0 \le \theta \le \dfrac{\pi}{2}\right\}$.

$\text{Volume} = \displaystyle\iint_R f\cdot dA = \int_{\theta=0}^{\pi/2} \int_{r=1}^{2} \frac{1}{1 + r^2}\cdot r\cdot dr\cdot d\theta$

$$\int_{r=1}^{2} \frac{1}{1+r^2} \cdot r \, dr = \frac{1}{2}\ln(1+r^2)|_{r=1}^{r=2} = \frac{1}{2}\ln\left(\frac{5}{2}\right) \text{ and } \int_{\theta=0}^{\pi/2} \frac{1}{2}\ln\left(\frac{5}{2}\right) d\theta = \frac{\pi}{2} \cdot \frac{1}{2}\ln\left(\frac{5}{2}\right) \approx 0.72 \cdot$$

19. $\int_0^1 \int_0^{\pi} r \, d\theta \, dr = \frac{\pi}{2}$

21. $\int_0^1 \int_0^{\pi/2} r \cdot \sin(\theta) \cdot r \, d\theta \, dr = \int_0^1 r^2 \, dr = \frac{1}{3}$

23. $\int_0^{\pi/2} \int_0^1 e^{-r^2} \cdot r \, dr \, d\theta = \int_0^{\pi/2} \frac{1}{2}\left(1-\frac{1}{e}\right) d\theta = \frac{\pi}{4}\left(1-\frac{1}{e}\right)$

25. $f(x,y) = 7 + 3x + 2y$ with $R = \{(x,y): x^2 + y^2 \leq 9\}$. (from Problem 5)

   Area of R = 9π and $\iint_R f \, dA = 63\pi$ (from Problem 5) so the average value of f on R is 7.

27. $f(x,y) = 5 + 2x + 3y$ with $R = \{(x,y): x^2 + y^2 \leq 16\}$. (from Problem 13)

   Area of R = 16π and $\iint_R f \, dA = 80\pi$ (from Problem 5) so the average value of f on R is 5.

29. A sprinkler (located at the origin) sprays water so after one hour the depth at location (x,y) feet is

   $f(x,y) = K \cdot e^{-(x^2+y^2)}$ feet.

   (a) How much water reaches the annulus $2 \leq r \leq 4$ and the annulus $8 \leq r \leq 10$ in one hour?
   From Practice 2 we know that the total amount of the water in a circle of radius A is
   $K \cdot \left(1 - e^{-A^2}\right)\pi$. So the amount in the annulus $2 \leq r \leq 4$ is
   $K \cdot \left(1 - e^{-16}\right)\pi - K \cdot \left(1 - e^{-4}\right)\pi = K \cdot \left(e^{-4} - e^{-16}\right)\pi$ and the amount in the annulus
   $8 \leq r \leq 10$ is $K \cdot \left(e^{-64} - e^{-100}\right)\pi$.

   (b) The area of the $2 \leq r \leq 4$ annulus is $(4^2 - 2^2)\pi = 12\pi$ square feet, and the area of the $8 \leq r \leq 10$
   annulus is 36π square feet. The average depth for the first annulus is $\dfrac{K \cdot \left(e^{-4} - e^{-16}\right)\pi}{12\pi} \approx 0.00153K$

   and it is $\dfrac{K \cdot \left(e^{-64} - e^{-100}\right)\pi}{36\pi} \approx 4.5 \cdot 10^{-30} \cdot K$ (almost no water) for the second annulus.

14.4 Odd Answers

Author confession: Many of these integrals are very messy and take a long time. I used Maple to evaluate them.

1. $\text{Area} = \iint_R 1 \, dA = \int_{x=0}^{2} \int_{y=0}^{4} 1 \, dy \, dx = \int_{x=0}^{2} 4 \, dx = 8 = M$

   $M_x = \iint_R y \cdot \delta \, dA = \int_{x=0}^{2} \int_{y=0}^{4} y \cdot (1) \, dy \, dx = \int_{x=0}^{2} 8 \, dx = 16$

Chapter 14: Odd Answers   Contemporary Calculus

$$M_y = \iint_R x \cdot \delta \ dA = \int_{x=0}^{2} \int_{y=0}^{4} x \cdot (1) \ dy \ dx = \int_{x=0}^{2} 4x \ dx = 4, \text{ and } \bar{x} = 1, \ \bar{y} = 2$$

3. $$\text{Area} = \iint_R 1 \ dA = \int_{x=0}^{2} \int_{y=0}^{x^2} 1 \ dy \ dx = \int_{x=0}^{2} x^2 \ dx = \frac{8}{3}$$

$$M = \iint_R \delta \ dA = \int_{x=0}^{2} \int_{y=0}^{x^2} (1+x) \ dy \ dx = \int_{x=0}^{2} x^3 + x^2 \ dx = \frac{20}{3}$$

$$M_x = \iint_R y \cdot \delta \ dA = \int_{x=0}^{2} \int_{y=0}^{x^2} y \cdot (1+x) \ dy \ dx = \int_{x=0}^{2} \frac{1}{2}(x^4 + x^5) \ dx = \frac{128}{15}$$

$$M_y = \iint_R x \cdot \delta \ dA = \int_{x=0}^{2} \int_{y=0}^{x^2} x \cdot (1+x) \ dy \ dx = \int_{x=0}^{2} (x^3 + x^4) \ dx = \frac{52}{5}$$

$$\bar{x} = \frac{M_y}{M} = \frac{39}{25} \text{ and } \bar{y} = \frac{M_x}{M} = \frac{32}{25}$$

5. $\text{Area} = \frac{15}{2}$, $M = 24$, $M_x = \frac{341}{8}$, $M_y = \frac{405}{8}$, $\bar{x} = \frac{135}{64}$, $\bar{y} = \frac{341}{192}$

7. $$\text{Area} = \iint_R 1 \ dA = \int_{\theta=0}^{\pi/3} \int_{r=0}^{\sin(3\theta)} 1 \ r \ dr \ d\theta = \frac{2}{3} = M,$$

$$M_x = \iint_R y \cdot \delta \ dA = \int_{\theta=0}^{\pi/3} \int_{r=0}^{\sin(3\theta)} r \cdot \sin(\theta) \ r \ dr \ d\theta = \frac{9}{70}$$

$$M_y = \iint_R x \cdot \delta \ dA = \int_{\theta=0}^{\pi/3} \int_{r=0}^{\sin(3\theta)} r \cdot \cos(\theta) \ r \ dr \ d\theta = \frac{9}{70}\sqrt{3}$$

$$\bar{x} = \frac{27}{140}\sqrt{3}, \ \bar{y} = \frac{27}{140}$$

9. $\text{Area} = \pi$, $M = \frac{3}{4}\pi$, $M_x = \frac{3}{4}$, $M_y = \frac{5}{8}\pi$, $\bar{x} = \frac{5}{6}$, $\bar{y} = \frac{16}{9\pi}$

11. $M = 8$, $I_x = \frac{128}{3}$, $R_x = \frac{4}{3}\sqrt{3}$ 13. $M = 16$, $I_x = 128$, $R_x = 2\sqrt{2}$

15. The density of the bar is 4 kg/m. The force between the point mass and a small Δx piece of the bar at location x is $f_i = GM \frac{4 \cdot \Delta x}{x^2}$. Forming the usual Riemann sum of the little forces and taking the limit,

total force = $G \cdot 10 \cdot \int_{2}^{4} \frac{4}{x^2} \ dx$,

17. The density is $k = 5$ kg/m for the first bar and $K = 3$ kg/m for the second bar. The force between a small Δx piece of the first bar at location x and a small Δy piece of the second bar at location y is $f_{ij} = G \cdot \dfrac{(5\Delta x)(3\Delta y)}{(y-x)^2}$.

Total force $= G \cdot \displaystyle\int_4^7 \int_0^2 \dfrac{15}{(y-x)^2}\, dx\, dy$

14.5 Odd Answers

1. $SA = \displaystyle\int_0^2 \int_0^{2x} \sqrt{1+(2x)^2+1}\, dy\, dx = \int_0^2 2x\sqrt{4x^2+2}\, dx = \dfrac{26}{3}\sqrt{2}$

3. $SA = \displaystyle\int_0^4 \int_0^{y/2} \sqrt{1+(4)^2+(2y)^2}\, dx\, dy = \int_0^4 \dfrac{y}{2}\sqrt{4y^2+17}\, dx = \dfrac{243}{8} - \dfrac{17}{24}\sqrt{17}$

5. $SA = \displaystyle\int_{-2}^2 \int_0^{4-x^2} \sqrt{26}\, dy\, dx = \int_{-2}^2 (4-x^2)\sqrt{26}\, dx = \dfrac{32}{3}\sqrt{26}$

7. $SA = \displaystyle\iint_R \sqrt{26}\, dA = 9\pi \cdot \sqrt{26}$ since R is a circle of radius 3

   or $SA = \displaystyle\int_{\theta=0}^{2\pi} \int_{r=0}^{3} \sqrt{26} \cdot r\, dr\, d\theta = \int_{\theta=0}^{2\pi} \dfrac{9}{2}\sqrt{26}\, d\theta = 9\sqrt{26} \cdot \pi$

9. $8\pi\sqrt{3}$

11. R is an annulus ($1 \le r \le 3$). $SA = \displaystyle\int_{\theta=0}^{2\pi} \int_{r=1}^{3} \sqrt{2} \cdot r\, dr\, d\theta = 9\pi\sqrt{2} - 1\pi\sqrt{2} = 8\pi\sqrt{2}$

13. $SA = \displaystyle\int_{-2}^2 \int_{x^2}^4 \sqrt{1+(y^2)^2+(2xy)^2}\, dy\, dx$

15. $SA = \displaystyle\int_0^2 \int_0^3 \sqrt{1+\cos^2(x)+\sin^2(y)}\, dy\, dx$

14.6 Odd Answers

1. $\displaystyle\int_{x=0}^{4} \int_{y=0}^{2} \int_{z=0}^{4-2y} f\, dz\, dy\, dx$

3. $\displaystyle\int_{x=-2}^{2} \int_{y=-\sqrt{4-x^2}}^{\sqrt{4-x^2}} \int_{z=\sqrt{x^2+y^2}}^{4} f\, dz\, dy\, dx$

5. $\displaystyle\int_{x=0}^{2} \int_{y=0}^{4-x^2} \int_{z=0}^{16-4x^2-y^2} f\, dz\, dy\, dx$

7. $7/6$ 9. -5 11. 24

13. $(e^3 - 1)/6$ 15. 16/3   17. 144

19. int(2*x+3, y=0..2*x, x=0..2, z=0..3);   21. int(x*y*z, z=0..8-2*y, y=x..4, x=0..4);

14.7 Odd Answers

1. $\frac{\pi}{3}(2\sqrt{2} - 1)$ 3. $\frac{23}{24}\pi$ 5. $\frac{1}{54}\pi^5$ 7. 8π

9. $\int_0^4 \int_{\pi/4}^{\pi/2} \int_0^1 r \cdot e^{-r^2} \, dr \cdot d\theta \cdot dz = \frac{\pi}{2}\left(1 - \frac{1}{e}\right)$ 11. $\int_0^1 \int_0^{\pi/2} \int_0^{4-r\sin(\theta)} r \, dz \cdot d\theta \cdot dr = \pi - \frac{1}{3}$

13. $\iiint_R f \, dV = \int_0^{2\pi} \int_0^3 \int_3^5 r \cdot r \, dz \cdot dr \cdot d\theta = 36\pi$

15. $\iiint_R f \, dV = \int_0^{2\pi} \int_0^{\sqrt{7}} \int_0^{1+r^2} (e^z) \cdot r \, dz \cdot dr \cdot d\theta = 2\pi\left(\frac{1}{2}e^8 - \frac{1}{2}e - \frac{7}{2}\right)$

17. $\iiint_R f \, dV = \int_0^{\pi/2} \int_0^3 \int_{r^2}^{36-3r^2} r \, dz \cdot dr \cdot d\theta = \frac{81}{2}\pi$

19. 0   21. $\frac{5}{3}\pi$ 23. $\frac{7}{3}\pi$ 25. $\frac{31}{160}\pi$

14.8 Odd Answers

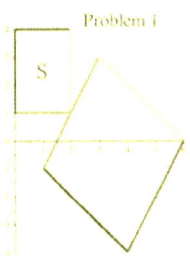
Problem 1

1. $S = \{(u,v): 0 \le u \le 2, 1 \le v \le 4\}$ under $x = u + v$ and $y = 2u - v$.

   $J(u,v) = \begin{vmatrix} 1 & 1 \\ 2 & -1 \end{vmatrix} = -3, \quad J(x,y) = -\frac{1}{3}$

   $\iint_R f(x,y) \, dx \, dy = \int_1^4 \int_0^2 f(u+v, 2u-v)(3) \, du \, dv$

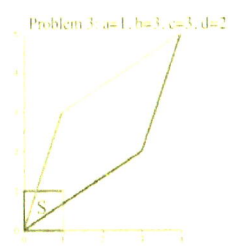
Problem 3: a=1, b=3, c=3, d=2

3. $S = \{(u,v): 0 \le u \le 1, 0 \le v \le 1\}$ under

   $x = au + bv$ and $y = cu + dv$.

   $J(u,v) = \begin{vmatrix} a & b \\ c & d \end{vmatrix} = ad - bc, \quad J(x,y) = \frac{1}{ad - bc}$

   $\iint_R f(x,y) \, dx \, dy = \int_0^1 \int_0^1 f(au+bv, cu+dv) \, |ad - bc| \, du \, dv$

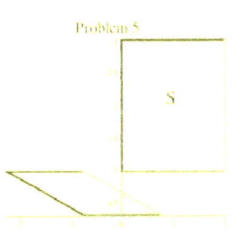
Problem 5

5. $S = \{(x,y): 0 \le x \le 2, 1 \le y \le 3\}$ under $u = \frac{3x - 3y}{4}$ and $v = \frac{y}{3}$.

   Then x=4u/3+3v and y=3v. $J(x,y) = \begin{vmatrix} \frac{3}{4} & -\frac{3}{4} \\ 0 & \frac{1}{3} \end{vmatrix} = \frac{1}{4}, \quad J(u,v) = 4$

7. $S = \{(x,y): 0 \le x \le 1, 0 \le y \le 1\}$ under $u = x^2 - y^2$ and $v = 2xy$.

   $J(x,y) = \begin{vmatrix} 2x & -2y \\ 2y & 2x \end{vmatrix} = 4x^2 + 4y^2$, $J(u,v) = \dfrac{1}{4x^2 + 4y^2}$

   (It is difficult/impossible to solve for x and y.)

   Problem 7

   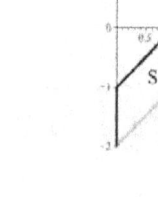

9. R = {trapezoid with vertices $(x,y) = (1,0), (2,0), (0,-2)$ and $(0,-1)$}

   under $u = x - y$ and $v = x + y$. Then $x = \dfrac{u+v}{2}$ and $y = \dfrac{-u+v}{2}$.

   $J(u,v) = \begin{vmatrix} 1 & -1 \\ 1 & 1 \end{vmatrix} = 2$, $J(x,y) = \dfrac{1}{2}$

   Problem 9

11. $\displaystyle\iint\limits_R 1\ dV = \iint\limits_S 1 \cdot |J(u,v)|\ du\ dv = \iint\limits_S 7\ du\ dv = 7(14) = 98$.

13. $J(x,y) = 3$ so $J(u,v) = \dfrac{1}{3}$. Then

   $\displaystyle\iint\limits_R 1\ dV = \iint\limits_S 1 \cdot |J(u,v)|\ dV \iint\limits_S \dfrac{1}{3}\ dV = \dfrac{1}{3}(15) = 5$.

15. The new uv domain is a triangle with vertices (0,), (1,0) and (0,1). $J(u,v) = \begin{vmatrix} 2 & 1 \\ 1 & 2 \end{vmatrix} = 3$.

   $\displaystyle\int_0^1 \int_0^{1-u} [(2u+v) + 3(u+2v)] \cdot 3\ dv\ du = 6$.

17. Using the substitution from Fig. 4, u=x+y, v=x-2y, so $J(x,y) = \begin{vmatrix} 1 & 1 \\ 1 & -2 \end{vmatrix} = -3$ and $J(u,v) = -\dfrac{1}{3}$.

   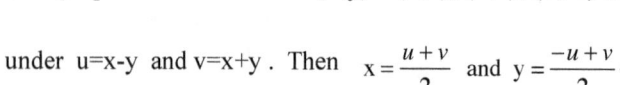 (Note: The substitution u=(x+y)/3 and v=(x-2y) from example

   leads to the same integral value since with this substitution u goes from 0 to 2 and v from 1 to 2.)

19. u=y-x v=2x+y so $J(x,y) = \begin{vmatrix} 1 & 1 \\ 1 & -2 \end{vmatrix} = -3$ and $J(u,v) = -\dfrac{1}{3}$.

   x=(v-u)/3 y=(2u+v)/3 so 3x+6y=3u+3v.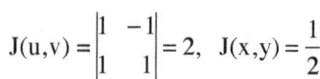

21. This is from Problem 8: $x = u/v$ and $y = uv$

   $J(u,v) = \begin{vmatrix} \dfrac{1}{v} & -\dfrac{u}{v^2} \\ v & u \end{vmatrix} = \dfrac{2u}{v}$ Then $u = \sqrt{xy}$, $v = \sqrt{\dfrac{y}{x}}$ with u=1 to 2 and v=1 to 2.

   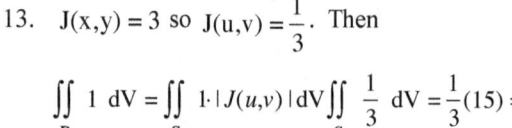

   (Note: See the Two Transformations, Same Result page at the end of the problem answers.)

23. R is the region bounded by the ellipse. If $x=au$ and $y=bv$ then the uv shape is $u^2+v^2 \leq 1$, a circle of radius 1 and area π. $J(u,v) = \begin{vmatrix} a & 0 \\ 0 & b \end{vmatrix} = ab$ so $\iint_{\text{ellipse}} 1\, dx\, dy = \iint_{\text{circle}} ab\, du\, dv = ab \cdot \iint_{\text{circle}} 1\, du\, dv = ab\pi$

15.0 Introduction to Vector Calculus

This may seem like a strange title since we have already been doing calculus with vectors, but the topics in this chapter extend the main calculus ideas into fields of vectors such as the one illustrated in Fig. 1 (from **Modeling Waves and Currents Produced by Hurricanes Katrina, Rita, and Wilma** by Lie-Yauw Oey and Dong-Ping Wang) . First we will examine some examples of vector fields. Then on to line integrals (integrals along paths in 2D or 3D) that will enable us to calculate the work done moving an object along a path in a wind field or magnetic field. Finally we will consider surface integrals and then some important theorems that generalize the Fundamental Theorem of Calculus to vector fields.

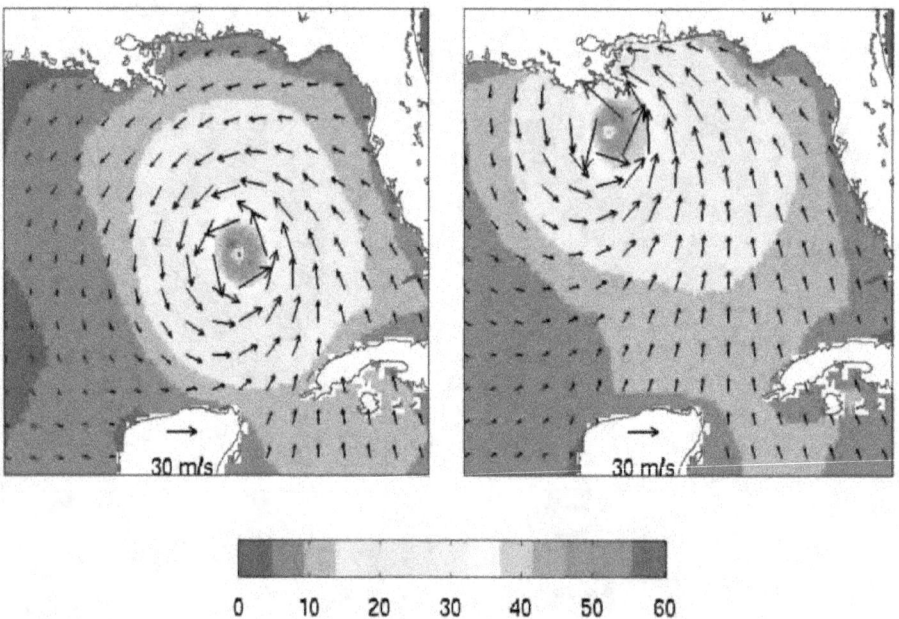

Fig. 1: Hurricane Katrina wind vectors (in m/s) on August 29, 2005

Vector fields contain a great deal of information and calculus can help us use that information. But even without calculus we can answer some questions about such fields.

Example 1: (a) Which way, clockwise or counterclockwise, where the winds blowing?
    (b) Were the strongest winds greater in the left picture or in the right picture?
    (c) Approximately what is the maximum wind speed in the right picture?
    (d) In which direction was the eye of the hurricane traveling?

Solution: (a) Counterclockwise. (b) Right picture
(c) Using the "30 m/s" key at the bottom of each picture, the maximum wind speed was about 50-60 m/s (about 120 miles/hour) . (d) To the NNW.

Other common vector fields are water velocity vectors indicating currents (Fig. 2) and force fields indicating the strength and direction of attractive or repulsive forces either in 2 or 3 dimensions.

(Note: Excellent, animated wind velocity fields for various parts of the world are at http://www.wunderground.com/maps/us/WindSpeed.html)

Problems

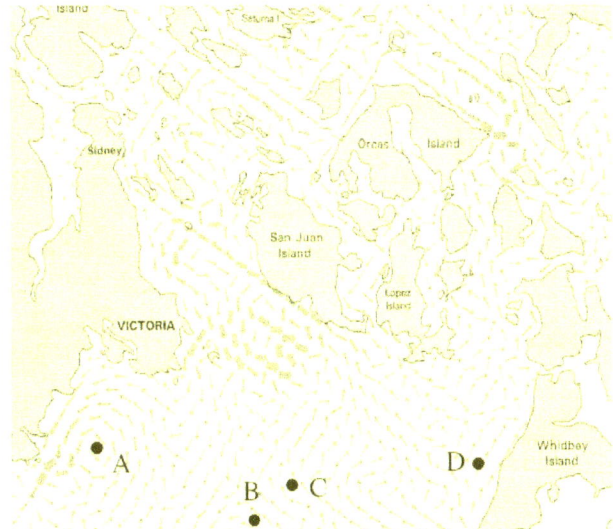

Fig. 2: Currents in the San Juan Islands

Problems 1 to 4 refer to Fig. 2. This picture uses wider arrows to represent stronger current.

1. What is happening to the current at point A?

2. If we drop small pieces of wood into the water at points A and B, do they stay close or do they drift apart?

3. Is it easier to row from point D to San Juan Island or from San Juan Island to point D?

4. Is it easier to row from point C to point D or from point D to point C?

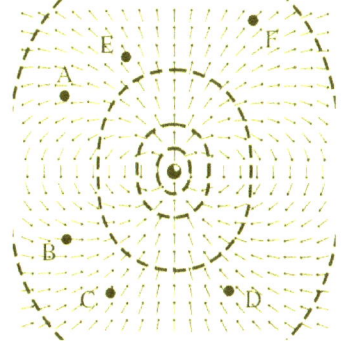

Fig. 3: Lines of Force for a bar magnet

Problems 5 to 8 refer to Fig. 3. The dotted curves each represent places of equal force. Moving from the center, the force at each dotted curve is 1/10 the force at the next closer dotted curve.

5. Is it easier to move from A to B or from B to A?

6. Is it easier to move from C to D or from D to C?

7. Is it easier to move from E to F or from F to E?

8. Imagine that this is a water current and that you dropped a cork into the water at A. Sketch the path of the cork.

15.1 Vector Fields

2D Vector Fields

> A **2D vector field** is a function **F** that assigns a 2D vector **F**(x,y) to each point (x,y) in the domain of the field. Since F(x,y) is a 2D vector we can write
>
> $$\mathbf{F}(x,y) = P(x,y)\mathbf{i} + Q(x,y)\mathbf{j} = \langle P(x,y), Q(x,y) \rangle \text{ or simply } \mathbf{F} = P\mathbf{i} + Q\mathbf{j}$$

Unlike our earlier work with vectors, these F(x,y) vectors have assigned locations. The vector field consists of an infinite number of vectors (one at each point) so we typically just graph enough of these vectors to make the pattern clear. The convention is to put the tail of the **F**(x,y) vector at the point (x,y).

Example 1: Plot the vectors **F**(x,y)=< y, -x> at the points (1,0), (1,1), (0,1), (-1,1), (-1,0) and (-1,-1).

Solution: **F**(1,0)=<0,-1>, **F**(1,1)=<-1,1>, **F**(0,1)=<1,0>, **F**(-1,1)=<1,1>, **F**(-1,0)=<0,1>, **F**(-1,-1)=<-1,1>. These are shown in Fig. 1.

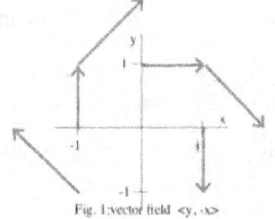
Fig. 1: vector field <y, -x>

It is very tedious to draw a vector field by hand, and some computer programs can do a very nice job. Fig. 2 is the same vector field **F**(x,y)=< y, -x> drawn by the program Maple with the command

with(plots): fieldplot([y, -x], x = -2 .. 2, y = -2 .. 2, arrows = THIN, color=red, thickness=2,grid=[11,11],title = "Vector Field <y, -x>", titlefont = ["ROMAN", 18]);

Maple automatically scaled the arrows to fit into the figure.

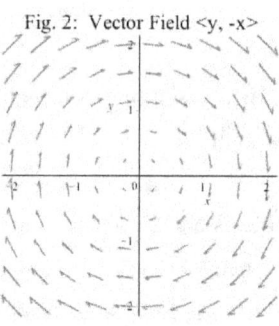
Fig. 2: Vector Field <y, -x>

Practice 1: Plot the vectors **F**(x,y)=< x, y/2> at the same points as in Example 1.

A few vector fields are very common and you should be able to recognize them after plotting just a few representative vectors.

Radial Fields <x,y> and <-x,-y>. These are shown in Fig. 3a and 3b. These vectors all point toward the origin or all vectors point away from the origin (except at the origin).

Fig. 3a <x, y>

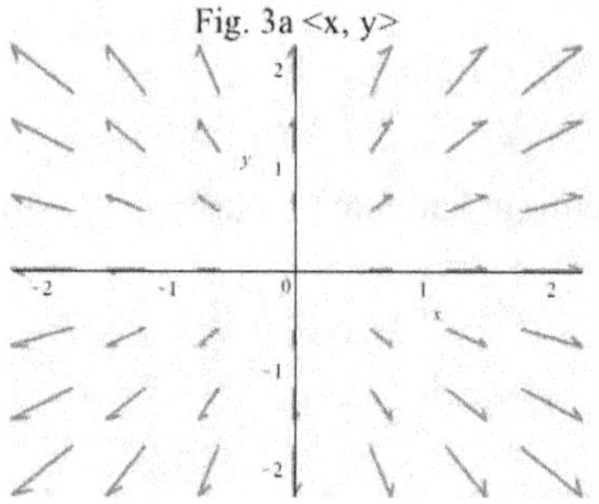

Fig. 3b <-x, -y>

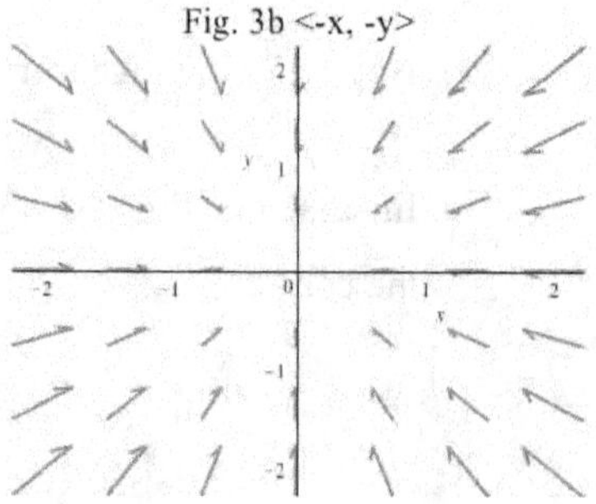

Rotational Fields: $\langle -y, x \rangle$, $\langle y, -x \rangle$ and $\left\langle \dfrac{y}{\sqrt{x^2+y^2}}, \dfrac{-x}{\sqrt{x^2+y^2}} \right\rangle$. These are shown in Fig. 4.

The first and second of these fields have vectors that increase in magnitude the farther they are from the origin. The third field has vectors of constant magnitude 1.

Fig. 4a <-y,x>

Fig. 4b <y,-x>

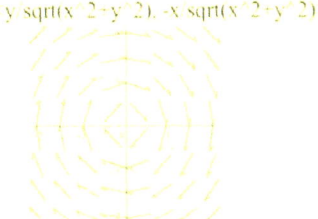
Fig. 4c <y/sqrt(x^2+y^2), -x/sqrt(x^2+y^2)>

The chapter Appendix gives the Maple commands for creating 2D and 3D vector fields.

3D Vector Fields

A 3D vector field **F** assigns a 3D vector at each point in the 3D domain of **F**:

$F(x,y,z) = \langle P(x,y,z), Q(x,y,z), R(x,y,z) \rangle = P\mathbf{i} + Q\mathbf{j} + R\mathbf{k}$ where P, Q and R are scalar-valued functions.

Since we live in a 3D world, 3D vector (force) fields are very common in applications, but they are much more difficult to create by hand and to visualize on a 2D page. Fortunately, some software can do the work for us.

Radial Fields: $F(x,y,z) = \langle x,y,z \rangle$ and $\langle -x,-y,-z \rangle$.

These are shown in Fig. 5a and 5b. In Fig. 5c all of the vectors of $F(x,y,z) = \langle x,y,z \rangle$ have been normalized to have the same length. The vectors in all three plots point toward or away from the origin (except at the origin).

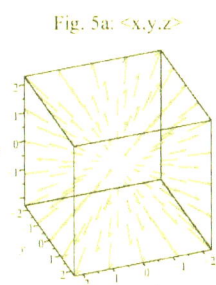
Fig. 5a: <x,y,z>

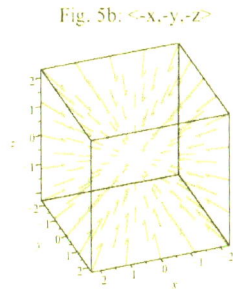
Fig. 5b: <-x,-y,-z>

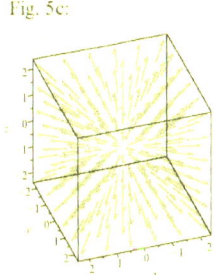
Fig. 5c:

Rotational Fields: These are more difficult to visualize because the rotation can be around any of the axes, some other line, or simply around the origin. Three views of $F(x,y,z) = \langle -y, x, 0 \rangle$ are shown in Fig. 6.

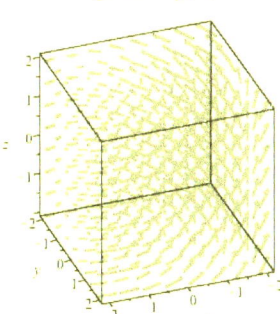
Fig. 6a <-y,x,0>

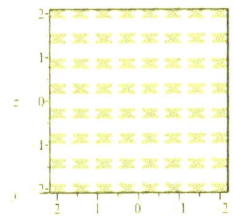
Fig. 6b <-y,x,0>

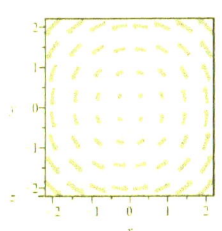
Fig. 6c <-y,x,0>

Fig. 7 shows the "swirl" field for
F(x,y,z) = <y,z,x>, and Fig. 8 is the simple field
F(x,y,z) = <1,3,2> in which all of the vectors have
the same magnitude and direction.

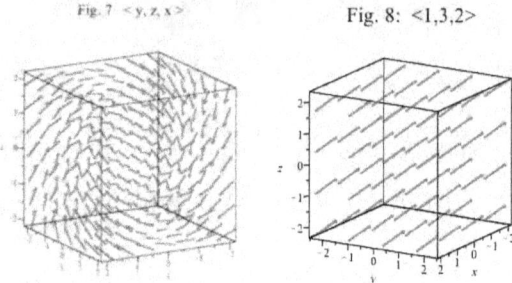

Fig. 7 <y,z,x>   Fig. 8: <1,3,2>

Gravitation and Electric Force Fields

If, as is the situation for gravitational and electric fields, the magnitude of the force is inversely proportional to the square of the distance between the objects, then $\mathbf{F}(x,y,z) = \dfrac{k}{x^2+y^2+z^2}$ where k is a positive or negative constant. The field will look like Fig. 5a or 5b.

Gravitational Field: Newton's Law of Gravitation says that k=GMm where m and M are the masses of the objects and G is the gravitational constant: $\mathbf{F} = \dfrac{GMm}{x^2+y^2+z^2}$. Suppose the object with mass M is located at the origin, and let $\mathbf{r} = \langle x,y,z \rangle$ be the position vector of the object of mass m. Then $|\mathbf{r}|^2 = x^2+y^2+z^2$ and the force is directed toward the origin so the direction of **r** is $-\dfrac{\mathbf{r}}{|\mathbf{r}|}$. Putting this together, the **gravitational field** is $\mathbf{F}(x,y,z) = \mathbf{F}(\mathbf{r}) = \dfrac{GMm}{|\mathbf{r}|^2}\left(-\dfrac{\mathbf{r}}{|\mathbf{r}|}\right) = \left(\dfrac{-GMm}{|\mathbf{r}|^3}\right)\mathbf{r}$. These vectors behave like those in Fig. 5b in which the vector at each point is directed toward the origin.

Electrical Field: Coulomb's Law says that the force **F** is inversely proportional to the square of the distance between the two charges. If an electric charge Q is located at the origin, and a charge q is located at (x,y,z) then $|\mathbf{F}| = \dfrac{eqQ}{|\mathbf{r}|^2}$ where $\mathbf{r} = \langle x,y,z \rangle$ and e is a constant. Finally, $\mathbf{F}(\mathbf{r}) = \left(\dfrac{eqQ}{|\mathbf{r}|^3}\right)\mathbf{r}$ with qQ<0 for unlike charges (attracting each other) and qQ>0 for like charges (repelling each other). If we consider the force per unit q of charge then we have the **electric field** $\mathbf{E}(\mathbf{r}) = \dfrac{1}{q}\mathbf{F}(\mathbf{r}) = \left(\dfrac{eQ}{|\mathbf{r}|^3}\right)\mathbf{r}$.

Gradient Fields

In section 13.5 the gradient vector, $\nabla f(x,y) = \left\langle \dfrac{\partial f}{\partial x}, \dfrac{\partial f}{\partial y} \right\rangle$, was introduced, and this very naturally assigns a vector at each (x,y) location where f has partial derivatives.

Example 2: For $z = f(x,y) = x^2 + y^2$, what is the gradient vector field for f?

Solution: $\frac{\partial f}{\partial x} = 2x$ and $\frac{\partial f}{\partial y} = 2y$ so the vector field is $\mathbf{F} = \langle 2x, 2y \rangle$ .which looks similar to Fig. 3a.

Practice 2: $z = f(x,y) = xy$. Determine the gradient vector field for f(x,y) and sketch several vectors from this field.

One of the important properties of the gradient vector $\nabla f(x,y)$ is that it is perpendicular to the level curve of f(x,y) at the point (x,y) and points "uphill." So if the level curves of z=f(x,y) are known, then it is easy to sketch the gradient vector field. Fig. 9 shows level curves and gradient vectors of the paraboloid $z = f(x,y) = x^2 + y^2$.

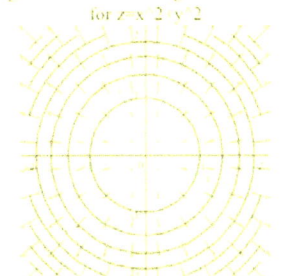

Fig. 9 level curves and gradient vectors for $z = x^2 + y^2$

Fig. 10a shows the surface $f(x,y) = \frac{-5x}{1 + x^2 + y^2}$. Fig. 10b shows level curves for f(x,y) as well as gradient vectors. The gradient vectors have been drawn to all have the same length to better show their directions and that they are perpendicular to the level curves.

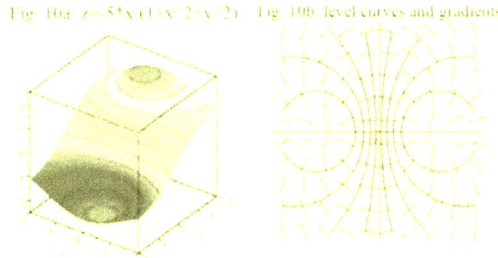

Fig. 10a: $z = -5x/(1+x^2+y^2)$ Fig. 10b: level curves and gradients

Practice 3: Fig. 11 shows the surface and the level curves for the saddle $f(x,y) = x^2 - y^2$. Sketch the gradient field.

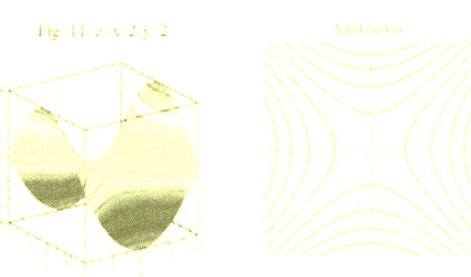

Fig. 11: $z = x^2 - y^2$ level curves

In a later section (give section number later) we will start with a vector field $\mathbf{F}(x,y)$ and determine if that field $\mathbf{F}$ is the gradient field of some function z=f(x,y). Such a field $\mathbf{F}$ is called **conservative field** and the function z=f(x,y) that gives rise to the field is called the **potential function** for $\mathbf{F}$. Conservative fields have a number of important properties.

Problems

In problems 1 to 8, plot vectors from the given field $\mathbf{F}(x,y)$ at several locations (x,y) for integer values for x and y with $-2 \le x \le 2$ and $-2 \le y \le 2$.

1. $\mathbf{F} = \langle 2, 1 \rangle$
2. $\mathbf{F} = \langle 1, x \rangle$
3. $\mathbf{F} = \langle x, x \rangle$
4. $\mathbf{F} = \langle y, y \rangle$
5. $\mathbf{F} = \langle x^2, y \rangle$
6. $\mathbf{F} = \langle 1, x \cdot y \rangle$
7. $\mathbf{F} = \langle x, -y \rangle$
8. $\mathbf{F} = \langle y, 1 - x \rangle$

In problems 9 to 11, match the vector field F to the 3D plot.

9. (a) $\mathbf{F} = \langle 0,0,1 \rangle$

   (b) $\mathbf{F} = \langle -y,x,0 \rangle$

   (c) $\mathbf{F} = \langle 1,1,0 \rangle$

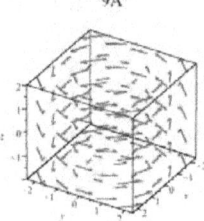
9A

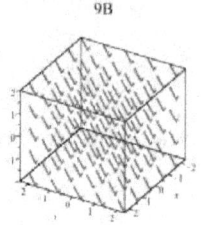
9B

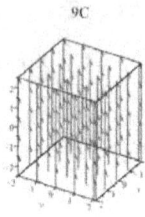
9C

10. (a) $\mathbf{F} = \langle 0,z,-y \rangle$

    (b) $\mathbf{F} = \langle -x,-y,-z \rangle / \sqrt{x^2 + y^2 + z^2}$

    (c) $\mathbf{F} = \langle -x,-y,-z \rangle$

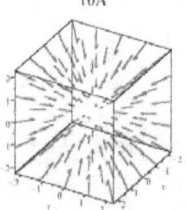
10A

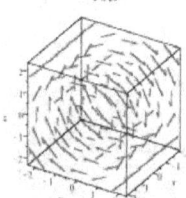
10B

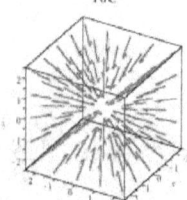
10C

11. (a) $\mathbf{F} = \langle x,y,z \rangle$

    (b) $\mathbf{F} = \langle -z,0,x \rangle$

    (c) $\mathbf{F} = \langle 1,0,0 \rangle$

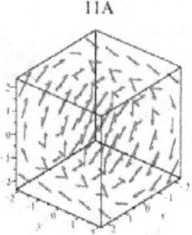
11A

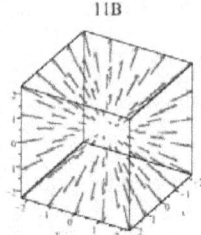
11B

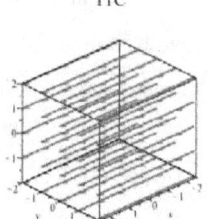
11C

In problems 12 to 21, find the gradient field for the given functions.

12. $f(x,y) = x^2 + y^2$

13. $f(x,y) = x^2 - y^2$

14. $f(x,y) = x/y$

15. $f(x,y) = x \cdot y^2 - x^2 \cdot y$

16. $f(x,y) = x \cdot \sin(y) + e^{x \cdot y}$

17. $f(x,y,z) = 3x + 2y - z$

18. $f(x,y,z) = x \cdot y \cdot z$

19. $f(x,y,z) = \sqrt{x^2 + y^2 + z^2}$

20. $f(x,y,z) = x^3 \cdot y + 2y \cdot z + 5$

21. $f(x,y,z) = \sin(x \cdot y \cdot z)$

In problems 22 to 27, some level curves of z=f(x,y) are given. On the each figure sketch the gradient vector field for this function. (Do not put heads on the arrows -- just use short line segments.)

22.

23.

24.

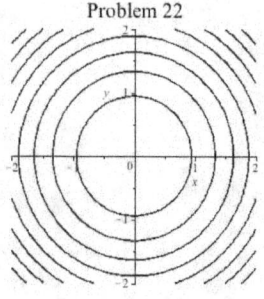
Problem 22

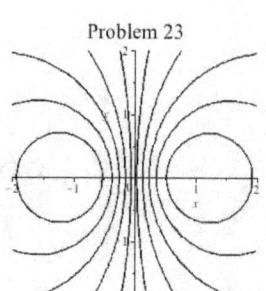
Problem 23

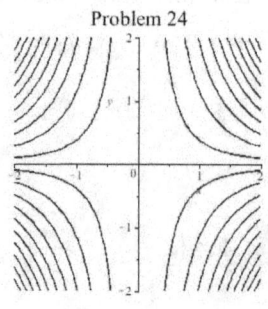
Problem 24

25.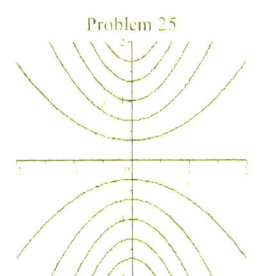

26.

27.

Practice Answers

Practice 1: See Fig. P1. Fig. P1 Maple shows the $<x, y/2>$ field for $-2 \leq x \leq 2$ and $-2 \leq y \leq 2$.

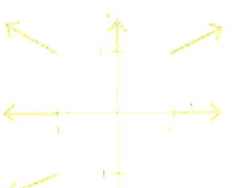

Fig. P1 Vector field $<x, y/2>$

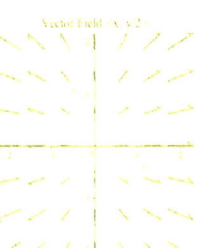

Fig. P1 Maple. $<x, y/2>$

Practice 2: $z = f(x,y) = xy$. $\nabla f(x,y) = \langle y, x \rangle$.

The field is shown in Fig. P2.

Practice 3: Fig. P3 shows gradient vectors for $f(x,y) = x^2 - y^2$. They have all been drawn the same length. The vectors closer to the origin should be shorter.

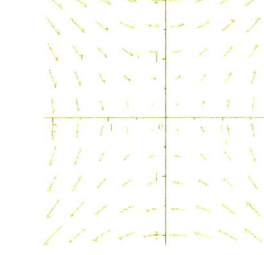

Appendix: Maple commands for plotting 2D and 3D vector fields

2D field <y,x>   with(plots): fieldplot([y, x], x = -2..2, y = -2 .. 2);
This variation adds more details:

    with(plots): fieldplot([y, x], x = -2..2, y = -2 .. 2, arrows = THIN, color = red, thickness=1,
    grid=[10,10], scaling=constrained, title = "Fig. P2: <y, x>", titlefont = ["ROMAN", 18]);

3D field "swirl" <y,z,x>   with(plots): fieldplot3d([y,z,x], x = -2 .. 2, y = -2 .. 2, z = -2 .. 2);
With more details:

    with(plots): fieldplot3d([y,z,x], x = -2 .. 2, y = -2 .. 2, z = -2 .. 2, arrows = THIN, color=red,
    thickness=1, grid=[8,8,8], title = "Fig. 6: <y,z,x>", titlefont = ["ROMAN", 18],orientation=[70,60,0]);

Note: Maple automatically centers each vector **F**(x,y) at the location (x,y).

Level curves and gradient vectors together

with(plots):

CP:=contourplot(x^2-y^2,x = -2..2, y = -2..2, color=blue, thickness=1, contours=8):

Gfld:=fieldplot([2*x,-2*y], x = -2..2, y = -2..2, arrows = THIN, color = red, thickness=1,grid=[10,10]):

display(Gfld,CP);

15.2 Del Operator and 2D Divergence and Curl

The divergence and curl of a vector field F describe two characteristics of the field at each point in the field. They will be extended into 3D in later sections, but they are more easily understood graphically and intuitively in 2D, so we start there.

Divergence: div F

Assume that the vector field F describes the flow of a liquid in 2D, perhaps water contained between two close-together parallel plates. The **divergence** is the rate per unit area that the water dissipates (**departs**, leaves) at the point P. A positive value for the divergence means that more water is leaving at P than is entering at P. But we can't really see a point so imagine a small circle C centered at P, and consider whether more water s leaving or entering this circle. Fig. 1(a) shows more water leaving than entering the circle around P so div $F(P) > 0$, Fig. 1(b) has more entering than leaving so div $F(P) < 0$, and Fig. 1(c) shows the same amount leaving as entering so div $F(P) = 0$.

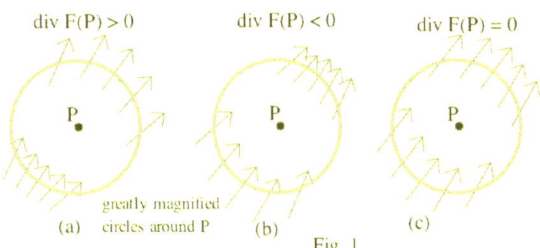

Fig. 1

Example 1: Estimate whether div F(P) is positive, negative or zero when P=A, B and C in Fig. 2.

Solution: (a) div F > 0 (b) div F = 0 (c) div F < 0

Practice 1: Estimate whether div F(P) is positive, negative or zero when P=A, B and C in Fig. 3.

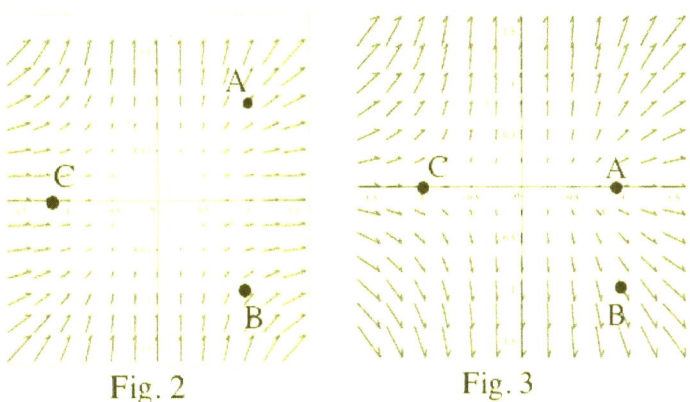

Fig. 2   Fig. 3

It turns out that div F(P) is very easy to calculate.

Definition: Divergence   div F

For a vector field $\mathbf{F}(x,y) = M\mathbf{i} + N\mathbf{j}$ with continuous partial derivatives,

the divergence of F at Point P is $\text{div } \mathbf{F}(P) = \dfrac{\partial M}{\partial x} + \dfrac{\partial N}{\partial y}$.

(In 3D with $\mathbf{F}(x,y,z) = M\mathbf{i} + N\mathbf{j} + P\mathbf{k}$, $\text{div } \mathbf{F}(P) = \dfrac{\partial M}{\partial x} + \dfrac{\partial N}{\partial y} + \dfrac{\partial P}{\partial z}$.)

We will justify this definition for divergence in section 15.10, but, for now, we will simply use it.

Example 2: The vector field in Fig. 2 is $\mathbf{F} = \langle x^2, y^2 \rangle$. Calculate div $\mathbf{F}(P)$ at $P=A=(1,1)$, $P=B=(1,-1)$ and $P=C=(-1,0)$.

Solution: $\text{div } \mathbf{F} = \dfrac{\partial M}{\partial x} + \dfrac{\partial N}{\partial y} = 2x+2y$. (a) at A div $\mathbf{F} = 4$, at B div $\mathbf{F} = 0$, at C div $\mathbf{F} = -2$.

Practice 2: The vector field in Fig. 3 is $\mathbf{F} = \langle x^2, 2 \cdot y \rangle$. Calculate div $\mathbf{F}(P)$ at $P=A=(1,0)$, $P=B=(1,-1)$ and $P=C=(-1,0)$.

For the radial field $\mathbf{F} = \langle x, y \rangle$, div $\mathbf{F} = 2$ at every point. For the radial field $\mathbf{F} = \langle -x, -y \rangle$, div $\mathbf{F} = -2$ at every point. And for the rotational field $\mathbf{F} = \langle -y, x \rangle$, div $\mathbf{F} = 0$ at every point. (Fig. 4)

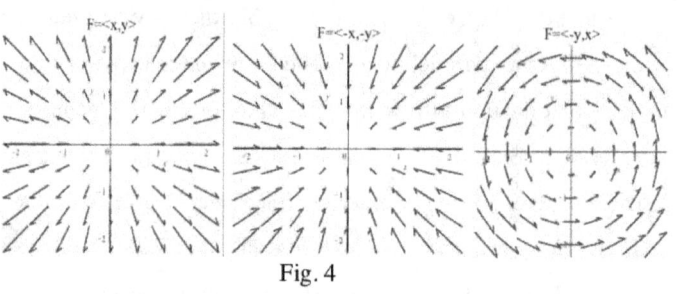

Fig. 4

Curl: curl F

The curl $\mathbf{F}$ (P) measures the rotation of the vector field $\mathbf{F}$ at the point P. Again picture a small circle centered at P, but this time imagine an axle at P and small paddles on the circle (Fig. 5). If the water vectors would rotate the little wheel counterclockwise (Fig. 6a) we say that curl $\mathbf{F}$ (P) > 0, if the rotation is clockwise (Fig. 6b) curl $\mathbf{F}$(P) < 0, and if there is no rotation then curl $\mathbf{F}$ (P) = 0 (Fig. 6c)

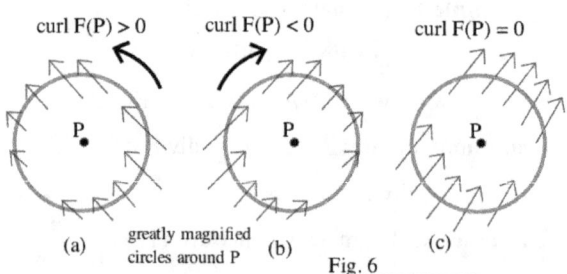

Fig. 6

Example 3: Estimate whether curl $\mathbf{F}(P)$ is positive, negative or zero when P=A, B and C in Fig. 7.

Solution: (a) at A curl $\mathbf{F}$ > 0, at B curl $\mathbf{F}$ < 0, at C curl $\mathbf{F}$ > 0

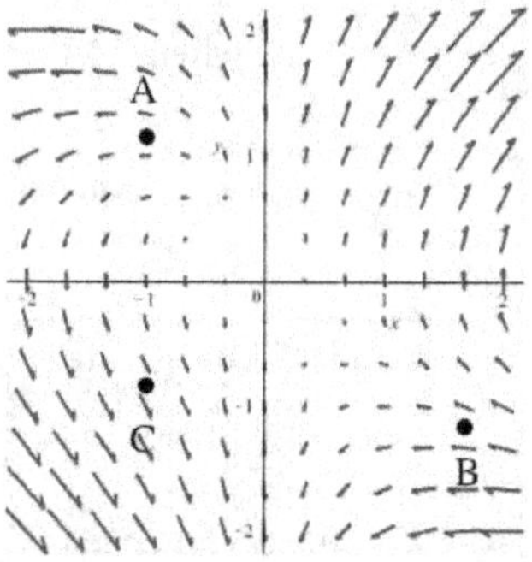

Definition: Curl in 2D   curl F

For a vector field $\mathbf{F}(x,y) = M\mathbf{i} + N\mathbf{j}$ with continuous partial derivatives, then the curl of F at Point P is $\text{curl } \mathbf{F}(P) = \dfrac{\partial N}{\partial x} - \dfrac{\partial M}{\partial y}$.

Fig. 7

Note: Later we will define curl F to be a vector quantity in 3D (still measuring a rotation) as

$$\text{curl } \mathbf{F} = \left(\dfrac{\partial P}{\partial y} - \dfrac{\partial N}{\partial z}\right)\mathbf{i} + \left(\dfrac{\partial M}{\partial z} - \dfrac{\partial P}{\partial x}\right)\mathbf{j} + \left(\dfrac{\partial N}{\partial x} - \dfrac{\partial M}{\partial y}\right)\mathbf{k}$$

For now we will just need and use the k component of this 3D vector. That will tell us about the rotation in the xy-plane,

Example 4: The vector field in Fig. 7 is $\mathbf{F} = \langle xy, x+y \rangle$ with A=(–1,1), B=(1.8,–1) and C=(–1,–1).

Calculate curl **F**(P) at each point.

Solution: curl $\mathbf{F} = \dfrac{\partial N}{\partial x} - \dfrac{\partial M}{\partial y} = 1 - x$. At A curl **F** = 2, at B curl **F** = –0.8, at C curl **F** = 2.

Practice 3: Calculate curl **F** for the radial field $\mathbf{F} = \langle x, y \rangle$ and the rotational field $\mathbf{F} = \langle -y, x \rangle$.

If curl F=0 at all points in the field, then F is called **irrotational**.

The divergence and the curl measure completely different characteristics of the field at a point, and knowing the sign of one does not tell us anything about the sign of the other. For example, the curl of $\mathbf{F} = \langle x^2 y + y, y^2 \rangle$ is curl $\mathbf{F} = \dfrac{\partial N}{\partial x} - \dfrac{\partial M}{\partial y} = (0) - (x^2 + 1) < 0$ at all points, but div $\mathbf{F} = \dfrac{\partial M}{\partial x} + \dfrac{\partial N}{\partial y} = 2xy + 2y$ which can be positive, negative or zero depending on the location (x,y).

Practice 4: Find a vector field **F** so that div **F** > 0 at all points but curl **F** can be positive, negative or zero depending on the location (x,y).

del operator $\nabla = \left\langle \dfrac{\partial}{\partial x}, \dfrac{\partial}{\partial y}, \dfrac{\partial}{\partial z} \right\rangle$

A mathematical operator is like a function but it typically operates on functions or other advanced objects. For example, differentiation $\dfrac{d}{dx}[\ ]$ and integration $\int [\ ] dx$ are operators that do things to whatever function is put into the brackets. Similarly, the del operator does things to functions and even vector fields. And the del notation is a compact way to represent complicated operations (and an easy way to remember them).

in 2D: $\nabla f = \left\langle \dfrac{\partial f}{\partial x}, \dfrac{\partial f}{\partial y} \right\rangle$ which we met earlier and called the 2D gradient of f

$\nabla \bullet \mathbf{F} = \left\langle \dfrac{\partial f}{\partial x}, \dfrac{\partial f}{\partial y} \right\rangle \bullet \langle M, N \rangle = \dfrac{\partial M}{\partial x} + \dfrac{\partial N}{\partial y}$ which is the 2D divergence, div **F**

$\nabla \times \mathbf{F} = \begin{vmatrix} \mathbf{i} & \mathbf{j} & \mathbf{k} \\ \dfrac{\partial}{\partial x} & \dfrac{\partial}{\partial y} & 0 \\ M & N & 0 \end{vmatrix} = \dfrac{\partial N}{\partial x} - \dfrac{\partial M}{\partial y}$ which is the 2D curl, curl **F**

In 3D: $\nabla f = \left\langle \dfrac{\partial f}{\partial x}, \dfrac{\partial f}{\partial x}, \dfrac{\partial f}{\partial x} \right\rangle$ the 3D gradient of f

$\nabla \bullet \mathbf{F} = \left\langle \dfrac{\partial f}{\partial x}, \dfrac{\partial f}{\partial y}, \dfrac{\partial f}{\partial z} \right\rangle \bullet \langle M, N, P \rangle = \dfrac{\partial M}{\partial x} + \dfrac{\partial N}{\partial y} + \dfrac{\partial P}{\partial z}$ the 3D divergence, div **F**

15.2 divergence, curl and del in 2D

$$\nabla \times F = \begin{vmatrix} i & j & k \\ \frac{\partial}{\partial x} & \frac{\partial}{\partial y} & \frac{\partial}{\partial z} \\ M & N & P \end{vmatrix} = \left(\frac{\partial P}{\partial y} - \frac{\partial N}{\partial y} \right) i + \left(\frac{\partial M}{\partial z} - \frac{\partial P}{\partial x} \right) j + \left(\frac{\partial N}{\partial x} - \frac{\partial M}{\partial y} \right) k \quad \text{the 3D curl of F}$$

You should recognize the z-component of $\nabla \times F$ as the 2D curl of F.

The del operator can be combined with itself to create new operations:

the Laplacian: $\nabla \bullet \nabla f = \left\langle \frac{\partial}{\partial x}, \frac{\partial}{\partial y}, \frac{\partial}{\partial z} \right\rangle \bullet \left\langle \frac{\partial f}{\partial x}, \frac{\partial f}{\partial y}, \frac{\partial f}{\partial z} \right\rangle = \frac{\partial^2 f}{\partial x^2} + \frac{\partial^2 f}{\partial y^2} + \frac{\partial^2 f}{\partial z^2}$

is important in mathematical physics.

And the del operator can help illuminate connections between objects that seem unrelated:

div curl $F = \nabla \bullet (\nabla \times F) = \left\langle \frac{\partial}{\partial x}, \frac{\partial}{\partial y}, \frac{\partial}{\partial z} \right\rangle \bullet \left(\frac{\partial P}{\partial y} - \frac{\partial N}{\partial y} \right) i + \left(\frac{\partial M}{\partial z} - \frac{\partial P}{\partial x} \right) j + \left(\frac{\partial N}{\partial x} - \frac{\partial M}{\partial y} \right) k = \ldots = 0$

You can fill in the " ..." by taking the dot product and then recognizing that the various mixed partial derivatives are equal (e.g, $\frac{\partial^2 f}{\partial x \partial y} = \frac{\partial^2 f}{\partial y \partial x}$).

Wrap up

The justifications for the definitions of the divergence and curl of a vector field at a point will come later in this chapter, but the ideas and calculations in 2D are relatively easy and they will help with the ideas and calculations when we get to Green's Theorem.

Problems

1. Estimate whether the div **F** is positive, negative or zero at A, B and C in Fig. 8.

2. Estimate whether the div **F** is positive, negative or zero at A, B and C in Fig. 9.

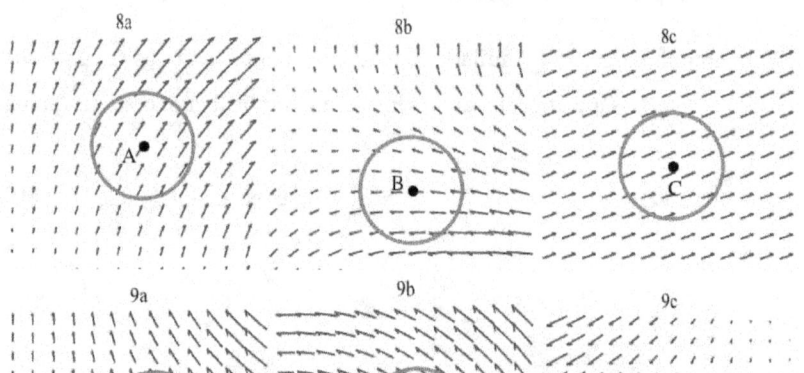

In problems 3 to 6, calculate div **F** at the given points.

3. $F = \left\langle x^2 + 3y, 2y + x \right\rangle$ at

   A=(1, 3), B=(2, −1) and C=(−1, 3).

4. $F = \left\langle x \cdot y^2, x^3 \cdot y + 3 \right\rangle$ at A=(1, 1), B=(2, −1) and C=(0, −2).

5. $F = \left\langle 5x - 3y, x + 2y \right\rangle$ at A=(3, 2), B=(0, 3) and C=(1, 4).

6. $\mathbf{F} = \langle x^2 - y^2, x^2 + y^2 \rangle$ at A=(2, 3), B=(-2, 2) and C=(3, -4).

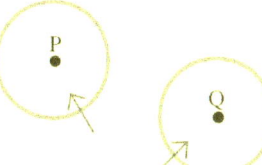

7. In Fig. 10 add additional vectors so div $\mathbf{F}(P) > 0$.

8. In Fig. 10 add additional vectors so div $\mathbf{F}(Q) < 0$.

Fig. 10

9. Estimate whether the curl $\mathbf{F}$ is positive, negative or zero at A, B and C in Fig. 8.

10. Estimate whether the curl $\mathbf{F}$ is positive, negative or zero at A, B and C in Fig. 9.

In problems 11 to 14, calculate curl $\mathbf{F}$ at the given points.

11. $\mathbf{F} = \langle x^2 + 3y, 2y + x \rangle$ at A=(1, 3), B=(2, -1) and C=(-1, 3).

12. $\mathbf{F} = \langle x \cdot y^2, x^3 \cdot y + 3 \rangle$ at A=(1, 1), B=(2, -1) and C=(0, -2).

13. $\mathbf{F} = \langle 5x - 3y, x + 2y \rangle$ at A=(3, 2), B=(0, 3) and C=(1, 4).

14. $\mathbf{F} = \langle x^2 - y^2, x^2 + y^2 \rangle$ at A=(2, 3), B=(-2, 2) and C=(3, -4).

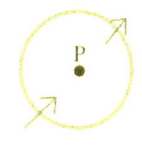

15. In Fig. 11 add additional vectors so curl $\mathbf{F}(P) > 0$.

Fig. 11

16. In Fig. 11 add additional vectors so curl $\mathbf{F}(Q) < 0$.

17. Show that the div and curl are always 0 for a constant field $\mathbf{F} = \langle a, b \rangle$.

18. Show that the div and curl are constant for a linear field $\mathbf{F} = \langle ax + by, cx + dy \rangle$.

19. If a rotational field is flowing counterclockwise, does that mean the curl at each point is positive. Justify your conclusion.

20. If the y-component of $\mathbf{F}$ is always 0, what can you conclude about the divergence?

21. If the y-component of $\mathbf{F}$ is always 0, what can you conclude about the curl?

Practice Answers

Practice 1: (a) div F > 0 (b) div F = 0 (c) div F < 0

Practice 2: div $\mathbf{F} = \dfrac{\partial M}{\partial x} + \dfrac{\partial N}{\partial y} = 2x + 2$. (a) at A div $\mathbf{F} = 4$, at B div $\mathbf{F} = 4$, at C div $\mathbf{F} = 0$.

Practice 3: For $\mathbf{F} = \langle x, y \rangle$ curl $\mathbf{F} = \dfrac{\partial N}{\partial x} - \dfrac{\partial M}{\partial y} = 0$ at every point. No rotation anywhere.

For $\mathbf{F} = \langle -y, x \rangle$ curl $\mathbf{F} = \dfrac{\partial N}{\partial x} - \dfrac{\partial M}{\partial y} = (1) - (-1) = 2$ at every point. The rotation is counterclockwise at every point.

Practice 4: You need a field $\mathbf{F} = \langle M, N \rangle$ so that $\operatorname{div} \mathbf{F} = \dfrac{\partial M}{\partial x} + \dfrac{\partial N}{\partial y} > 0$ everywhere but so $\operatorname{curl} \mathbf{F} = \dfrac{\partial N}{\partial x} - \dfrac{\partial M}{\partial y}$ contains variables. $\mathbf{F} = \langle x^3 + x + y^2, y^3 \rangle$ works.

$\operatorname{div} \mathbf{F} = \dfrac{\partial M}{\partial x} + \dfrac{\partial N}{\partial y} = 3x^2 + 1 + 3y^2 > 0$ and $\operatorname{curl} \mathbf{F} = \dfrac{\partial N}{\partial x} - \dfrac{\partial M}{\partial y} = (0) - (2y)$.

15.3 Line Integrals

A curtain is hanging from a very bent rod (Fig. 1). If we have an equation $z=f(x,y)$ or $(x(t), y(t), z(t))$ for the rod, how can we calculate the area of the curtain between the rod and the floor (the xy-plane). And if the rod has different densities, $\delta(x,y,z)$ at locations (x,y,z), along its length, how can we calculate the total mass of the rod?

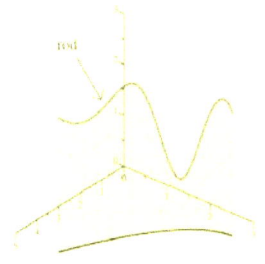

Fig. 1: Curtain on a bent rod

These are two silly questions, but their solutions illustrate the main idea of this section: an integral along a curve, a **line integral**. One of the main applications of line integrals is to determine the work done to move an object in a force field in two or three dimensions, and there are others.

Curtain area solution: In beginning calculus we created integral of $f(x)$ on the interval $a \leq x \leq b$ on the x-axis (Fig. 2) by partitioning the interval, picking a representative point x^* in each subinterval, calculating the area of the little rectangle above the subinterval as $f(x^*) \cdot \Delta x$, forming the Riemann sum $\sum f(x^*) \cdot \Delta x$ of these little areas, and finally taking the limit as $\Delta x \to 0$ to get an integral: $\int_a^b f(x)\,dx$.

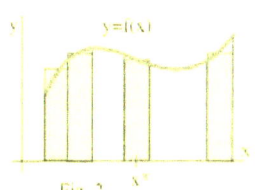

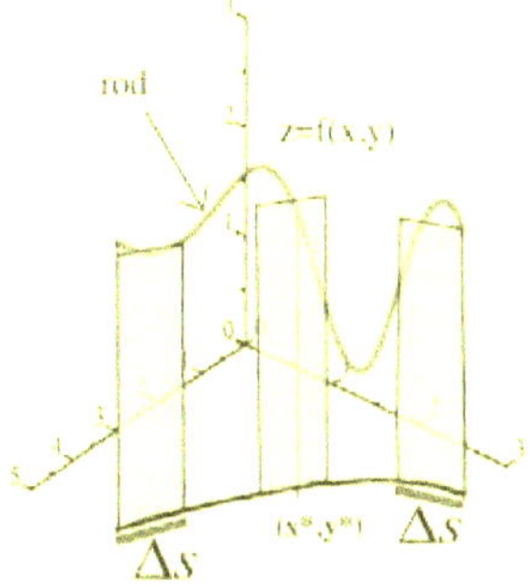

Fig. 2

The same strategy works for the shower curtain, except here we partition the curve C in the xy-plane (Fig. 3), pick a representative point (x^*, y^*) in each subinterval, and find the area of each little rectangle as $f(x^*,y^*) \cdot \Delta s$ where Δs is the length of the subinterval. Then as $\Delta s \to 0$ we get that $\sum f(x^*,y^*) \cdot \Delta s \to \int f(x,y)\,ds$. If our curve in the xy-plane is parameterized by t, $\mathbf{r}(t) = \langle x(t), y(t) \rangle$ then $z(t)=f(x(t), y(t))$ and $\Delta s = \sqrt{(\Delta x)^2 + (\Delta y)^2}$. The Riemann sum becomes

Fig. 3: Partitioned curtain

$$\sum f(x^*(t), y^*(t)) \cdot \Delta s = \sum f(x^*(t), y^*(t)) \cdot \frac{\sqrt{(\Delta x)^2 + (\Delta y)^2}}{\Delta t} \cdot \Delta t = \sum z(t^*) \cdot \sqrt{\left(\frac{\Delta x}{\Delta t}\right)^2 + \left(\frac{\Delta y}{\Delta t}\right)^2} \cdot \Delta t$$

Taking the limit of this as $\Delta s \to 0$, we get that

$$\{\text{area between f(x,y) and xy - plane}\} = \int_C f\,ds = \int_{t=a}^{t=b} z(t) \cdot \sqrt{\left(\frac{dx}{dt}\right)^2 + \left(\frac{dy}{dt}\right)^2} \cdot dt$$

Example 1: Suppose the curve C is parameterized by $\mathbf{r}(t) = \langle x(t), y(t) \rangle$ with $x(t) = 1+t^2$ and $y(t) = 3-t$, and that $f(x,y) = 2 + \sin(3xy) = 2 + \sin(3(1+t^2)(3-t))$. This is the situation in Fig. 1. Write an integral in terms of t for the area between the curve C in the xy-plane and f(x,y) for t from 0 to 2. If the units of x and y are meters and the t units are seconds, then what are the units of the result.

Solution: area between f(x,y) and the xy-plane $= \int_0^2 \left(2 + \sin(3(t^2+1)(3-t))\right)\sqrt{(2t)^2 + (-1)^2}\, dt$

Unfortunately we can not evaluate this integral by hand, but a calculator can: area=10.754 m².

Fig. 4: <2cos(t), 2sin(t), 2+cos(5t)>

Practice 1: Represent the area between the curve C parameterized by
$\mathbf{r}(t) = \langle 2\cos(t),\ 2\sin(t)\rangle$ $f(\mathbf{r}(t))=2+\cos(5t)$ (Fig. 4) as a definite integral and evaluate the integral.

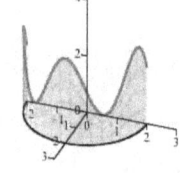

General case for line integral of a scalar function f over a path C : $\int_C f\, ds$

Suppose the path C is parameterized by $\mathbf{r}(t) = \langle x(t), y(t) \rangle$ where x(t) and y(t) are smooth (differentiable) functions of t, and that t varies from t=a to t=b. Then

$$\int_C f\, ds = \int_{t=a}^b f(\mathbf{r}(t)) \cdot \left|\frac{d\mathbf{r}}{dt}\right| dt = \int_{t=a}^b f(\mathbf{r}(t)) \cdot |\mathbf{r}'(t)|\, dt$$

Note: $ds = |\mathbf{r}'(t)|\, dt$ simply says that {change in position}={speed}·{change in time}: distance =rate·time.

Example 2: C1 is the semicircular path from (2,0) to (-2,0) parameterized by $\mathbf{r}(t) = \langle 2\cos(t),\ 2\sin(t)\rangle$ for t= 0 to π, and f(x,y)= 5+2x+4y (Fig. 5). C2 is the straight line path from (2,0) to (-2,0) parameterized by the path $\mathbf{r}(t) = \langle 2-4t,\ 0\rangle$ for t=0 to t=1.

Evaluate $\int_{C1} f\, ds$ and $\int_{C2} f\, ds$.

Solution: Along C1 $|\mathbf{r}'(t)| = \sqrt{(-2\sin(t))^2 + (2\cos(t))^2} = 2$ and
$f(\mathbf{r}(t)) = 5 + 2\cdot 2\cdot\cos(t) + 4\cdot 2\cdot\sin(t)$ so

$$\int_{C1} f\, ds = \int_0^\pi (5 + 4\cos(t) + 8\sin(t))(2)\, dt = 32 + 10\pi \approx 63.4 \ .$$

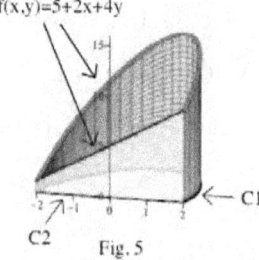

Fig. 5

Along C2 $|\mathbf{r}'(t)| = \sqrt{(-4)^2 + (0)^2} = 4$ and $f(\mathbf{r}(t)) = 5 + 2\cdot(2-4t) + 4\cdot(0) = 9 - 8t$ so

$$\int_{C2} f\, ds = \int_0^1 (9-8t)(4)\, dt = 20 \ .$$

Even though C1 and C2 begin and end at the same points, the values of the line integrals are different.

Note: We can always parameterize the straight line from A to B by $\mathbf{P}(t) = (1-t)\cdot A + t\cdot B$ for t=0 to t=1. Circles are typically parameterized using variations of $x = r\cdot\cos(t)$ and $y = r\cdot\sin(t)$ that take directions and shifts into account.

15.3 Line Integrals

Units: In the previous example, suppose f(x,y) is the density (in kg/m) of the curve **r** at the location (x,y), that the location (x,y) is in meters, m, and that time is in seconds, s. Then the units of the integral are

$$\int_C f \, ds = \int_{t=a}^{b} f(\mathbf{r}(t))|\mathbf{r}'(t)| \, dt = \text{kg}$$

$$\left(\frac{kg}{m}\right)\left(\frac{m}{s}\right)(s)$$

Practice 2: C is the straight line path from (0,1) to (4,2) and f(x,y)=2x+y. Evaluate $\int_C f \, ds$.

If the units of f are pounds, the x and y units are feet, and the t units are minutes, then what are the units of this line integral?

3D Mass of a Rod

Suppose a curve $C(t) = (t, t^2, 1+t^2)$ meters (Fig. 6) has linear density $\delta(x,y,z) = x - y + z - 1$ g/m. We can represent the mass of the curve from t=1 to t=3 as an integral in t by proceeding as before and starting with a partition of the curve into segments of lengths $\Delta s = \sqrt{(\Delta x)^2 + (\Delta y)^2 + (\Delta z)^2} = \sqrt{1 + (2t)^2 + (2t)^2} = \sqrt{1+8t^2}$.

The density at each location (x*, y*, z*) on the curve is

$\delta(x,y,z) = x - y + z - 1 = (t) - (t^2) + (1+t^2) - 1 = t$ so the mass along each

segment is {segment mass} = density · length = $t \cdot \sqrt{1+8t^2}$. Finally, the total

mass of the rod is $\int_{t=1}^{3} t \cdot \sqrt{1+8t^2} \, dt = \frac{1}{24}(1+8t^2)^{3/2} \Big|_1^3 = \frac{73}{24}\sqrt{73} - \frac{9}{8} \approx 24.86$ g.

Fig. 6

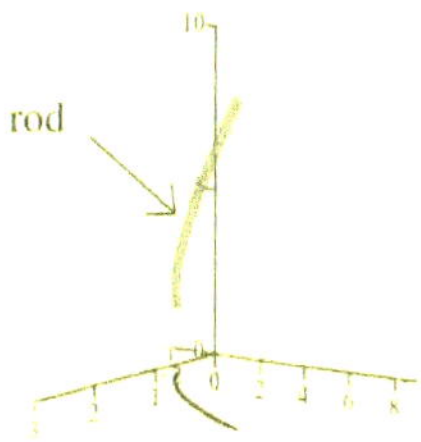

Generalizing this approach to any continuous function f(x,y,z) along a smooth curve C parameterized by $\mathbf{r}(t) = \langle x(t), y(t), z(t) \rangle$ from t=a to t=b, we again have

$$\int_C f \, ds = \int_{t=a}^{t=b} f(\mathbf{r}(t)) \cdot |\mathbf{r}'(t)| \cdot dt .$$

Practice 3: Represent the total mass of the curve $C(t) = (2t, t^2, t)$ that has density $\delta(x,y,z) = x + z$ from t=1 to t=4 as an integral and evaluate the integral.

Note: The various application formulas for first and second moments and centers of mass in section 14.6 also apply here, but we replace the triple integrals $\iiint_R$ with $\int_C$ and the dV with ds.

Work moving along a curve

Previously in this section the function f was a scalar-valued function, but very interesting situations arise when we move along a curve in a vector field.

In section 11.4 we saw that the elementary idea that "work=force times distance" could be extended to "work=(force in the direction of movement) · (displacement) = **F**•**D**," the dot product of **F** and **D**, where **F** is a force vector and **D** is the displacement vector. This is exactly the idea we need to calculate the work moving an object along a curve in two or three dimensions.

Line integral of a vector-valued function F over a path C

Work: Suppose C is a smooth curve in 3D parameterized by $\mathbf{r}(t) = \langle x(t), y(t), z(t) \rangle$ for $a \leq t \leq b$ and that $\mathbf{F}(x,y,z)$ is a 3D force vector field. If we partition C (Fig. 7) into small time increments, then at location $P^* = (x(t^*), y(t^*), z(t^*))$ the displacement is tangent to the curve and has length Δs so the displacement is $\Delta s \cdot \mathbf{T}(t^*)$ where $\mathbf{T}(t^*)$ is the unit tangent vector at P^* (Fig. 8). The work for **F** to move the object along that small Δs segment of the curve is $\mathbf{F} \cdot \mathbf{T} \cdot \Delta s$. Then the total work is approximately

$$\sum \mathbf{F}(x(t^*), y(t^*), z(t^*)) \cdot \mathbf{T}(x(t^*), y(t^*), z(t^*)) \cdot \Delta s$$

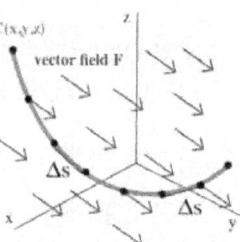

Fig. 7: Curve C and vector field F

As $\Delta s \to 0$ the approximations become better and better, and the work to move an object along C is defined to be

$$\text{work} = \int_C \mathbf{F} \cdot \mathbf{T} \, ds = \int_C \mathbf{F}(x(t), y(t), z(t)) \cdot \mathbf{T}(x(t), y(t), z(t)) \, ds$$

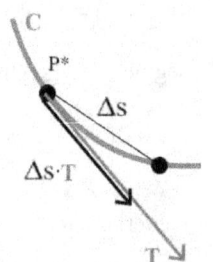

Fig. 8: work on one segment

But $\mathbf{T}(t) = \dfrac{\mathbf{r}'(t)}{|\mathbf{r}'(t)|}$ and $ds = |\mathbf{r}'(t)| \, dt$ so

Fig. 9: C=(cos(t),sin(t),t) F=<1,1,z>

$$\text{work} = \int_C \mathbf{F} \cdot \mathbf{T} \, ds = \int_{t=a}^{b} \mathbf{F}(\mathbf{r}(t)) \cdot \mathbf{r}'(t) \, dt$$

Note: The units for work are the units of **F** times the units of length, ds.

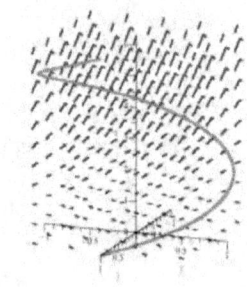

Example 2: Represent the work done in the field $\mathbf{F}(x,y,z) = \langle 1, 1, z \rangle$ to move an object along the helix $\mathbf{r}(t) = \langle \cos(t), \sin(t), t \rangle$ for t=0 to t=2π (Fig. 9) as an integral and then evaluate the integral.

Solution: $\mathbf{F}(\mathbf{r}(t)) \cdot \mathbf{r}'(t) = \langle 1, 1, z \rangle \cdot \langle -\sin(t), \cos(t), 1 \rangle = -\sin(t) + \cos(t) + t$ so $\text{work} = \int_{t=0}^{2\pi} (-\sin(t) + \cos(t) + t) \, dt = 2\pi^2$

15.3 Line Integrals

Practice 4: Represent the work done in the constant field $F(x,y,z) = \langle 3,2,1 \rangle$ to move an object along the curve $r(t) = \langle t, t^2, t^3 \rangle$ for t=0 to t=2 as an integral and then evaluate the integral.

Definition

If C is a smooth curve given by the vector function $\mathbf{r}(t)$ for $a \leq t \leq b$, and $\mathbf{F}$ is a continuous vector field defined on C, then the **line integral of F along C** is

$$\int_C \mathbf{F}(\mathbf{r}(t)) \bullet d\mathbf{r} = \int_{t=a}^{b} \mathbf{F}(\mathbf{r}(t)) \bullet \mathbf{r}'(t) \, dt = \int_C \mathbf{F} \bullet \mathbf{T} \, ds$$

Note: This pattern will occur often in future sections, and you should be familiar with all three notations. The middle integral is usually the easiest to use for computations.

Note: If $\mathbf{F} = \langle M, N, P \rangle$ and $\mathbf{T} = \left\langle \dfrac{dx}{ds}, \dfrac{dy}{ds}, \dfrac{dz}{ds} \right\rangle$ then $\mathbf{F} \bullet \mathbf{T} ds = M dx + N dy + P dz$. Units = (F units)(ds units).

You should also recognize what is happening geometrically (Fig. 10). If the angle θ between $\mathbf{F}$ and $\mathbf{r}'$ is 90° then $\mathbf{F} \bullet \mathbf{T} = |\mathbf{F}| |\mathbf{T}| \cos(\theta) = 0$, if the $\mathbf{F}$ and $\mathbf{r}'$ angle is acute ($-90° < \theta < 90°$) then $\mathbf{F} \bullet \mathbf{T} = |\mathbf{F}| |\mathbf{T}| \cos(\theta) > 0$, and if the $\mathbf{F}$ and $\mathbf{r}'$ angle is obtuse ($90° < \theta < 270°$) then $\mathbf{F} \bullet \mathbf{T} = |\mathbf{F}| |\mathbf{T}| \cos(\theta) < 0$.

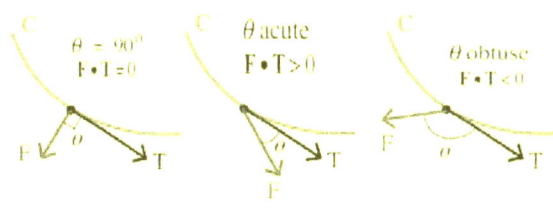

Fig. 10: Geometry of $\mathbf{F} \bullet \mathbf{T}$

Practice 5: C is the semicircle given by $\mathbf{r}(t) = \langle \cos(t), \sin(t) \rangle$ for $0 \leq t \leq \pi$, and four vector fields are $\mathbf{F1} = \langle 1, 0 \rangle$, $\mathbf{F2} = \langle -1, 0 \rangle$, $\mathbf{F3} = \langle 0, 1 \rangle$ and $\mathbf{F4} = \langle x, y \rangle$. Use sketches of the path C and a few vectors from the field $\mathbf{F}$ to estimate whether each line integral $\int_C \mathbf{F}(\mathbf{r}(t)) \bullet d\mathbf{r}$ is positive, negative or zero.

Work in Gravitational, Electrical and Magnetic Fields: In all three of these situations the force between points is inversely proportional to the square of the distance between the points and acts along the straight line connecting the points. If one point is at the origin and the other is at (x,y,z) then the magnitude of $\mathbf{F}$ is

$$|\mathbf{F}| = \frac{k}{|\mathbf{r}|^2} = \frac{k}{(x^2 + y^2 + z^2)} \quad \text{and the direction of } \mathbf{F} \text{ is } \frac{\mathbf{r}}{|\mathbf{r}|} = \frac{\langle x,y,z \rangle}{(x^2+y^2+z^2)^{1/2}} \quad \text{so} \quad \mathbf{F(r)} = \frac{k\langle x,y,z \rangle}{(x^2+y^2+z^2)^{3/2}} = \frac{k\mathbf{r}}{|\mathbf{r}|^3}.$$

Example 3: Find the work to moving an object along the path from (1,1,1) in a straight line to (c,c,c) where c>1.

Solution: The line can be parameterized by $\mathbf{r}(t) = \langle t, t, t \rangle$ as t goes from 1 to c, and

$$\mathbf{F}(\mathbf{r}(t)) \bullet \mathbf{r}'(t) = \frac{k\langle t,t,t \rangle}{|\langle t,t,t \rangle|^3} \bullet \langle 1,1,1 \rangle = \frac{k3t}{(3t^2)^{3/2}} = \frac{k}{\sqrt{3}} \frac{1}{t^2} \quad \text{so the work done is}$$

$$\text{work} = \int_{t=1}^{c} \mathbf{F}(\mathbf{r}(t)) \bullet \mathbf{r}'(t)\, dt = \frac{k}{\sqrt{3}} \int_{t=1}^{c} \frac{1}{t^2}\, dt = \frac{k}{\sqrt{3}} \left(-\frac{1}{t}\right)\bigg|_{t=1}^{c} = \frac{k}{\sqrt{3}}\left(1 - \frac{1}{c}\right)$$. As the object is moved farther and farther away, as $c \to \infty$, the work approaches the finite value $k/\sqrt{3}$.

Flow along a curve C in a vector field F

> **Flow:** If a smooth curve C is parameterized by r(t) in a continuous vector field F, then the **flow along C** from t=a to t=b is $\text{flow} = \int_C \mathbf{F} \bullet \mathbf{T}\, ds$.

F might represent the velocity field of a fluid in a region of space (water in a river channel, air in a wind tunnel), then the integral of $\mathbf{F} \bullet \mathbf{T}$ along a curve C in that space is the "flow" along that curve. In this case, if the units of F are m/sec, then the units for flow are m^2/sec.

If C is a closed loop, then $\int_C \mathbf{F} \bullet \mathbf{T}\, ds$ is called the **circulation** around C.

Flow is calculated in the same way as work.

Flux across a closed curve C in 2D

Just as flow measured the accumulation of a vector field along a curve C, the flux measures the accumulation as the vector field crosses perpendicular to the curve so we need the velocity of the field in the direction of the normal vector to the curve at each point. And then we want to accumulate all of those little values, an integral.

> {Flux of F across closed curve C} $= \int_C \mathbf{F} \bullet \mathbf{n}\, ds$ where **n** is the (outward) unit normal vector to C and the curve C is traversed exactly once in the counterclockwise direction.

Flow is the line integral of the scalar component $\mathbf{F} \bullet \mathbf{T}$ of $\mathbf{F}$ in the direction of the unit tangent vector to C.
Flux is the line integral of the scalar component $\mathbf{F} \bullet \mathbf{n}$ of $\mathbf{F}$ in the direction of the unit normal vector to C.

If $\mathbf{F}(x,y,z) = \langle M(x,y), N(x,y) \rangle = M\mathbf{i} + N\mathbf{j}$ is a vector field in 2D and C is parameterized in the xy-plane in the counterclockwise direction by $\mathbf{r}(t) = \langle x(t), y(t) \rangle$ then the unit normal vector is $\mathbf{n} = \mathbf{T} \mathbf{x} \mathbf{k}$ where $\mathbf{T}$ is the unit tangent vector and $\mathbf{k} = \langle 0,0,1 \rangle$ (Fig. 11). Then

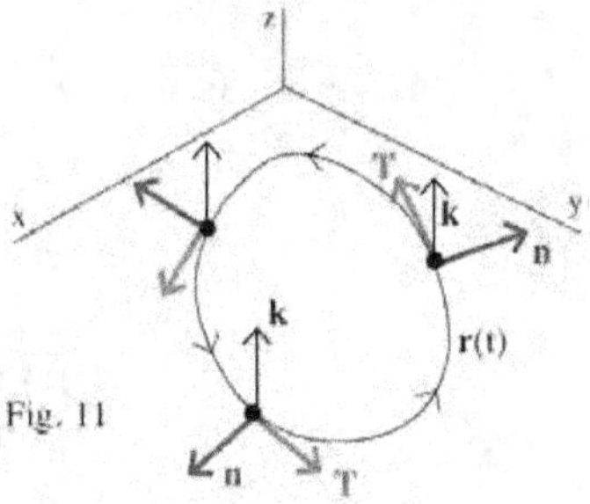

Fig. 11

15.3 Line Integrals

$$\mathbf{n} = \mathbf{T} \times \mathbf{k} = \begin{vmatrix} \mathbf{i} & \mathbf{j} & \mathbf{k} \\ \frac{dx}{ds} & \frac{dy}{ds} & 0 \\ 0 & 0 & 1 \end{vmatrix} = \left\langle \frac{dy}{ds}, -\frac{dx}{ds}, 0 \right\rangle \text{ so}$$

$$\mathbf{F} \bullet \mathbf{n} = \langle M, N \rangle \bullet \left\langle \frac{dy}{ds}, -\frac{dx}{ds} \right\rangle = M\frac{dy}{ds} - N\frac{dx}{ds} \text{ and}$$

$$\text{flux} = \int_C \mathbf{F} \bullet \mathbf{n} \, ds = \int_C \left(M\frac{dy}{ds} - N\frac{dx}{ds} \right) ds = \int_C M\,dy - N\,dx.$$

Fig. 12: Flux across C

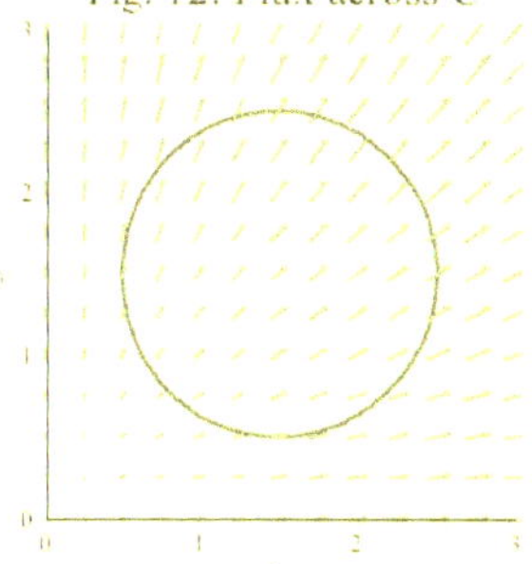

Example 4: Calculate the flux across the circle C parameterized by
$\mathbf{r}(t) = \langle 1.5 + \cos(t), 1.5 + \sin(t) \rangle$ for the vector field $\mathbf{F}(x,y) = \langle x, y \rangle = x\mathbf{i} + y\mathbf{j}$. (Fig. 12)

Solution: M=1.5+cos(t), N=1.5+sin(t), dx=-sin(t) dt, dy=cos(t) dt, $0 \le t \le 2\pi$. This path traverses the circle in the counterclockwise direction. Then

$$\text{flux} = \int_C M\,dy - N\,dx = \int_{t=0}^{2\pi} (1.5 + \cos(t))(\cos(t)\,dt) - (1.5 + \sin(t))(-\sin(t)\,dt)$$

$$= \int_{t=0}^{2\pi} \left(1.5\cos(t) + \cos^2(t) + 1.5\sin(t) + \sin^2(t) \right) dt = 2\pi$$

The net outward flow across the circle is positive – more is leaving than is entering the circular region.

Practice 6: What flux do you expect for this curve C if the field is reversed to become $\mathbf{F}(x,y) = \langle -x, -y \rangle$? Calculate the flux over this C for the field.

If the simple closed curve C encloses a source of water, then the flux across C will be positive and equal to the rate of water input from the source. If C contains a sink (a drain), then the flux across C will be negative. If C contains both sources and sinks, then the flux across C will be the signed sum of the sources (counted as positive) and the sinks (counted as negative).

Wrap up

This section has examined line integrals and a variety of their applications, but there are really only two mathematical situations: when the function is scalar-valued and when the function is vector-valued.

If f is a **scalar-valued** function then:

$$\int_C f \, ds = \int_{t=a}^{b} f(\mathbf{r}(t)) \cdot \left| \frac{d\mathbf{r}}{dt} \right| dt = \int_{t=a}^{b} f(\mathbf{r}(t)) \cdot |\mathbf{r}'(t)| \, dt$$

$$\text{area} = \int_C f(\mathbf{r}(t))\,ds = \int_{t=a}^{b} f(\mathbf{r}(t)) \cdot |\mathbf{r}'(t)| \, dt$$

$$\text{mass} = \int_C \delta(\mathbf{r}(t))\,ds = \int_{t=a}^{b} \delta(\mathbf{r}(t)) \cdot |\mathbf{r}'(t)| \, dt$$

If **F** is a **vector-valued** function then:

$$\int_C \mathbf{F} \bullet \mathbf{T}\, ds = \int_C \mathbf{F}(\mathbf{r}(t)) \bullet d\mathbf{r} = \int_{t=a}^{b} \mathbf{F}(\mathbf{r}(t)) \bullet \mathbf{r}'(t)\, dt = \int_C (M dx + N dy + P dz)$$

$$\text{work} = \int_C \mathbf{F} \bullet \mathbf{T}\, ds$$

$$\text{flow} = \int_C \mathbf{F} \bullet \mathbf{T}\, ds$$

$$\text{flux} = \int_C \mathbf{F} \bullet \mathbf{n}\, ds = \int_C (M dy - N dx) \quad \text{(C a simple, closed 2D curve oriented counterclockwise)}$$

Later in this chapter we will consider 3D electric and magnetic vector fields and their flows and fluxes.

Problems

1. Determine the area between the curve $\mathbf{r}(t) = \langle 2t+1, 3+t^2 \rangle$ for t from 0 to 3 in the xy-plane and $f(x,y) = x^2 - 4y + 11$.

2. Determine the area between the curve $\mathbf{r}(t) = \langle \sin(t), 1+\cos(t) \rangle$ for t from 0 to π in the xy-plane and $f(x,y) = 2 + xy$.

3. Create an integral to calculate the area between $\mathbf{r}(t) = \langle x(t), y(t), z(y) \rangle$ (x,y,z$\geq$0) and the xz-plane.

4. Create an integral to calculate the area between $\mathbf{r}(t) = \langle x(t), y(t), z(y) \rangle$ (x,y,z$\geq$0) and the yz-plane.

5. A pipe is parameterized by $\mathbf{r}(t) = \langle 3\cos(t), 3\sin(t) \rangle$ for $0 \leq t \leq \pi/2$ and the density of the pipe is $\delta(x,y) = 1 + x + 2y$ at location (x,y). Find the total mass of the pipe.

6. Find the mass of the pipe in Problem 5 if $\delta(x,y) = 1 + 3x$.

7. A pipe is parameterized by $\mathbf{r}(t) = \langle 1+t, 3t, 2+t \rangle$ for $0 \leq t \leq 2$ and has density $\delta(x,y,z) = y + z$ at (x,y,z). Find the mass of the pipe.

8. Find the mass of the pipe in Problem 7 if $\delta(x,y,z) = x + 2y + 3z$.

In problems 9 to 14, evaluate $\int_C f\, ds$ for the given function f on the curve C.

9. $f(x,y) = 2x + y$ on C given by $\mathbf{r}(t) = \langle 3t+2, 5-4t \rangle$ for $0 \leq t \leq 2$.

10. $f(x,y) = x - 2y$ on C given by $\mathbf{r}(t) = \langle 12t+1, 5t-4 \rangle$ for $1 \leq t \leq 4$.

11. $f(x,y) = x^2 y + y$ on C given by $\mathbf{r}(t) = \langle 3, t^2 + 1 \rangle$ for $0 \leq t \leq 3$.

12. $f(x,y) = xy$ on C given by $\mathbf{r}(t) = \langle 4t^2, 3t^2 + 3 \rangle$ for $0 \leq t \leq 1$.

13. $f(x,y) = x + y$ on C given by $\mathbf{r}(t) = \langle 2 - \sin(t), 1 + \cos(t) \rangle$ for $0 \leq t \leq \pi$.

14. $f(x,y) = x - 2y$ on C given by $\mathbf{r}(t) = \langle 2 + 5\cos(t), 1 + 2\cos(t) \rangle$ for $0 \leq t \leq 2\pi$.

15. In Fig. 13 is the work along each path A and B positive, negative or zero?

16. In Fig. 13 is the work along each path C and D positive, negative or zero?

17. In Fig. 14 is the flow along each path A and B positive, negative or zero?

18. In Fig. 14 is the flow along each path C and D positive, negative or zero?

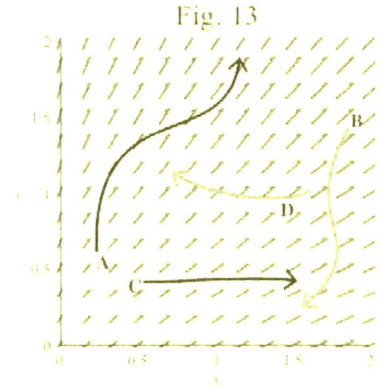

Fig. 13

19. Calculate the work to move an object along the path
    $r(t) = \langle 2+3t, 4t \rangle$ for $0 \le t \le 3$ in the field $\mathbf{F} = \langle x, x+y \rangle$.

20. Calculate the work to move an object along the path
    $r(t) = \langle t^2, t \rangle$ for $0 \le t \le 2$ in the field $\mathbf{F} = \langle -y, x \rangle$

21. Calculate the work to move an object along the path
    $r(t) = \langle \cos(t), t, \sin(t) \rangle$ for $0 \le t \le \pi$ in the field
    $\mathbf{F} = \langle 1, 2, 3 \rangle$.

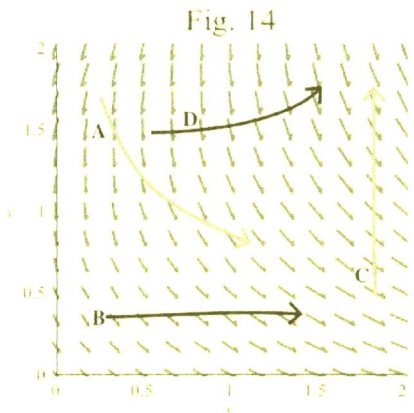

Fig. 14

22. Calculate the work to move an object along the path
    $r(t) = \langle \cos(t), t, \sin(t) \rangle$ for $0 \le t \le \pi$ in the field
    $\mathbf{F} = \langle x, y, z \rangle$.

23. Calculate the flow along the path $r(t) = \langle t, 4t, t^2 \rangle$ for $1 \le t \le 2$ in the field $\mathbf{F} = \langle z, 2y, x \rangle$.

24. Calculate the flow along the path $r(t) = \langle t^2, 3+t \rangle$ for $0 \le t \le 5$ in the field $\mathbf{F} = \langle -y, 2x \rangle$.

25. Calculate the flux around the closed path $r(t) = \langle \cos(t), 2 \cdot \sin(t) \rangle$ for $0 \le t \le 2\pi$ in the field
    $\mathbf{F} = \langle 2x, 1+2y \rangle$.

26. Calculate the flux around the closed path $r(t) = \langle \cos(t), 2 \cdot \sin(t) \rangle$ for $0 \le t \le 2\pi$ in the field
    $\mathbf{F} = \langle 3x, 1+2y \rangle$.

27. If the circulation is 0 around a closed path in a vector field **F**, can the flux be positive, negative, zero? (Think about the unit circle path in a radial vector field.)

28. If the work along a path C in a vector field is positive, what can be said about the flow along that path?

In problems 29 to 34, assume that the orientation of the path is reversed. Is the original value changed or not?

29. What happens to the area? Why?

30. What happens to the mass? Why?

31. What happens to the work? Why?

32. What happens to the flow? Why?

33. What happens to the circulation? Why?

34. What happens to the flux? Why?

Practice Answers

Practice 1: $$\text{area} = \int_{t=-\pi/2}^{\pi/2} (2+\cos(5t)) \cdot \sqrt{(-2\sin(t))^2 + (2\cos(t))^2} \cdot dt = \int_{t=-\pi/2}^{\pi/2} ((2+\cos(5t))) \cdot 2 \cdot dt$$
$$= 2\left(2t + \frac{1}{5}\sin(5t)\right)\Big|_{-\pi/2}^{\pi/2} = 4\pi + \frac{4}{5} \approx 13.37$$

Practice 2: C is parameterized by $r(t) = \langle (1-t)\cdot 0 + t\cdot 4, (1-t)\cdot 1 + t\cdot 2 \rangle = \langle 4t, 1+t \rangle$ for t=0 to t=1. Then $|r'(t)| = \sqrt{17}$ and $f(r(t)) = 2(4t) + (1+t) = 9t+1$ so

$$\int_C f\, ds = \int_{t=0}^{1} f(r(t))\left|\frac{dr}{dt}\right| dt = \int_{t=0}^{1} (9t+1)(\sqrt{17})\, dt = \frac{11}{2}\sqrt{17} .$$ The units are $(\$)\left(\frac{\text{feet}}{\text{min}}\right)(\text{min}) = \$ \cdot \text{feet}$

Practice 3: total mass $= \int_{t=1}^{4} 3t\cdot\sqrt{5+4t^2}\, dt = \frac{1}{4}(5+4t^2)^{3/2}\Big|_1^4 = \frac{69}{4}\sqrt{73} - \frac{27}{4} \approx 136.54$

Practice 4: $F(r(t))\bullet r'(t) = \langle 3,2,1\rangle \bullet \langle 1, 2t, 3t^2\rangle = 3+4t+3t^2$ and work $= \int_{t=0}^{2} (3+4t+3t^2)\, dt = 22$

Practice 5: C and the vector fields are shown in Fig. P5. The line integral is negative for F1, positive for F2, and 0 for F3 (by symmetry) and F4.

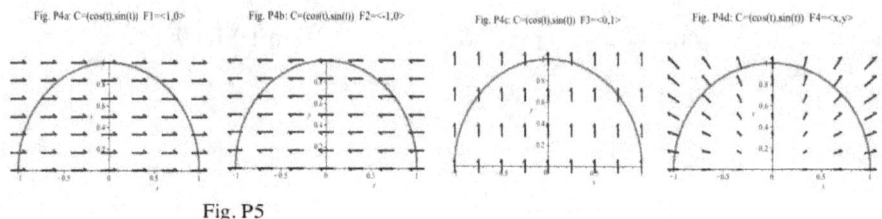

Fig. P5

Practice 6: M=-(1.5+cos(t)), N=-(1.5+sin(t)), dx=-sin(t) dt, dy=cos(t) dt, $0 \le t \le 2\pi$.

$$\text{flux} = \int_C M\,dy - N\,dx = \int_{t=0}^{2\pi} -(1.5+\cos(t))(\cos(t)\,dt) + (1.5+\sin(t))(-\sin(t)\,dt)$$

$$= -\int_{t=0}^{2\pi} \left(1.5\cos(t) + \cos^2(t) + 1.5\sin(t) + \sin^2(t)\right) dt = -2\pi,$$ the opposite of the Example 4 value.

15.4 The Fundamental Theorem of Line Integrals and Potential Functions

Section 15.2 introduced the line integral along a curve in a vector field F, $\int_C \mathbf{F} \bullet \mathbf{T} \, ds$, discussed some applications of these integrals and showed how to calculate them. But something curious is going on.

Let $\mathbf{F}(x,y) = \langle x, y+x \rangle$. If we take three different paths from (0,0) to (2,4) in this vector field we get three different values. Along the curve C_1 given by $\mathbf{r}_1(t) = \langle 2t, 4t \rangle$

$$\int_{C_1} \mathbf{F} \bullet \mathbf{T} \, ds = \int_{C_1} \mathbf{F}(\mathbf{r}_1(t)) \bullet \mathbf{r}'_1(t) \, dt = \int_{t=0}^{1} \langle 2t, 4t+2t \rangle \bullet \langle 2, 4 \rangle \, dt = \int_{t=0}^{1} 28t \, dt = 14.$$

But along the curve C_2 given by $\mathbf{r}_2(t) = \langle 2t, 4t^2 \rangle$, $\int_{C_2} \mathbf{F} \bullet \mathbf{T} \, ds = 46/3$.

And along the curve C_3 given by $\mathbf{r}_3(t) = \langle 2t, 4t^5 \rangle$, $\int_{C_3} \mathbf{F} \bullet \mathbf{T} \, ds = 50/3$.

In this vector field the work to move an object from (0,0) to (2,4) depends on the path of the object.

However, if we change the vector field slightly, to $\mathbf{F}(x,y) = \langle x, y \rangle$, and calculate the work along the same three paths from (0,0) to (2, 4), the results are always the same:

$$\int_{C_1} \mathbf{F} \bullet \mathbf{T} \, ds = \int_{t=0}^{1} \langle 2t, 4t \rangle \bullet \langle 2, 4 \rangle \, dt = \int_{t=0}^{1} 20t \, dt = 10,$$

$$\int_{C_2} \mathbf{F} \bullet \mathbf{T} \, ds = \int_{t=0}^{1} \langle 2t, 4t^2 \rangle \bullet \langle 2, 8t \rangle \, dt = \int_{t=0}^{1} 32t^3 + 4t \, dt = 10,$$

and $\int_{C_3} \mathbf{F} \bullet \mathbf{T} \, ds = \int_{t=0}^{1} \langle 2t, 4t^5 \rangle \bullet \langle 2, 20t^4 \rangle \, dt = \int_{t=0}^{1} 80t^9 + 4t \, dt = 10$.

And the result will be 10 no matter what other smooth paths we take from (0, 0) to (2, 4) in this field.

This field $\mathbf{F}(x,y) = \langle x, y \rangle$ is a special type of vector field called a **gradient field** or a **potential field** or a **conservative field**, and it has the wonderful property that the value of the line integral does not depend on the path. This property is called **path independence**.

Definition: $\int_C \mathbf{F} \bullet \mathbf{T} \, ds$ is **independent of path** if $\int_{C_1} \mathbf{F} \bullet \mathbf{T} \, ds = \int_{C_2} \mathbf{F} \bullet \mathbf{T} \, ds$ for any two paths C_1 and C_2 with the same initial and ending points.

15.4 Potential Fields and Line Integrals

Conservative Fields

But what are the conservative vector fields and how can we determine if a given field is conservative?

> **Definitions: Conservative Field and Potential Function**
> A vector field **F** is called a **conservative** (or gradient, or potential) **field**
> on a region R in 2D or 3D if there is a scalar function f on R so that $\nabla f = \mathbf{F}$.
> This scalar function f is called the **potential function** for the conservative field **F**.

Example 1: Each of these scalar-valued functions generates a conservative vector field. Determine the vector field
F for $f_1(x,y) = xy$, $f_2(x,y) = \sqrt{x^2 + y^2}$ and $f_3(x,y,z) = x^2 y + 2yz$.

Solution: All that is needed is the gradient of each function: $\mathbf{F}_1(x,y) = \langle y, x \rangle$,

$$\mathbf{F}_2(x,y) = \left\langle \frac{x}{\sqrt{x^2+y^2}}, \frac{y}{\sqrt{x^2+y^2}} \right\rangle \text{ and } \mathbf{F}_3(x,y,z) = \langle 2xy, x^2 + 2z, 2y \rangle.$$

Each of these is a conservative vector field since each $\mathbf{F} = \nabla f$ for a differentiable function f.

Practice 1: Determine the conservative vector fields generated by $f_1(x,y) = 3x - 2y$, $f_2(x,y) = \sin(xy)$ and
$f_3(x,y,z) = \dfrac{k}{x^2 + y^2 + z^2}$.

> **Fundamental Theorem of Line Integrals**
> If there is a scalar function f on an open, connected region R in 2D or 3D so
> that $\nabla f = \mathbf{F}$ (**F** is a conservative field), and C is any piecewise smooth curve in R
> from point A (when t=a) to point B (when t=b)
> then $\displaystyle\int_C \mathbf{F} \bullet \mathbf{T} \, ds = \int_{t=a}^{b} \mathbf{F}(\mathbf{r}(t)) \bullet \mathbf{r}'(t) \, dt = \int_{t=a}^{b} \mathbf{F} \bullet d\mathbf{r} = f(B) - f(A)$
> and the line integral does not depend on the path C (path independence).

This is considered one of the four fundamental theorems of vector calculus (with Green's and Stokes's and the Divergence theorems). The meanings of "open," and "connected" are given in the chapter Appendix.

Proof: Suppose $\mathbf{F}(x,y,z) = \nabla f(x,y,z) = f_x(x,y,z)\mathbf{i} + f_y(x,y,z)\mathbf{j} + f_z(x,y,z)\mathbf{k}$ for some smooth
scalar-valued function f. Then

$$\mathbf{F} \bullet \mathbf{r}' = \nabla f \bullet \mathbf{r}' = \left\langle \frac{\partial f}{\partial x}, \frac{\partial f}{\partial y}, \frac{\partial f}{\partial z} \right\rangle \bullet \left\langle \frac{dx}{dt}, \frac{dy}{dt}, \frac{dz}{dt} \right\rangle = \frac{\partial f}{\partial x} \cdot \frac{dx}{dt} + \frac{\partial f}{\partial y} \cdot \frac{dy}{dt} + \frac{\partial f}{\partial x} \cdot \frac{dz}{dt} = \frac{df}{dt} \quad \text{(by the Chain Rule)}$$

so $\int_{t=a}^{b} \mathbf{F}(\mathbf{r}(t)) \bullet \mathbf{r}'(t)\ dt = \int_{t=a}^{b} \dfrac{df(\mathbf{r}(t))}{dt}\ dt = f(\mathbf{r}(b)) - f(\mathbf{r}(a)) = f(B) - f(A)$.

Example 2: $\mathbf{F}(x,y) = \langle y, x \rangle$. Evaluate $\int_C \mathbf{F} \bullet \mathbf{T}\ ds$ for a curve C that starts at A=(1,2) and ends at B=(4,3).

Solution: If we recognize (from Example 1) that $\mathbf{F} = \nabla f$ for $f(x,y) = xy$ then by the Fundamental Theorem of Line Integrals, $\int_C \mathbf{F} \bullet \mathbf{T}\ ds = f(B) - f(A) = f(4,3) - f(1,2) = 12 - 2 = 10$

along any smooth path C from A to B. You might want to check this value by using the path $\mathbf{r}(t) = \langle 1 + 3t, 2 + t \rangle$ for t from 0 to 1 and explicitly calculating $\int_{t=0}^{1} \mathbf{F}(\mathbf{r}(t)) \bullet \mathbf{r}'(t)\ dt$.

If we know a scalar function f whose gradient is the vector function $\mathbf{F}$, then a difficult calculus problem becomes an easy arithmetic problem. You should recognize the similarity of this calculation with the way integrals were evaluated in beginning calculus. In beginning calculus if f was an antiderivative of F (Df=F), then $\int_a^b F(x)\ dx = f(b) - f(a)$. Here if f is a potential function of $\mathbf{F}$ ($\nabla f = \mathbf{F}$), then $\int_{t=a}^{b} \mathbf{F} \bullet d\mathbf{r} = f(B) - f(A)$.

Practice 2: $\mathbf{F}(x,y,z) = \langle 2xy, x^2 + 2z, 2y \rangle$. Evaluate $\int_C \mathbf{F} \bullet \mathbf{T}\ ds$ for the curve C that starts at A=(1,0,4) and ends at B=(4,3,2). (Suggestion: Look at the answers in Example 1.)

Example 3: Fig. 1 shows the level curves for a smooth function z=f(x,y) and the vector field $\mathbf{F} = \nabla f$. Calculate the amount of work done moving an object from point A to point B along the two given paths.

Solution: work $= \int_C \mathbf{F} \bullet \mathbf{T}\ ds = f(B) - f(A) = 10$ along each path.

Practice 3: Using the same figure, calculate the amount of work done moving an object from point C to point D along the given curve. What is the total work to go from C to D and then back to C?

Fig. 1: Level curves of z=f(x,y)

Theorem: $\int_C \mathbf{F} \bullet \mathbf{T}\ ds$ is **independent of path** on open, simply-connected region R

if and only if $\int_C \mathbf{F} \bullet \mathbf{T}\ ds = 0$ for every closed path C in R.

Finding potential functions

Finding potential functions is a lot like finding antiderivatives but now we need to match two antiderivatives. If $\mathbf{F} = \langle M(x,y), N(x,y) \rangle$, then the potential function f must satisfy both $f_x = \frac{\partial f}{\partial x} = M$ and $f_y = \frac{\partial f}{\partial y} = N$. So typically two antiderivatives are needed, one with respect to x and one with respect to y.

Example 4: Find a potential function f for $\mathbf{F} = \langle 4y, 4x+5 \rangle$.

Solution: $\frac{\partial f}{\partial x} = 4y$ so (taking an antiderivative with respect to x) $f(x,y) = 4xy + g(y)$ (since $\frac{\partial g(y)}{\partial x} = 0$).

But $\frac{\partial}{\partial y}(4xy + g(y)) = 4x + g'(y)$ must equal $N = 4x+5$, so $g'(y) = 5$ and $g(y) = 5y$ (typically the "+k" is omitted). Then $f(x,y) = 4xy + 5y$. A quick check shows that $f_x = 4y$ and $f_y = 4x+5$ so $\nabla f = \mathbf{F}$.

Note: We could have first taken the antiderivative of $N = 4x+5$ with respect to y.

Practice 4: Find a potential functions f for $\mathbf{F} = \langle y \cdot e^{xy} + 2, x \cdot e^{xy} + 3 \rangle$ and $\mathbf{F} = \langle 2xy + \cos(x), x^2 \rangle$.

But the vector field $\mathbf{F}(x,y) = \langle x, y+x \rangle$ at the beginning of this section was not path independent so it is not a conservative field and does not have a potential function. There is a very easy way to determine if a vector field has a potential function and is conservative.

> **Theorem: Test for a conservative field**
>
> If R is an open, connected, and simply-connected region then
>
> 2D: $\mathbf{F} = \langle M(x,y), N(x,y) \rangle$ is a conservative field on R if and only if $\frac{\partial M}{\partial y} = \frac{\partial N}{\partial x}$.
>
> 3D: $\mathbf{F} = \langle M(x,y), N(x,y), P(x,y) \rangle$ is a conservative field on R if and only if
>
> $\frac{\partial M}{\partial y} = \frac{\partial N}{\partial x}$, $\frac{\partial M}{\partial z} = \frac{\partial P}{\partial x}$ and $\frac{\partial N}{\partial z} = \frac{\partial P}{\partial y}$.

Proof: These follow from Clairaut's Theorem which says that mixed partial derivatives are equal: $f_{xy} = f_{yx}$.

If has a potential function f, then $f_{xy} = \frac{\partial}{\partial y}\left(\frac{\partial f}{\partial x}\right) = \frac{\partial M}{\partial y}$ and $f_{yx} = \frac{\partial}{\partial x}\left(\frac{\partial f}{\partial y}\right) = \frac{\partial N}{\partial x}$ so $\frac{\partial M}{\partial y} = \frac{\partial N}{\partial x}$.

The proof for the 3D case follows from Clairaut's Theorem that $f_{xy} = f_{yx}$, $f_{xz} = f_{zx}$ and $f_{yz} = f_{zy}$.

The proof in 2D that $M_y = N_x$ implies that $\mathbf{F} = \langle M, N \rangle$ is conservative requires Green's Theorem which appears later.

For the vector field $\mathbf{F}(x,y) = \langle x, y+x \rangle$, $\dfrac{\partial M}{\partial y} = 0$ and $\dfrac{\partial N}{\partial x} = 1$ so the field does not have a potential function.

Practice 5: Which of these fields are conservative: $\mathbf{F}_1 = \langle 2xy^2 + 3, 2x^2y \rangle$, $\mathbf{F}_2 = \langle -y, x \rangle$, $\mathbf{F}_3 = \langle yz, xz, xy + 2x \rangle$ and $\mathbf{F}_4 = \langle yz, xz, xy + 2 \rangle$?

Potential or Gradient fields are also called Conservative fields because the total energy, kinetic plus potential, is constant at each point in the field, energy is conserved. The derivation of this result is shown in an Appendix after the Practice Answers.

Summary of results

If R is an open, connected, simply-connected region, then the following are equivalent:

(1) $\mathbf{F}$ is a conservative field.

(2) There is a potential function f so that $\mathbf{F} = \nabla f$ ..

(3) $\displaystyle\int_C \mathbf{F} \bullet \mathbf{T} \ ds = \int_{t=a}^{b} \mathbf{F} \bullet d\mathbf{r} = f(B) - f(A)$ for all points A and B in R and for all smooth curves C.

(4) $\displaystyle\int_C \mathbf{F} \bullet \mathbf{T} \ ds = \int_{t=a}^{b} \mathbf{F} \bullet d\mathbf{r} = 0$ for all simple, closed, smooth curves in R.

Problems

In problems 1 to 6, determine the conservative field generated by the given potential function.

1. $f(x,y) = x^2 + 3y^2$
2. $f(x,y) = 3x^2 - 4y^2$
3. $f(x,y) = \sin(3x + 2y)$
4. $f(x,y) = x^2y^3 + xy^2$
5. $f(x,y) = \ln(2x + 5y) + e^y$
6. $f(x,y) = \tan(x) - \sec(y)$

In problems 7 to 12, determine the work to move an object in the given field from point A to point B.

7. $\mathbf{F} = \langle 2x, 2y \rangle$, A = (1,2), B = (5,1)
8. $\mathbf{F} = \langle y, x \rangle$, A = (1,3), B = (3,5)
9. $\mathbf{F} = \langle x, x \rangle$, A = (0,2), B = (3,6)
10. $\mathbf{F} = \langle 3x^2y, 3x^2 \rangle$, A = (1,0), B = (3,1)
11. $\mathbf{F} = \langle yz, xz, xy \rangle$, A = (1,0,0), B = (4,2,1)
12. $\mathbf{F} = \langle -x, -y, -z \rangle$, A = (0,0,0), B = (2,4,6)

In problems 13 to 16, determine the work to move the object along the given paths from A to B and from C to D. Each figure shows the level curves of z=f(x,y) for some smooth function f.

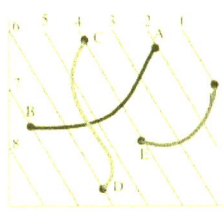
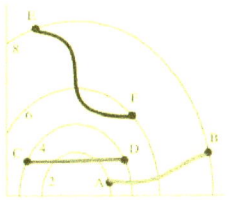
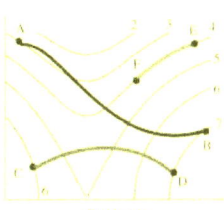

Problem 13 Problem 14 Problem 15 Problem 16

15.4 Potential Fields and Line Integrals

In problems 17 to 24, find a potential function for the given vector field F or show that F does not have a potential function.

17. $\langle 3x^2 + 4, 6 \rangle$

18. $\left\langle \dfrac{2}{2x+3y}, \dfrac{3}{2x+3y} + 3y^2 \right\rangle$

19. $\langle y \cdot \cos(xy) + 3x^2 y, x \cdot \cos(xy) + x^3 \rangle$

20. $\langle xy^3 + 3x^2 y, 3xy^2 + x^2 \rangle$

21. $\langle \sin(y), x \cdot \cos(y) \rangle$

22. $\langle 2y + 3z, 2x, 3x \rangle$

23. $\langle yz \cdot \cos(xyz), xz \cdot \cos(xyz) + z, xy \cdot \cos(xyz) \rangle$

24. $\langle y \cdot \cos(xy), x \cdot \cos(xy) + z, y \rangle$

25. (a) For $f(x,y) = \arctan(y/x)$ show that $\nabla f = \left\langle \dfrac{-y}{x^2 + y^2}, \dfrac{x}{x^2 + y^2} \right\rangle = \langle M, N \rangle$.

   (b) Show that $M_y = N_x$.

   (c) Show that for the closed circle $r(t) = (\cos(t), \sin(t))$ for $0 \le t \le 2\pi$, $\displaystyle\int_C \mathbf{F} \bullet \mathbf{T} \, ds = \int_{t=a}^{b} \mathbf{F} \bullet d\mathbf{r} = 2\pi$.

   (d) Why do these three results not violate the equivalent statements in the Summary?

Practice Answers

Practice 1: $\mathbf{F}_1(x,y) = \langle 3, -2 \rangle$, $\mathbf{F}_2(x,y) = \langle y \cdot \cos(xy), x \cdot \cos(xy) \rangle$ and

$$\mathbf{F}_3(x,y,z) = \left\langle \dfrac{-2kx}{(x^2+y^2+z^2)^2}, \dfrac{-2ky}{(x^2+y^2+z^2)^2}, \dfrac{-2kz}{(x^2+y^2+z^2)^2} \right\rangle = \dfrac{2k}{(x^2+y^2+z^2)^2} \langle -x, -y, -z \rangle$$

Note: $\mathbf{F}_3$ is a radial field with $|\mathbf{F}_3| = |2k|$ and it always points toward the origin.

Practice 2: From Example 1, $\mathbf{F} = \nabla f$ for $f(x,y) = x^2 y + 2yz$ so $\displaystyle\int_C \mathbf{F} \bullet \mathbf{T} \, ds = f(4,3,2) - f(1,0,4) = 72$.

Practice 3: work $= \displaystyle\int_C \mathbf{F} \bullet \mathbf{T} \, ds = f(D) - f(C) = 6$ along the given path, and, in fact, along every piecewise smooth path from C to D.

The work to go from C to D and then back to C is 0: the work from C to D is f(D)-f(C), and the work from D to C is f(C)-f(D) so the total work is $\{f(D)-f(C)\} + \{f(C)-f(D)\} = 0$.

Practice 4: $\nabla(e^{xy} + 22 + 3y) = \langle y \cdot e^{xy} + 2, x \cdot e^{xy} + 3 \rangle$ and $\nabla(x^2 y + \sin(x)) = \langle 2xy + \cos(x), x^2 \rangle$

Practice 5: $\mathbf{F}_1$ and $\mathbf{F}_4$ are conservative. $\mathbf{F}_2$ and $\mathbf{F}_3$ are not conservative.

Appendix: A bit of vocabulary about Curves and Regions

A curve C parameterized by r(t) for $a \leq t \leq b$ is **simple** if
$r(t_1) \neq r(t_2)$ for $a < t_1 < t_2 < b$. (Fig. A1)

A curve C parameterized by r(t) for $a \leq t \leq b$ is **closed** if $r(a) = r(b)$.

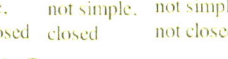

simple, closed   simple, not closed   not simple, closed   not simple, not closed

Fig. A1: Curves

A region R is **open** if every point in the R is inside a circle that lies completely in R.

$R = \{(x,y): x^2 + y^2 < 1\}$ is **open**.

$R = \{(x,y): x^2 + y^2 \leq 1\}$ is **not open** since every circle around a boundary point contains points not in R.

A region R is **connected** if any two points in R can be joined by a continuous curve that lies in R.

not connected, not simply-connected   connected, not simply-connected

simply-connected, not connected   open, connected, simply-connected

Fig. A3: Regions

A region R is **simply-connected** if every closed path in R can be contracted to a path without leaving R. A region with any holes (even just missing a single point) is not simply-connected.

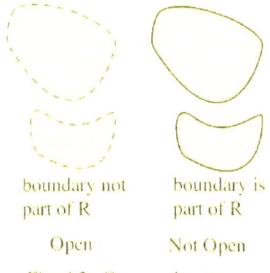

boundary not part of R   boundary is part of R
Open   Not Open

Fig. A2: Open and not open

Appendix Problems

For problems A1-A4, give examples of curves with the specified properties.

A1. closed, not simple   A2. not closed, not simple

A3. closed, simple   A4. not closed, simple

For problems A5-A10, give examples of regions with the specified properties. Give an example of a region that is

A5. open, connected and not simply-connected.   A6. open, simply-connected and not connected.

A7. connected, simply-connected, and not open.   A8. open, not connected and not simply-connected.

A9. connected, not open and not simply-connected.   A10. simply-connected, not open and not connected.

Appendix: Conservation of Energy

We need Newton's Second Law, $\mathbf{F} = m\mathbf{a}$ where $\mathbf{a}$ is the acceleration, so $\mathbf{F}(\mathbf{r}(t)) = m \cdot \mathbf{a}(t) = m \cdot \mathbf{r}''(t)$.

We also need that for any vector $\mathbf{w}$,

$$\frac{d}{dt}(\mathbf{w} \bullet \mathbf{w}) = \mathbf{w} \bullet \frac{d\mathbf{w}}{dt} + \frac{d\mathbf{w}}{dt} \bullet \mathbf{w} = 2\frac{d\mathbf{w}}{dt} \bullet \mathbf{w} \quad \text{so} \quad \frac{d\mathbf{w}}{dt} \bullet \mathbf{w} = \frac{1}{2}\frac{d}{dt}(\mathbf{w} \bullet \mathbf{w}),$$

so if $\mathbf{w} = \mathbf{r}'$ then $\mathbf{r}'' \bullet \mathbf{r}' = \frac{d\mathbf{r}'}{dt} \bullet \mathbf{r}' = \frac{1}{2}\frac{d}{dt}(\mathbf{r}' \bullet \mathbf{r}')$.

Kinetic energy: work to move from A to B

$$\text{work} = \int_C \mathbf{F} \bullet d\mathbf{r} = \int_{t=a}^{b} \mathbf{F}(\mathbf{r}(t)) \bullet \mathbf{r}'(t)\, dt = \int_{t=a}^{b} m \cdot \mathbf{r}''(t) \bullet \mathbf{r}'(t)\, dt$$

$$= \int_{t=a}^{b} m \cdot \frac{1}{2}\frac{d}{dt}(\mathbf{r}' \bullet \mathbf{r}')\, dt = \frac{m}{2}\mathbf{r}' \bullet \mathbf{r}' \Big|_{t=a}^{b} = \frac{m}{2}|\mathbf{r}'(t)|^2 \Big|_{t=a}^{b} = \frac{m}{2}|\mathbf{v}(t)|^2 \Big|_{t=a}^{b} = \frac{m}{2}|\mathbf{v}(b)|^2 - \frac{m}{2}|\mathbf{v}(a)|^2$$

But kinetic energy $K = \frac{1}{2}mv^2$ so work $= K(B) - K(A)$.

Potential energy: work to move from A to B

If f is a conservative field so $\mathbf{F} = -\nabla f$ for a potential function f, then

$$\text{work} = \int_C \mathbf{F} \bullet d\mathbf{r} = \int_C -\nabla f \bullet d\mathbf{r} = (-f(\mathbf{r}(t))\Big|_{t=a}^{b} = -f(\mathbf{r}(b)) + f(\mathbf{r}(a)) = f(A) - f(B)$$

Putting both type of energy together, $K(B) - K(A) = \text{work} = f(A) - f(B)$ so $K(B) + f(B) = K)A) + f(A)$.

The total energy, kinetic plus potential, at A is the same as at B for any two points A and B in R. In a conservative field the total energy is conserved.

Gravitational, electrical and magnetic fields are conservative fields.

15.5 Theorems of Green, Stokes and Gauss: An Introduction

The final sections of this text deal with the last three fundamental theorems of calculus, the theorems of Green, Stokes and Gauss. Each of these theorems extends the ideas of our earlier fundamental theorem of calculus to situations for vector-valued functions, and each has important applications to fields of physics and even to Maxwell's equations for magnetism and electricity. These theorems are technically sophisticated and difficult to prove, but the main ideas behind them are remarkably geometric and straightforward. The goal of this section is to approach these theorems geometrically and to illustrate why the ideas behind them are "easy and natural." The theorems will be clearly and precisely presented in the following sections and partial proofs will be given; the presentation in this section can help you understand what the theorems are saying and perhaps help you to remember them.

Introduction to Green's Theorem

Calculus deals with infinite collections of points, but sometimes a finite situation can give insight into the infinite.

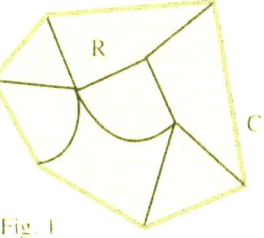
Fig. 1

In the following, R is a simple, simply-connected region consisting of a finite collection of cells. The boundary of R is a simple closed curve C (C consists of only the exterior edges of R). (Fig. 1)

Version 1 Green's Theorem: Divergence and Flux

Suppose water flows through the region R. Let's attach in-out flows to the edges of each cell. If we define the divergence of a cell to be the net outward flow of the cell then we can calculate the net outward flow of the collection of cells along the boundary of the collection – let's call this the flux across C.

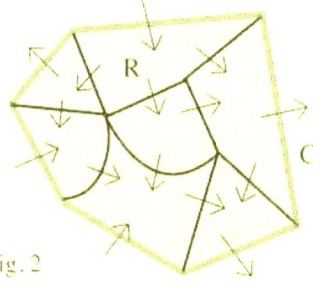
Fig. 2

We can calculate this flux in Fig. 2 in two ways. One way is to go around the boundary and add up the outward flows (counted as positive) and the inward flows (counted as negative). But if we add up the divergences for each cell in Fig. 2 we get the same net outward flow for the collection, the flux across C. This will always be true since for each inside boundary between cells, the outward flow from one cell becomes the inward flow into the next cell (Fig. 3) so the sum of those two flows will be zero, and that is the case for every shared edge inside the collection. Then the sum of all of the individual cell divergences is equal to just the sum of the flows on the outside edges (Fig. 4). This can be stated as

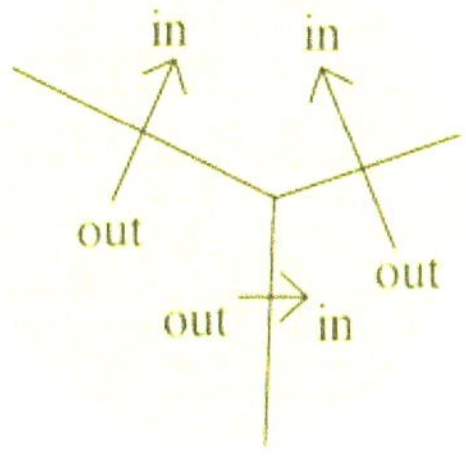

Fig. 3

In words: The sum of all of the individual cell divergences is equal
           to the sum of the flows on the outside edges (Fig. 4).

Finite version (divergence-flux form Green's Theorem):

$$\sum_R \{\text{divergence of cell}\} = \sum_C \{\text{flow across each outside edge}\} = \{\text{flux across C}\}.$$

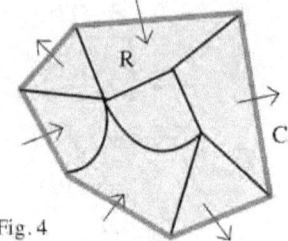

Fig. 4

Integral version (divergence-flux form Green's Theorem): For $\mathbf{F} = \langle M(x,y), N(x,y) \rangle$

$$\iint_R \text{div } \mathbf{F} \, dA = \oint_C \mathbf{F} \cdot \mathbf{n} \, ds = \text{flux across C}$$

Version 2 Green's Theorem: Curl and Circulation

Instead of looking at flow across edges of the cells, consider flow along each edge of a cell as in Fig. 5, and define the curl of each cell to be the sum of the flows around the edges of the cell counting flows in the counterclockwise direction to be positive and flows in the clockwise direction to be negative. On each inside shared edge (Fig. 6) the flow gets counted once as positive and once as negative so the sum of those two flows is 0. But this happens along every inside edge. If we add all of the curls together, the only flows that are not cancelled out in this way are the flows along the exterior edges of the collection, the flows along C (Fig. 7). This total flow around the boundary C of the collection is called the circulation.

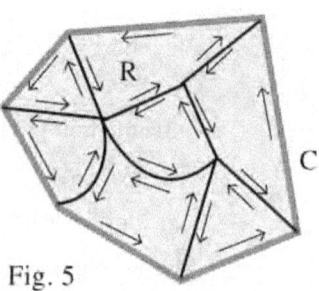

Fig. 5

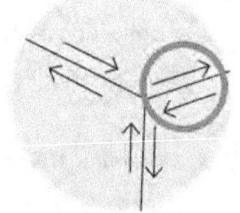

Fig. 6

In words: The circulation around the boundary C equals
           the sum of the circulations (curls) on the cells of R.

Finite version (Green's Theorem): $\sum_C \{\text{flow along outside edge}\} = \sum_R \text{curls } dA$

Integral version (curl-circulation form Green's Theorem): For $\mathbf{F} = \langle M(x,y), N(x,y) \rangle$

$$\oint_C \mathbf{F} \cdot \mathbf{T} \, ds = \iint_R \text{curl } \mathbf{F} \, dA = \iint_R \left(\frac{\partial N}{\partial x} - \frac{\partial M}{\partial y} \right) dA$$

If we could just take limits as all of the cells got smaller and smaller (it is not so easy) we would have both versions Green's Theorem which is discussed in Section 15.5:

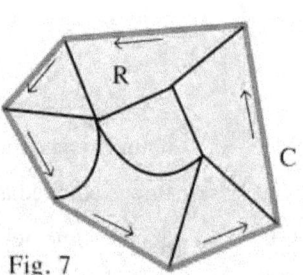

Fig. 7

$$\iint_R \left(\frac{\partial M}{\partial x} + \frac{\partial N}{\partial y} \right) = \oint_C \mathbf{F} \cdot \mathbf{n} \, d = \int_C M \, dy - N \, dx = \text{flux}$$

$$\iint_R \left(\frac{\partial N}{\partial x} - \frac{\partial M}{\partial y} \right) = \oint_C \mathbf{F} \cdot \mathbf{T} \, ds = \int_C M \, dx + N \, dy = \text{circulation}$$

Problems

1. For Fig. 8 verify that the sum of the divergences of all of the cells is equal to the flux across the boundary of the region.

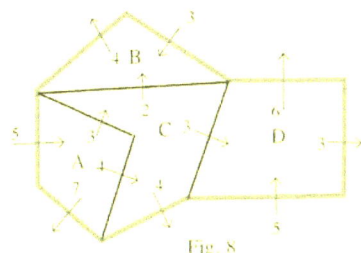

Fig. 8

2. For Fig. 9 verify that the sum of the curls of all of the cells is equal to the circulation around the boundary of the region.

Answer 1: A divergence = 9, B div = –1, C div = 2, D div = 1, so $\sum \text{div} = 11$.

Flux across the boundary = (7)+(4)+(–5)+(3)+(6)+(–3)+(4)+(–5)=11.

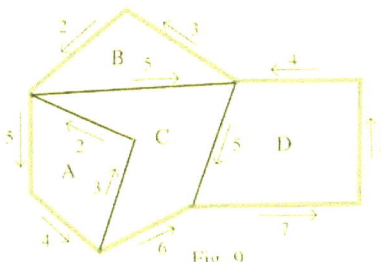

Fig. 9

Answer 2: A curl = 14, B curl = 10,

C curl =(6)+(–5)+(–5)+(–2)+(–3)= –9 , D curl = 19, so $\sum \text{curl} = 34$.

Circulation around the boundary = 34.

The following theorems of Stokes and Gauss extend Green's Theorem to higher dimensions.

Stokes' Theorem: Curl and Circulation

In Green's Theorem R was a planar region with boundary curve C (Fig. 10). Now suppose that the region R is a soap film and the boundary C is a rigid wire. If we gently blow on R to create an oriented, smooth surface with the same boundary C then each cell in the region R becomes a cell on

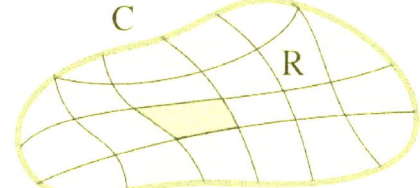

Fig. 10: Flat region R with boundary curve C

the surface S. Just like in Green's Theorem, on each inside shared edge (Fig. 11) the flow gets counted once as positive and once as negative so the sum of those two flows is 0. But this happens along every inside edge. If we add all of the curls together, the only flows that are not cancelled out in this way are the flows along the exterior edges of the collection, the flows along the boundary C. This total flow around the boundary C is called the circulation.

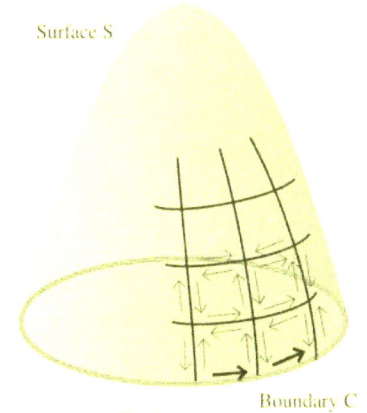

Fig. 11

In words: The circulation around the boundary C equals
the sum of the circulations (curls) on the cells of surface S.

Finite version (Stokes' Theorem):

$$\sum_C \{\text{flow along outside edge}\} = \sum_R \text{curls } dA$$

Integral version (Stokes' Theorem): $\oint_C \mathbf{F} \cdot \mathbf{T} \, ds = \oint_C \mathbf{F} \cdot d\mathbf{r} = \iint_S \text{curl } \mathbf{F} \cdot \mathbf{n} \, dS$

Gauss/Divergence Theorem: Flux and Divergence

In Green's Theorem R was a planar region with boundary curve C, and the sum of the internal cell divergences was equal to the flux across the boundary C. Now suppose that instead of a region R in 2D there is a solid region E in 3D, a volume, and that the boundary (skin) of E is a surface S. Also imagine that E is partitioned into little 3D cells, and that each of these internal cells has a divergence, a net outward flow.

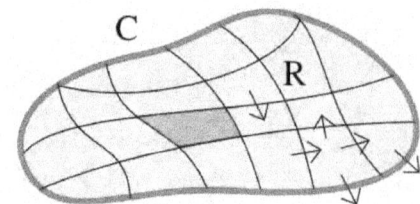

Fig. 12: Flat region R with boundary curve C

Just like in Green's Theorem, on each inside shared cell face the flow gets counted once as positive and once as negative so the sum of those two flows is 0. But this happens along every inside cell face. If we add all of the flows (divergences) for each cell together, the only flows that are not cancelled out in this way are the flows across the exterior faces of the collection, the flows across the boundary surface S. This total flow across the boundary S of the solid E is called the flux across S.

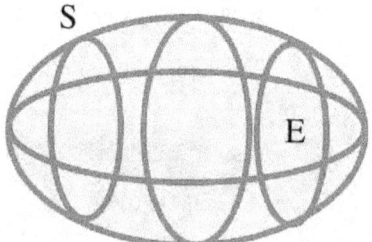

Fig. 13: E is the 3D region enclosed by the 2D boundary (skin) S

In words: The flux across the boundary S equals
the sum of the divergences on the cells of solid E.

Finite version (Gauss/Divergence Theorem):

flux across S = $\sum_S \mathbf{F} \bullet \mathbf{n} \, dA = \sum_E \text{divs} \, dV$

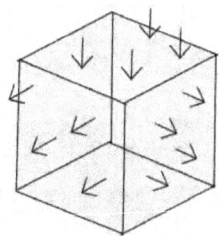

Fig. 14: cell in E

Integral Version: flux across S = $\iint_S \mathbf{F} \bullet \mathbf{n} \, dA = \iiint_E \text{div } \mathbf{F} \, dV$

Wrap up

In the following sections these theorems will be more carefully presented and partially proved, and we will actually do calculations using them. These are the final big three theorems of calculus, and they are both beautiful and very useful.

15.6 Green's Theorem

Green's Theorem makes statements about the equivalence between what happens on the boundary of a 2D region with what is happening on the inside of the region. It says, in two ways, that a single integral around the boundary is equal to a double integral over the region enclosed by the boundary. In this way Green's Theorem is similar to the Fundamental Theorem of Calculus which relates the area above an interval to the values of the antiderivatives at the boundary (endpoints) of the interval.

As with some other theorems in mathematics, Green's Theorem allows us to trade one calculation for another one that may be easier. And it shows connections between ideas that do not seem to be related.

Green's Theorem:

If   C is a simple, closed, and piecewise smooth curve (Fig. 1) and

$\mathbf{F}(x,y) = \langle M(x,y), N(x,y) \rangle = M\mathbf{i} + N\mathbf{j}$ where M and N have continuous first partial derivatives in the region R enclosed by C,

then $\oint_C \mathbf{F} \cdot \mathbf{T}\, ds = \oint_C M\, dx + N\, dy = \iint_R \left(\frac{\partial N}{\partial x} - \frac{\partial M}{\partial y} \right) dA$ Circulation-Curl Form

and $\oint_C \mathbf{F} \cdot \mathbf{n}\, ds = \oint_C M\, dy - N\, dx = \iint_R \left(\frac{\partial M}{\partial x} + \frac{\partial N}{\partial y} \right) dA$.   Flux-Divergence Form

---

Notation: The little circle on the integral sign indicates that C is a closed path parameterized in the counterclockwise direction..

The first conclusion says that the circulation is the accumulation of the interior curl. The second says that the flux is the accumulation of the interior divergence. It may help you make these connections by remembering that both circulation and curl deal with rotation while both flux and divergence deal with dissipation.

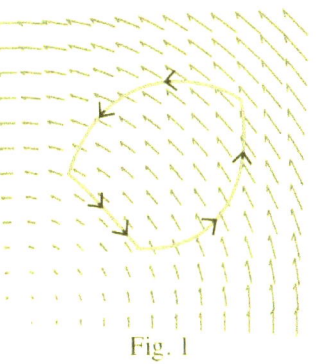

Fig. 1

A proof for simple regions is given in the Appendix of this section.

Example 1: For $\mathbf{F} = \langle x - y, x \rangle$ and $\mathbf{r}(t) = \langle 2 \cdot \cos(t), 2 \cdot \sin(t) \rangle$ (Fig. 2),

evaluate $\int_C (M\, dx + N\, dy)$ and $\iint_R \left(\frac{\partial N}{\partial x} - \frac{\partial M}{\partial y} \right) dA$.

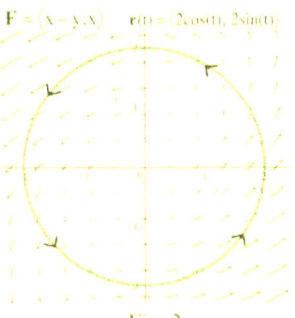

Fig. 2

Solution: First we need to write everything in terms of t:

$$M = 2\cos(t) - 2\sin(t), \quad N = 2\cos(t), \quad dx = -2\sin(t)\,dt, \quad dy = 2\cos(t)\,dt. \text{ Then}$$

$$\int_C (M\,dx + N\,dy) = \int_{t=0}^{2\pi} (2\cos(t) - 2\sin(t))(-2\sin(t)) + (2\cos(t))(2\cos(t))\,dt = \int_{t=0}^{2\pi} 4 - 4\cos(t)\sin(t)\,dt = 8\pi.$$

The circulation is 8π.

$\dfrac{\partial N}{\partial x} = 1$, $\dfrac{\partial M}{\partial y} = -1$ so $\dfrac{\partial N}{\partial x} - \dfrac{\partial M}{\partial y} = 2$ and $\iint_R (2)\,dA = 2(\text{area of circle of radius 2}) = 8\pi$.

As promised by the first conclusion of Green's Theorem, these two values are equal, and in this case the double integral was much easier to evaluate.

Practice 1: For $\mathbf{F} = \langle x - y, x \rangle$ and $\mathbf{r}(t) = \langle 2\cos(t), 2\sin(t) \rangle$, evaluate $\int_C (M\,dy - N\,dx)$ and $\iint_R \left(\dfrac{\partial M}{\partial x} + \dfrac{\partial N}{\partial y} \right) dA$.

Example 2: Evaluate $\int_C M\,dy - N\,dx$ and $\iint_R \left(\dfrac{\partial M}{\partial x} + \dfrac{\partial N}{\partial y} \right) dA$ for $\mathbf{F} = \langle -y, x \rangle$ and the triangular region R bounded by the x-axis, the line x=2 and the line y=x (Fig. 3).

Solution: $\dfrac{\partial M}{\partial x} = 0$ and $\dfrac{\partial N}{\partial y} = 0$ so we immediately have $\iint_R \left(\dfrac{\partial M}{\partial x} + \dfrac{\partial N}{\partial y} \right) dA = 0.$

The flux is 0.

The boundary of R consists of 3 line segments:

$C_1 = \langle 2t, 0 \rangle$, $C_2 = \langle 2, 2t \rangle$, and $C_3 = \langle 2 - 2t, 2 - 2t \rangle$ with $0 \le t \le 1$. We need that choice for C_3 in order for the orientation to be counterclockwise.

On C_1, $\int_{t=0}^{1} M\,dy - N\,dx = \int_{t=0}^{1} (-0)(2) - (2t)(2)\,dt = \int_{t=0}^{1} -4t\,dt = -2.$

On C_2, $\int_{t=0}^{1} M\,dy - N\,dx = \int_{t=0}^{1} (-2t)(2) - (2)(0)\,dt = \int_{t=0}^{1} -4t\,dt = -2.$

On C_3, $\int_{t=0}^{1} M\,dy - N\,dx = \int_{t=0}^{1} (2t - 2)(-2) - (2 - 2t)(-2)\,dt = \int_{t=0}^{1} 8 - 8t\,dt = 4.$

So $\int_{C_1} + \int_{C_2} + \int_{C_3} = (-2) + (-2) + (4) = 0.$ Certainly the double integral was easier.

Practice 2: Evaluate $\int_C (M\,dx + N\,dy)$ and $\iint_R \left(\dfrac{\partial N}{\partial x} - \dfrac{\partial M}{\partial y} \right) dA$ for $\mathbf{F} = \langle -y, x \rangle$ and the triangular region R bounded by the x-axis, the line x=2 and the line y=x.

15.6 Green's Theorem

Green's Theorem can also be used to evaluate line integrals.

Example 3: Evaluate $\oint_C x^2 y \, dy - y^2 \, dx$ where C is the boundary of the rectangle

$R = \{(x,y): 0 \le x \le 2, 0 \le y \le 1\}$ oriented counterclockwise.

Solution: If we can match the form of the line integral with one of the forms of Green's Theorem then we can evaluate one double integral instead of the four line integrals around R.

If we use the Flux-Divergence form $\oint_C M \, dy - N \, dx = \iint_R \left(\frac{\partial M}{\partial x} + \frac{\partial N}{\partial y} \right) dA$ then we need

$M = x^2 y$ and $N = y^2$ so $\oint_C x^2 y \, dy - y^2 \, dx = \iint_R (2xy + 2y) \, dA = \int_0^1 \int_0^2 (2xy + 2y) \, dx \, dy = \int_0^1 8y \, dy = 4$.

Practice 3: Use the Circulation-Curl form of Green's Theorem to evaluate the same line integral on the same region R.

In the previous examples we always traded a line integral for a double integral, but sometimes the opposite trade is useful.

Using Green's Theorem to Find Area

If the boundary of R is a simple closed curve C then $\text{area of } R = \iint_R 1 \, dA$. If we can find M and N so

that $\frac{\partial M}{\partial x} + \frac{\partial N}{\partial y} = 1$ then we can use $\iint_R 1 \, dA = \iint_R \left(\frac{\partial M}{\partial x} + \frac{\partial N}{\partial y} \right) dA = \oint_C M \, dy - N \, dx$. Putting

$M = \frac{x}{2}$ and $N = \frac{y}{2}$ works so $\iint_R 1 \, dA = \iint_R \left(\frac{\partial M}{\partial x} + \frac{\partial N}{\partial y} \right) dA = \oint_C M \, dy - N \, dx$.

If the boundary of R is a simple closed curve C, then $\text{area of } R = \frac{1}{2} \oint_C x \, dy - y \, dx$.

Example 4: Use this result to determine the area of the elliptical region $\frac{x^2}{4} + \frac{y^2}{25} \le 1$.

Solution: The boundary of this region can be parameterized by $\mathbf{r}(t) = \langle 2\cos(t), 5\sin(t) \rangle$.

Then $\text{area} = \frac{1}{2} \oint_C (2\cos(t))(5\cos(t)) - (5\sin(t))(-2\sin(t)) \, dt = \frac{1}{2} \int_0^{2\pi} 10 \, dt = 10\pi$.

Practice 4: Use the same method to determine the area of the general elliptical region $\dfrac{x^2}{a^2} + \dfrac{y^2}{b^2} \leq 1$.

Green's Theorem in More General Regions

The proof of Green's Theorem in the Appendix is valid for simple regions R (Fig. 4) in the plane for which any line parallel to an axis cuts the region R in at most 2 points or along an edge of R.

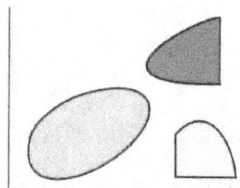

Fig. 4: Some simple regions

But Green's Theorem is true in much more complex regions if they can be decomposed into a union of simple regions.

If the region R is "bent," (Fig. 5) there are usually a finite number of cuts parallel to an axis so that R is the union of the some simple regions. Since the counterclockwise orientation on each simple region moves along each cut once in each direction (Fig. 6) the sum of those integral pieces is 0 and we are left with the integral around the original boundary of R.

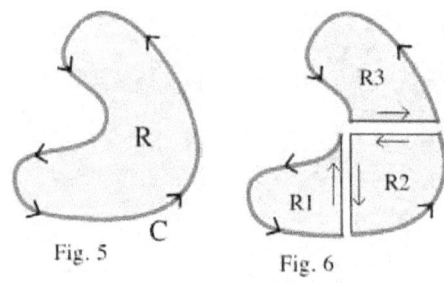

Fig. 5   Fig. 6

Practice 5: Decompose the region in Fig. 7 into several simple regions. Indicate the direction(s) of the paths along each cut.

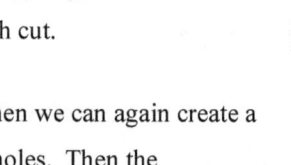

Fig. 7

Similarly, if R contains a finite number of holes (Fig. 8), then we can again create a single boundary for R by adding paths that connect to the holes. Then the integral along this new path will be the sum of the counterclockwise integrals around the outer boundary of R minus the sum of the counterclockwise integrals around the holes. The integrals along the added paths sum to 0 since they are traveled once in each direction. Remember, for a counterclockwise orientation of the curve C the region is always on our left hand side as we walk along C.

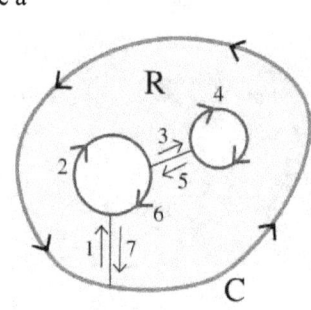

Fig. 8

Practice 6: Decompose the region in Fig. 9 into several simple regions. Indicate the directions and order of the paths along each cut.

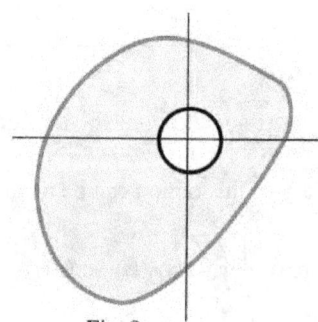

Fig. 9

Example 5: An interesting situation. $F(x,y) = \left\langle \dfrac{-y}{x^2+y^2}, \dfrac{x}{x^2+y^2} \right\rangle$.

Let C_1 be any simple closed curve that does not enclose the origin and let C_2 be a simple closed curve that does enclose the origin (Fig. 10). Use the Circulation-Curl form of Green's Theorem to calculate the circulations around C_1 and C_2.

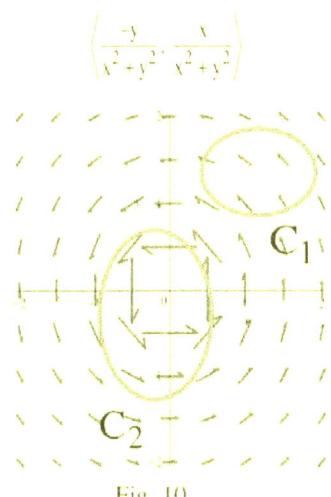

Fig. 10

Solution: On C_1 circulation

$$\oint_C \mathbf{F} \cdot \mathbf{T}\, ds = \oint_C M\, dx + N\, dy = \iint_R \left(\frac{\partial N}{\partial x} - \frac{\partial M}{\partial y} \right) dA.$$

$\dfrac{\partial M}{\partial y} = \dfrac{(x^2+y^2)(-1)-(-y)(2y)}{(x^2+y^2)^2} = \dfrac{y^2-x^2}{(x^2+y^2)^2}$, $\dfrac{\partial N}{\partial x} = \dfrac{(x^2+y^2)(1)-(x)(2x)}{(x^2+y^2)^2} = \dfrac{y^2-x^2}{(x^2+y^2)^2}$ so $\dfrac{\partial N}{\partial x} - \dfrac{\partial M}{\partial y} = 0$

so circulation $= \oint_C M\, dx + N\, dy = \iint_R \left(\dfrac{\partial N}{\partial x} - \dfrac{\partial M}{\partial y} \right) dA = \iint_R 0\, dA = 0$ on C_1.

Lets begin the "encloses the origin" by looking at the particular circle C that encloses the origin: $\mathbf{r}(t) = \langle h \cdot \cos(t), h \cdot \sin(t) \rangle$ for a positive value of h. In this case we can work with the line integral for circulation directly by putting everything in terms of t:

$$\text{circulation} = \oint_C M\, dx + N\, dy = \int_0^{2\pi} \left(\frac{-h \cdot \sin(t)}{h^2} \right)(-h \cdot \sin(t)) + \left(\frac{h \cdot \cos(t)}{h^2} \right)(h \cdot \cos(t))\, dt = \int_0^{2\pi} 1\, dt = 2\pi.$$

If C_2 is any simple closed curve that encloses the origin, we can take h small enough that the circle C is inside C_2. Then the region R bounded by D= the union of C_2 counterclockwise and C clockwise (Fig 10) does not contain the origin so $0 = \int_D = \int_{C_2} + \int_C = \int_{C_2} + (-2\pi)$

and $\{\text{circulation around } C_2\} = \oint_{C_2} M\, dx + N\, dy = 2\pi$.

For this vector field the circulation is 0 for any simple closed curve that does not surround the origin, and the circulation is always 2π for any simple, closed, positively-oriented curve that does surround the origin.

15.6 Green's Theorem

Problems

In problems 1 to 6 evaluate the line integral directly and by using Green's Theorem where C is positively oriented.

1. $\int_C x^2 y \, dx + 3y \, dy$ when C is the rectangle $0 \le x \le 2$, $0 \le y \le 1$.

2. $\int_C xy^2 \, dx + 5xy \, dy$ when C is the square $0 \le x \le 2$, $0 \le y \le 2$.

3. $\int_C 3xy \, dx + 2x^2 \, dy$ when C is the triangle with vertices (0,0), (1,0) and (1,2).

4. $\int_C x^2 \, dx + xy \, dy$ when C is the triangle with vertices (0,0), (0,2) and (2,2).

5. $\int_C ax \, dx + by \, dy$ when C is the circle $x^2 + y^2 = r^2$.

6. $\int_C ay \, dx + bx \, dy$ when C is the circle $x^2 + y^2 = r^2$.

In problems 7 to 10 use Green's Theorem to find the counterclockwise circulation and the outward flux for the field **F** and the curve C.

7. $\mathbf{F} = \langle x + 2y, y - x \rangle$, C is the square $0 \le x \le 2$, $0 \le y \le 2$.

8. $\mathbf{F} = \langle 3x + 2y, 4y - 5x \rangle$, C is the rectangle $0 \le x \le 3$, $0 \le y \le 1$.

9. $\mathbf{F} = \langle x^2 + y^2, x^2 - y^2 \rangle$, C is the triangle with vertices (0,0), (0,2) and (2,2).

10. $\mathbf{F} = \langle x^2 y, 3x + y^2 \rangle$, C is the triangle bounded by the lines $x=0$, $y=1$ and $x=2y$.

In problems 11 and 12 use Green's Theorem to find the area enclosed by the curve C.

11. C is given by $\mathbf{r}(t) = \langle t, t^2 \rangle$ as t goes from -2 to 2 and by $\mathbf{r}(t) = \langle t, 8 - t^2 \rangle$ as t goes from 2 to -2.

12. C is the curve bounded by the x-axis and the cycloid
$\mathbf{r}(t) = \langle A(t - \sin(t)), A(1 - \cos(t)) \rangle$.

13. $\mathbf{F} = \langle M, N \rangle$ and $\frac{\partial N}{\partial x} - \frac{\partial M}{\partial y} = 5$ on region R in Fig. P13. The area of R is 100, and $\int_{C_2} \mathbf{F} \cdot d\mathbf{r} = 20$. Use Green's Theorem to determine $\int_{C_1} \mathbf{F} \cdot d\mathbf{r}$.

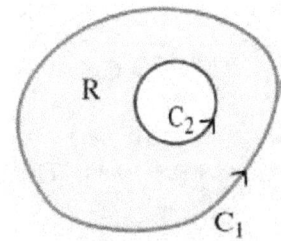

Fig. P13

14. $\mathbf{F} = \langle M, N \rangle$ and $\dfrac{\partial N}{\partial x} - \dfrac{\partial M}{\partial y} = 7$ on region R (inside C_1, outside C_2) in Fig.

    P14. If $\displaystyle\int_{C_2} \mathbf{F} \bullet d\mathbf{r} = 3\pi$, use Green's Theorem to determine $\displaystyle\int_{C_1} \mathbf{F} \bullet d\mathbf{r}$.

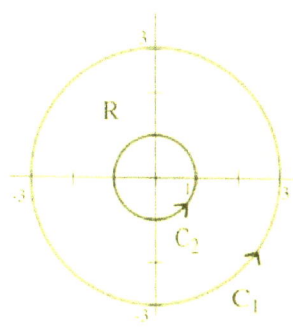

Fig. P14

15. $\mathbf{F} = \langle M, N \rangle$ and $\dfrac{\partial N}{\partial x} - \dfrac{\partial M}{\partial y} = 9$ on region R in Fig. P15,

    $\displaystyle\int_{C_2} \mathbf{F} \bullet d\mathbf{r} = 3\pi$ and $\displaystyle\int_{C_3} \mathbf{F} \bullet d\mathbf{r} = 4\pi$. Use Green's

    Theorem to determine $\displaystyle\int_{C_1} \mathbf{F} \bullet d\mathbf{r}$.

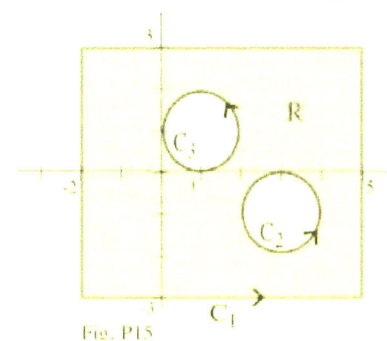

Fig. P15

16. $\mathbf{F} = \langle M, N \rangle$ and $\dfrac{\partial N}{\partial x} - \dfrac{\partial M}{\partial y} = 5$ on region R in Fig. P16, and

    $\displaystyle\int_{C_2} \mathbf{F} \bullet d\mathbf{r} = 2\pi$. Use Green's Theorem to determine $\displaystyle\int_{C_1} \mathbf{F} \bullet d\mathbf{r}$.

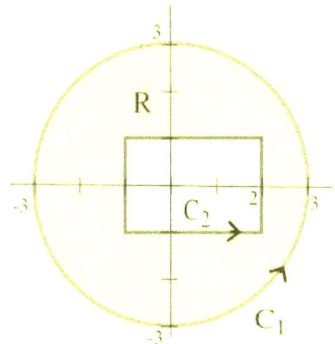

17. Show that the circulation and flux of a constant field $\mathbf{F} = \langle a, b \rangle$ are 0 over every simply connected region R.

18. Show that the flux across any counterclockwise oriented simple closed curve C of a linear vector field $\mathbf{F} = \langle ax + by, cx + dy \rangle$ is always a constant multiple of the area of the region enclosed by C. Find the constant.

Fig. P16

19. Show that the circulation around any counterclockwise oriented simple closed curve C of a linear vector field $\mathbf{F} = \langle ax + by, cx + dy \rangle$ is always a constant multiple of the area of the region enclosed by C. Find the constant.

Practice Answers

Practice 1: $M = 2\cdot\cos(t) - 2\cdot\sin(t)$, $N = 2\cdot\cos(t)$, $dx = -2\sin(t)\,dt$, $dy = 2\cdot\cos(t)\,dt$, $\dfrac{\partial M}{\partial x} = 1$, $\dfrac{\partial N}{\partial y} = 0$

$$\int_C (M\,dy - N\,dx) = \int_{t=0}^{2\pi} ((2\cdot\cos(t) - 2\cdot\sin(t))(2\cdot\cos(t)) - (2\cdot\cos(t))(-2\sin(t)))\,dt = \int_{t=0}^{2\pi} 4\cdot\cos^2(t)\,dt = 4\pi.$$

$$\iint_R \left(\frac{\partial M}{\partial x} + \frac{\partial N}{\partial y}\right)\,dA = \iint_R (1+0)\,dA = \iint_R (1)\,dA = 1(\text{area of circle of radius 2}) = 4\pi.$$

As promised by the second conclusion of Green's Theorem, these two values are equal.

Practice 2: $\dfrac{\partial N}{\partial x} = 1$ and $\dfrac{\partial M}{\partial y} = -1$ so $\iint_R \left(\dfrac{\partial N}{\partial x} - \dfrac{\partial M}{\partial y}\right)\,dA = \iint_R (2)\,dA = 2(\text{triangle area}) = 2(2) = 4$.

The circulation is 4.

$\mathbf{F} = \langle -y, x\rangle$. $C_1 = \langle 2t, 0\rangle$, $C_2 = \langle 2, 2t\rangle$, and $C_3 = \langle 2-2t, 2-2t\rangle$ with $0 \le t \le 1$.

On C_1, $\displaystyle\int_{t=0}^{1} M\,dx + N\,dy = \int_{t=0}^{1} (-0)(2) + (2t)(0)\,dt = \int_{t=0}^{1} 0\,dt = 0$.

On C_2, $\displaystyle\int_{t=0}^{1} M\,dx + N\,dy = \int_{t=0}^{1} (-2t)(0) + (2)(2)\,dt = \int_{t=0}^{1} 4\,dt = 4$.

On C_3, $\displaystyle\int_{t=0}^{1} M\,dx + N\,dy = \int_{t=0}^{1} (2t-2)(-2) + (2-2t)(-2)\,dt = \int_{t=0}^{1} 0\,dt = 0$.

$\displaystyle\int_{C_1} + \int_{C_2} + \int_{C_3} = (0) + (4) + (0) = 4$. Again the double integral was easier.

The flows along C_1 and C_3 make sense in terms of Fig. 3. Along C_1 and C_3 the vector field is perpendicular to the boundary so there is no flow.

Practice 3: $\displaystyle\oint_C x^2 y\,dy - y^2\,dx = \oint_C (M\,dx + N\,dy) = \iint_R \left(\dfrac{\partial N}{\partial x} - \dfrac{\partial M}{\partial y}\right)\,dA$ so $M = -y^2$ and $N = x^2 y$.

$\displaystyle\oint_C x^2 y\,dy - y^2\,dx = \iint_R (2xy + 2y)\,dA = \int_0^1 \int_0^2 (2xy + 2y)\,dx\,dy = \int_0^1 8y\,dy = 4$.

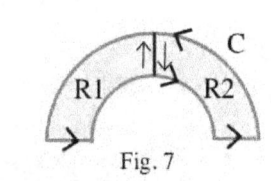

Fig. 7

Practice 4: Take $\mathbf{r}(t) = \langle a\cdot\cos(t), b\cdot\sin(t)\rangle$. Then

$$\text{area} = \frac{1}{2}\oint_C (a\cdot\cos(t))(b\cdot\cos(t)) - (b\cdot\sin(t))(-a\cdot\sin(t))\,dt = \frac{1}{2}\int_0^{2\pi} ab\,dt = ab\pi.$$

Practice 5: See Fig. 7

Practice 6: Fig. 9 shows one solution.

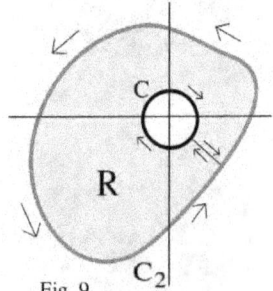

Fig. 9

Appendix A: Proof of Green's Theorem for Simple Regions

A simple region R is one in which lines parallel to an axis intersect the boundary of R in at most two places (Fig. A1).

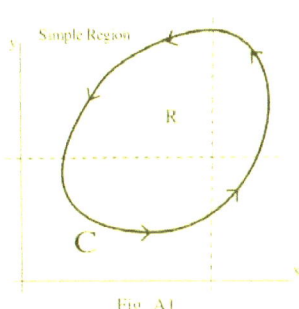

Fig. A1

Label Fig. A2 so $f_1(x) \leq y \leq f_2(x)$ for $a \leq x \leq b$. C_1 is the curve $y = f_1(x)$ oriented counterclockwise as x goes from a to b, and C_2 is $y = f_2(x)$ which has a counterclockwise orientation as x goes from b to a. The closed curve C is the union of C_1 and C_2. With labeling we will compute $\oint_C M\, dx$ and $\iint_R \frac{\partial M}{\partial y}\, dA$.

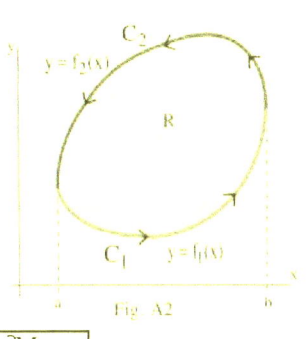

Fig. A2

$$\oint_C M\, dx = \oint_{C_1} M\, dx + \oint_{C_2} M\, dx = \int_a^b M(x,f_1)\, dx + \int_b^a M(x,f_2)\, dx$$

$$= \int_a^b M(x,f_1)\, dx - \int_a^b M(x,f_2)\, dx = \int_a^b M(x,f_1) - M(x,f_2)\, dx$$

$$\iint_R \frac{\partial M}{\partial y}\, dA = \int_a^b \int_{f_1}^{f_2} \frac{\partial M}{\partial y}\, dy\, dx = \int_a^b M(x,f_2) - M(x,f_1)\, dx \quad \text{so} \quad \boxed{\oint_C M\, dx = -\iint_R \frac{\partial M}{\partial y}\, dA} \quad (1)$$

To get the other part of the result we need, re-label the region R as in Fig. A3 so $x = g_1(y)$ for $c \leq y \leq d$. C_1 is the curve $x = g_1(x)$ oriented counterclockwise as y goes from c to d, and C_2 is $x = g_2(y)$ which has counterclockwise orientation as y goes from d to c. With this labeling we will compute $\oint_C N\, dy$ and $\iint_R \frac{\partial N}{\partial x}\, dA$.

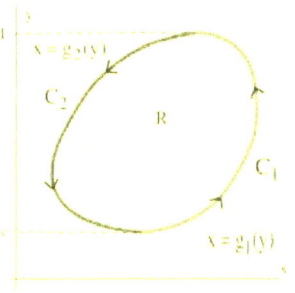

Fig. A3

$$\oint_C N\, dy = \oint_{C_1} N\, dy + \oint_{C_2} N\, dy = \int_c^d N(g_2,y)\, dy + \int_d^c N(g_1,y)\, dy$$

$$= \int_c^d N(g_2,y)\, dy - \int_c^d N(g_1,y)\, dy = \int_c^d N(g_2,y) - N(g_1,y)\, dy$$

$$\iint_R \frac{\partial N}{\partial x}\, dA = \int_c^d \int_{g_1}^{g_2} \frac{\partial N}{\partial x}\, dx\, dy = \int_c^d N(g_2,y) - N(g_1,y)\, dy \quad \text{so} \quad \boxed{\oint_C N\, dy = \iint_R \frac{\partial N}{\partial x}\, dA} \quad (2)$$

Adding result (1) and result (2)), we have $\boxed{\oint_C M\, dx + N\, dy = \iint_R \frac{\partial N}{\partial x} - \frac{\partial M}{\partial y}\, dA}$, the Circulation-Curl form of Green's Theorem.

Using a similar approach, you can show that $-\oint_C N\, dx = \iint_R \frac{\partial N}{\partial y}\, dA$ and $\oint_C M\, dy = \iint_R \frac{\partial M}{\partial x}\, dA$ so $\oint_C M\, dy - N\, dx = \iint_R \frac{\partial M}{\partial x} + \frac{\partial N}{\partial y}\, dA$, the Flux-Divergence form of Green's Theorem.

On a Modified Simple Region

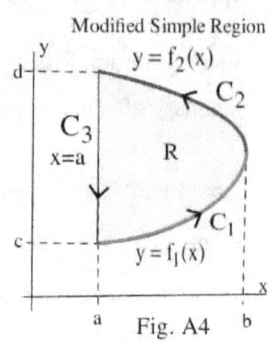
Modified Simple Region

Suppose the region R has one edge that if parallel to the y-axis (Fig. A4) labeled so $f_1(x) \leq y \leq f_2(x)$ for $a \leq x \leq b$. C_1 is the curve $y = f_1(x)$ oriented counterclockwise as x goes from a to b, and C_2 is $y = f_2(x)$ which has a counterclockwise orientation as x goes from b to a. C_3 is oriented counterclockwise with x=a as y goes from d to . The closed curve C is the union of C_1, C_2 and C_3.

$\oint_C M\,dx$ and $\iint_R \frac{\partial M}{\partial y}\,dA$.

$$\oint_C M\,dx = \oint_{C_1} M\,dx + \oint_{C_2} M\,dx + \oint_{C_3} M\,dx = \int_a^b M(x,f_1)\,dx + \int_b^a M(x,f_2)\,dx + \int_a^a M(x,f_2)\,dx$$

$$= \int_a^b M(x,f_1)\,dx + \int_b^a M(x,f_2)\,dx = \int_a^b M(x,f_1)\,dx - \int_a^b M(x,f_2)\,dx = \int_a^b M(x,f_1)\,dx - M(x,f_2)\,dx \cdot$$

$$\iint_R \frac{\partial M}{\partial y}\,dA = \int_a^b \int_{f_1}^{f_2} \frac{\partial M}{\partial y}\,dy\,dx = \int_a^b M(x,f_2) - M(x,f_1)\,dx \quad \text{so} \quad \oint_C M\,dx = -\iint_R \frac{\partial M}{\partial y}\,dA.$$

To get the other part of the result we need, label the region R as in Fig. A5 so C_1 is the curve $x = g(y)$ oriented counterclockwise as y goes from c to d, and C_2 is $x = a$ which has counterclockwise orientation as y goes from d to c. With this labeling we will compute $\oint_C N\,dy$ and $\iint_R \frac{\partial N}{\partial x}\,dA$.

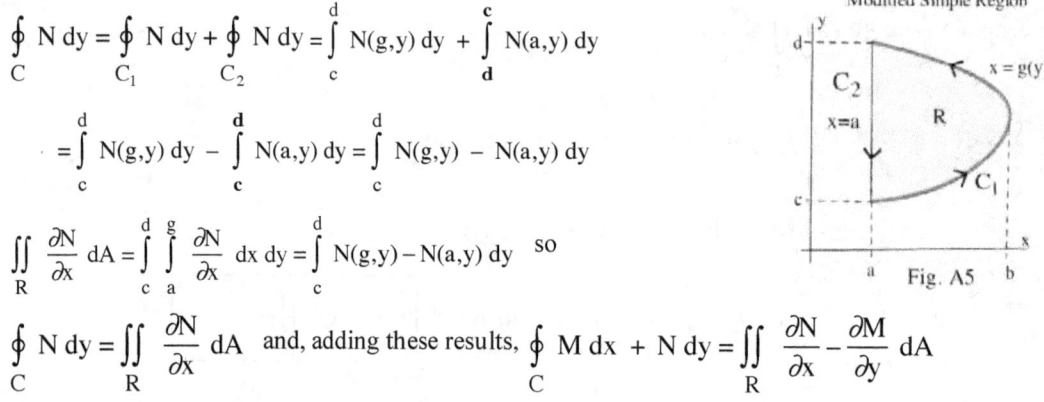

Fig. A5

Modified Simple Region

$$\oint_C N\,dy = \oint_{C_1} N\,dy + \oint_{C_2} N\,dy = \int_c^d N(g,y)\,dy + \int_d^c N(a,y)\,dy$$

$$= \int_c^d N(g,y)\,dy - \int_c^d N(a,y)\,dy = \int_c^d N(g,y) - N(a,y)\,dy$$

$$\iint_R \frac{\partial N}{\partial x}\,dA = \int_c^d \int_a^g \frac{\partial N}{\partial x}\,dx\,dy = \int_c^d N(g,y) - N(a,y)\,dy \quad \text{so}$$

$\oint_C N\,dy = \iint_R \frac{\partial N}{\partial x}\,dA$ and, adding these results, $\oint_C M\,dx + N\,dy = \iint_R \frac{\partial N}{\partial x} - \frac{\partial M}{\partial y}\,dA$

Other "simple" regions can be handled in similar ways.

The proof for general regions is difficult and is not included here.

If the region R lies on a plane in 3D (R is flat), then after a rotation of axes R can be made to lie in a new x'y'-plane and Green's Theorem applies.

15.7 Divergence and Curl in 3D

The divergence and curl were introduced in Section 15.2 for a 2D vector field F since it is easier in 2D to visualize what they measure and because we only needed the 2D versions for Green's Theorem in Section 15.5. Here we extend those definitions to a 3D vector field.

Divergence: div **F** in 3D

Assume that the vector field F describes the flow of a liquid in 3D. The **divergence** is the rate per unit volume that the water dissipates (**departs**, leaves) at the point P. A positive value for the divergence means that more water is leaving at P than is entering at P. But we can't really see a point so imagine a small sphere C centered at P, and consider whether more water s leaving or entering this sphere. Fig. 1(a) shows more water leaving than entering the sphere around P so div **F**(P) > 0, Fig. 1(b) has more entering than leaving so div **F**(P) < 0, and Fig, 1(c) shows the same amount leaving as entering so div **F** (P) = 0. This is much more difficult to visualize in 3D than in 2D, but the calculations are not any harder.

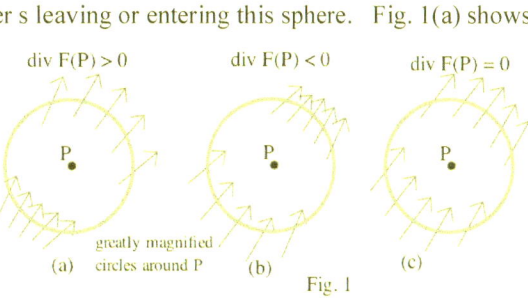

Fig. 1

Example 1: Fig. 2 shows the radial field $F(x,y,z) = \langle x,y,z \rangle$. Based on this figure, is div **F** positive, negative or zero at P=(1,2,0), Q=(1,0,2) and R=(0,0,0).

Solution: The vectors are increasing in magnitude as we move away from the origin so more are leaving little spheres at each of these points (in fact, at every point in 3D) so div **F** > 0 at every point in 3D.

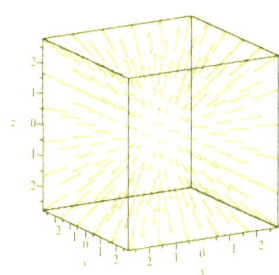

Fig. 2: $\langle x, y, z \rangle$

Definition: Divergence F in 3D   div **F**

For a vector field $\mathbf{F}(x,y,z) = M\mathbf{i} + N\mathbf{j} + P\mathbf{k}$ with continuous partial derivatives,

the divergence of F at Point P is $\operatorname{div} \mathbf{F}(P) = \dfrac{\partial M}{\partial x} + \dfrac{\partial N}{\partial y} + \dfrac{\partial P}{\partial z}$.   div **F** = $\nabla \bullet$ **F**

Example 2: Calculate div **F** for $\mathbf{F}(x,y,z) = \langle x,y,z \rangle$ at the points in Example 1.

Solution: $\operatorname{div} \mathbf{F} = \dfrac{\partial M}{\partial x} + \dfrac{\partial N}{\partial y} + \dfrac{\partial P}{\partial z} = 1+1+1 = 3$ at every point (x,y,z) so more water is leaving the is entering every tiny sphere in 3D.

Practice 1: Calculate div **F** for $\mathbf{F}(x,y,z) = \langle x^2, y^2, z \rangle$ at the points at (1,1,1), (2, –3,4) and (–0.5, 0,3).

Divergence of the inverse-square radial field $\mathbf{F} = \langle x, y, z\rangle/\sqrt{x^2+y^2+z^2}$

Radial inverse-square vector fields are very common in applications such as electricity and magnetism so it is worth calculating the divergence of such a field even if it is a bit tedious.

If $\mathbf{r} = \langle x, y, z\rangle$ then $|\mathbf{r}| = \sqrt{x^2+y^2+z^2}$ and the direction of r is $\dfrac{\mathbf{r}}{|\mathbf{r}|}$. If a field $\mathbf{F}$ has the same direction a r (a radial field) and follows an inverse-square law with magnitude $\dfrac{1}{|\mathbf{r}|^2}$ then $\mathbf{F} = \left(\dfrac{1}{|\mathbf{r}|^2}\right)\left(\dfrac{\mathbf{r}}{|\mathbf{r}|}\right) = \dfrac{\mathbf{r}}{|\mathbf{r}|^3}$.

$\dfrac{\partial |\mathbf{r}|^3}{\partial x} = \dfrac{\partial}{\partial x}\left(x^2+y^2+z^2\right)^{3/2} = 3x\cdot\left(x^2+y^2+z^2\right)^{1/2} = 3x|\mathbf{r}|$ and similarly for y and z, $\dfrac{\partial |\mathbf{r}|^3}{\partial y} = 3y|\mathbf{r}|$

and $\dfrac{\partial |\mathbf{r}|^3}{\partial z} = 3z|\mathbf{r}|$. But to calculate the divergence of $\dfrac{\mathbf{r}}{|\mathbf{r}|^3}$ we need the quotient rule for each partial

derivative. $\dfrac{\partial}{\partial x}\dfrac{x}{|\mathbf{r}|^3} = \dfrac{|\mathbf{r}|^3\left(\dfrac{\partial x}{\partial x}\right) - x\cdot\left(\dfrac{\partial |\mathbf{r}|^3}{\partial x}\right)}{\left(|\mathbf{r}|^3\right)^2} = \dfrac{|\mathbf{r}|^3\left(\dfrac{\partial x}{\partial x}\right) - x\cdot 3x|\mathbf{r}|}{\left(|\mathbf{r}|^3\right)^2} = \dfrac{1}{|\mathbf{r}|^3} - \dfrac{3x^2}{|\mathbf{r}|^5}$. Similarly

$\dfrac{\partial}{\partial x}\dfrac{y}{|\mathbf{r}|^3} = \dfrac{1}{|\mathbf{r}|^3} - \dfrac{3y^2}{|\mathbf{r}|^5}$ and $\dfrac{\partial}{\partial x}\dfrac{z}{|\mathbf{r}|^3} = \dfrac{1}{|\mathbf{r}|^3} - \dfrac{3z^2}{|\mathbf{r}|^5}$ so div $\mathbf{F}$ = div $\dfrac{\mathbf{r}}{|\mathbf{r}|^3} = \dfrac{3}{|\mathbf{r}|^3} - \dfrac{3(x^2+y^2+z^2)}{|\mathbf{r}|^5}$

$= \dfrac{3}{|\mathbf{r}|^3} - \dfrac{3|\mathbf{r}|^2}{|\mathbf{r}|^5} = 0$. The divergence of the inverse-square vector field $\mathbf{F}$ is 0 everywhere except at the origin where the field is not defined.

Curl: curl $\mathbf{F}$ in 3D

In section 15.2 the curl of a vector field F at a point P was introduced and described as a measure of the counterclockwise rotation of a small paddle wheel or circle at P caused by the vector field. (Fig. 3) We could also view this as vectors in the xy-plane causing rotation about an axis in the z direction. In 3D, instead of a small circle, imagine a small sphere that is fixed at its center and rotates about that center point (Fig. 4). The 3D curl is a vector with two important properties (these will proved in section 15.9 using Stoke's Theorem):

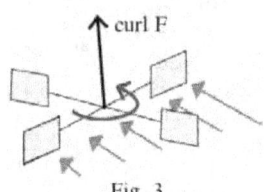

Fig. 3

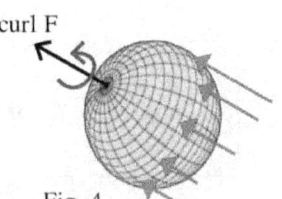

Fig. 4

* the magnitude of the curl gives the rate of the fluid's rotation, and
* the direction of the curl is normal to the plane of greatest circulation and points in the direction so that the circulation at the point has a right hand orientation (Fig. 5).

If we have a small paddle wheel at point P and tilt it in different directions, then the wheel will spin fastest with a right-hand orientation when the axis points in the direction of the curl vector.

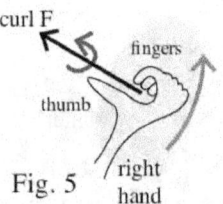

Fig. 5

15.7 Divergence and Curl in 3D

Definition: Curl F in 3D

For a vector field $\mathbf{F}(x,y,z) = M\mathbf{i} + N\mathbf{j} + P\mathbf{k}$ with continuous partial derivatives, then

the curl of F at Point P is $\quad \text{curl } \mathbf{F} = \left(\dfrac{\partial P}{\partial y} - \dfrac{\partial N}{\partial z}\right)\mathbf{i} + \left(\dfrac{\partial M}{\partial z} - \dfrac{\partial P}{\partial x}\right)\mathbf{j} + \left(\dfrac{\partial N}{\partial x} - \dfrac{\partial M}{\partial y}\right)\mathbf{k}$. $\quad$ curl $\mathbf{F} = \nabla \mathbf{x} \mathbf{F}$

Note: The k component of curl, $\dfrac{\partial N}{\partial x} - \dfrac{\partial M}{\partial y}$, is just the 2D curl F from section 15.2.

Example 3: Calculate curl $\mathbf{F}$ for $\mathbf{F}(x,y,z) = \langle y - z,\ z - 2x,\ x + 3z \rangle$.

Solution: $\quad \text{curl } \mathbf{F} = \nabla \mathbf{x} \mathbf{F} = \begin{vmatrix} \mathbf{i} & \mathbf{j} & \mathbf{k} \\ \dfrac{\partial}{\partial x} & \dfrac{\partial}{\partial y} & \dfrac{\partial}{\partial z} \\ y-z & z-2x & x+3z \end{vmatrix} = \langle 0-1,\ -(1+1),\ -2-1 \rangle = \langle -1,\ -2,\ -3 \rangle$.

The curl of this field is the same at every point in 3D, and the little paddle wheel will spin fastest if the axis of the wheel is oriented in the direction of $\langle -1,\ -2,\ -3 \rangle$.

Practice 2: Calculate curl $\mathbf{F}$ for $\mathbf{F}(x,y,z) = \langle 2y + z,\ x^2,\ 3z \rangle$.

Example 4: (a) Calculate the gradient vector field $\mathbf{F}$ of $f(x,y,z) = x^3 z + 3xy^2 + 4z$.

(b) Calculate curl $\mathbf{F}$.

Solution: (a) $\mathbf{F} = \nabla f = \langle 3x^2 z + 3y^2,\ 6xy,\ x^3 + 4 \rangle$.

(b) $\quad \text{curl } \mathbf{F} = \nabla \mathbf{x} \mathbf{F} = \begin{vmatrix} \mathbf{i} & \mathbf{j} & \mathbf{k} \\ \dfrac{\partial}{\partial x} & \dfrac{\partial}{\partial y} & \dfrac{\partial}{\partial z} \\ 3x^2 z + 3y^2 & 6xy & x^3 + 4 \end{vmatrix} = \langle 0-0,\ -(3x^2 - 3x^2),\ 6y - 6y \rangle = \langle 0,\ 0,\ 0 \rangle$.

The result in the previous example was not a lucky accident – the curl of every gradient field is 0 everywhere.

Theorem: If $\mathbf{F}$ is a conservative field ($\mathbf{F} = \nabla f$),

then curl $\mathbf{F} = \langle 0,\ 0,\ 0 \rangle$ at every point in 3D. curl $(\nabla f) = \mathbf{0}$

If curl $\mathbf{F} \neq \mathbf{0}$ at any point, then $\mathbf{F}$ is a not a conservative field.

Proof: If $\mathbf{F}$ is a gradient field then there is a potential function f so that $\mathbf{F} = \nabla f = \left\langle \dfrac{\partial f}{\partial x},\ \dfrac{\partial f}{\partial y},\ \dfrac{\partial f}{\partial z} \right\rangle$.

Then
$$\text{curl } \mathbf{F} = \nabla \times \mathbf{F} = \begin{vmatrix} \mathbf{i} & \mathbf{j} & \mathbf{k} \\ \frac{\partial}{\partial x} & \frac{\partial}{\partial y} & \frac{\partial}{\partial z} \\ \frac{\partial f}{\partial x} & \frac{\partial f}{\partial y} & \frac{\partial f}{\partial z} \end{vmatrix}.$$

The $\mathbf{i}$ component is $\frac{\partial}{\partial y}\left(\frac{\partial f}{\partial z}\right) - \frac{\partial}{\partial z}\left(\frac{\partial f}{\partial y}\right) = 0$ by Clairaut's Theorem of Mixed Partial Derivatives.

You can easily verify that the $\mathbf{j}$ and $\mathbf{k}$ components of this curl are also 0.

Practice 3: Show that $\mathbf{F} = \langle -y, x, z \rangle$ is not a conservative vector field.

Unfortunately, knowing that curl F = 0 is not sufficient to guarantee that F is a conservative field. However there is a partial converse to the previous theorem.

> **Theorem:** If $\mathbf{F}$ is defined and has continuous partial derivatives at every point in 3D and curl $\mathbf{F} = 0$
> then $\mathbf{F}$ is a conservative field.

The proof requires Stoke's Theorem and is not given here.

Curls of the radial and inverse-square radial fields

We could use the definition of curl to calculate the curls for $\mathbf{r} = \langle x, y, z \rangle$ and $\mathbf{F} = \left(\frac{1}{|\mathbf{r}|^2}\right)\left(\frac{\mathbf{r}}{|\mathbf{r}|}\right) = \frac{\mathbf{r}}{|\mathbf{r}|^3}$ but it is much easier to recognize that both $\mathbf{r}$ and $\mathbf{F}$ are gradient fields and then invoke the theorem. It is easy to check that $\mathbf{r} = \nabla\left[\frac{1}{2}(x^2 + y^2 + z^2)\right]$ and $\mathbf{F} = \nabla\sqrt{x^2 + y^2 + z^2}$ so curl $\mathbf{r} = 0$ and curl $\mathbf{F} = 0$ everywhere where each of them is defined. $\mathbf{r}$ is defined everywhere and is a conservative field. $\mathbf{F}$ is not defined at the origin and is not a conservative field in a domain that includes the origin.

> **Theorem:** If $\mathbf{F}$ has continuous partial second derivatives, then div (curl $\mathbf{F}$) = 0.

Proof: The proof is straightforward and the last step depends on Clairaut's Theorem.

$$\text{div (curl } \mathbf{F}) = \text{div}\left\{\left(\frac{\partial P}{\partial y} - \frac{\partial N}{\partial z}\right)\mathbf{i} + \left(\frac{\partial M}{\partial z} - \frac{\partial P}{\partial x}\right)\mathbf{j} + \left(\frac{\partial N}{\partial x} - \frac{\partial M}{\partial y}\right)\mathbf{k}\right\}$$

$$= \frac{\partial}{\partial x}\left(\frac{\partial P}{\partial y} - \frac{\partial N}{\partial z}\right) + \frac{\partial}{\partial y}\left(\frac{\partial M}{\partial z} - \frac{\partial P}{\partial x}\right) + \frac{\partial}{\partial z}\left(\frac{\partial N}{\partial x} - \frac{\partial M}{\partial y}\right)$$

$$= \frac{\partial^2 P}{\partial x \partial y} - \frac{\partial^2 N}{\partial x \partial z} + \frac{\partial^2 M}{\partial y \partial z} - \frac{\partial^2 P}{\partial y \partial x} + \frac{\partial^2 N}{\partial z \partial x} - \frac{\partial^2 M}{\partial z \partial y} = 0$$

Problems

In problems 1 to 6, calculate the divergence and curl of each vector field.

1. $\mathbf{F} = \langle x^2 y, xyz, xz^3 \rangle$
2. $\mathbf{F} = \langle yz, xz, xy + 2 \rangle$
3. $\mathbf{F} = \langle x \cdot e^z, z \cdot e^y, y \cdot e^x \rangle$
4. $\mathbf{F} = \langle y, z, x^3 \rangle$
5. $\mathbf{F} = \langle x + y, y + z, z + x \rangle$
6. $\mathbf{F} = \langle x^2 + y^2, y^2 + z^2, z^2 + x^2 \rangle$

In problems 7 to 10, F is a vector field and f is a scalar function in 3D. Determine whether the given calculation is meaningful. If it is not, explain why. If it is, determine whether the result is a scalar or a vector.

7. div f , div F , div (curl F) , gradient (curl F), curl F
8. curl f , curl (div F) , gradient (div F) , div (curl (gradient f))
9. gradient f , curl (curl F) , curl (div (gradient f))
10. gradient F , curl (gradient F) , div (div F)

In problems 11 to 16, determine if the vector field F is conservative. If F is conservative, find a function f so the F=gradient f.

11. $\mathbf{F} = \langle x^2 y, xyz, xz^3 \rangle$
12. $\mathbf{F} = \langle yz, xz, xy + 2 \rangle$
13. $\mathbf{F} = \langle y, z, x^3 \rangle$
14. $\mathbf{F} = \langle y^2, 2xy, 3z^2 \rangle$
15. $\mathbf{F} = \langle \sin(y \cdot z), x \cdot z \cdot \cos(y \cdot z), x \cdot y \cdot \cos(y \cdot z) + 2 \rangle$
16. $\mathbf{F} = \langle y^3, 3xy^2 + 5z, 5y \rangle$

17. Suppose curl $\mathbf{F}(1,2,3) = \langle 2, 4, 1 \rangle$. If you are at the location (5, 10, 5) and look towards (1,2,3) will you see the rotation at (1,2,3) to be clockwise or counterclockwise?

18. Suppose curl $\mathbf{F}(5,1,4) = \langle -1, 3, 1 \rangle$. If you are at the location (2, 10, 7) and look towards (5,1,4) will you see the rotation at (5,1,4) to be clockwise or counterclockwise?

19. Suppose curl $\mathbf{F}(6,3,2) = \langle 2, -1, 1 \rangle$. If you are at the location (4, 4, 3) and look towards (6,3,2) will you see the rotation at (6,3,2) to be clockwise or counterclockwise?

20. Suppose curl $\mathbf{F}(7,-4,5) = \langle -1, -2, 3 \rangle$. If you are at the location (9, 0, -1) and look towards (7,-4,5) will you see the rotation at (7,-4,5) to be clockwise or counterclockwise?

Practice Answers

Practice 1: div $\mathbf{F} = 2x + 2y + 1$ so div $\mathbf{F}(1,1,1) = 5$, div $\mathbf{F}(2,-3,4) = -1$ and div $\mathbf{F}(-0.5,0,3) = 0$.

Practice 2:
$$\text{curl } \mathbf{F} = \nabla \times \mathbf{F} = \begin{vmatrix} \mathbf{i} & \mathbf{j} & \mathbf{k} \\ \dfrac{\partial}{\partial x} & \dfrac{\partial}{\partial y} & \dfrac{\partial}{\partial z} \\ 2y+z & x^2 & 3z \end{vmatrix} = \langle 0-0, -(0-1), 2x-2 \rangle = \langle 0, 1, 2x-2 \rangle .$$

Practice 3: $\mathbf{F} = \langle -y, x, z \rangle$. $\text{curl } \mathbf{F} = \begin{vmatrix} \mathbf{i} & \mathbf{j} & \mathbf{k} \\ \dfrac{\partial}{\partial x} & \dfrac{\partial}{\partial y} & \dfrac{\partial}{\partial z} \\ -y & x & z \end{vmatrix} = (0-0)\mathbf{i} + (0-0)\mathbf{j} + (1-(-1))\mathbf{k} = \langle 0,0,2 \rangle \neq \mathbf{0}$

so $\mathbf{F}$ is not a conservative field.

15.8 Parametric Surfaces

In earlier work we extended the basic idea of a function of one variable, y = f(x), to parametric equations in which both x and y were functions of a single parameter t: x=x(t) and y=y(t) (in 3D, z=(t), also). Basically this mapped a 1-dimensional object (think of a piece of wire) into 2 or 3 dimensions (Fig. 1). And by treating this mapping as a vector-valued function, $\mathbf{r}(t) = \langle x(t), y(t), z(t) \rangle$, we could use the ideas and tools of vectors. In this section we will do something similar, except now we will map a 2-dimension object (think of a sheet of paper) into 3 dimensions by treating x, y, and z as functions of two parameters, u and v: x=x(u,v), y=y(u,v) and z=z(u,v). These graphs will be surfaces in 3D. By using parametric surfaces we can work with more complicated and more general shapes (Fig. 2). And by treating these surfaces as vector-valued functions, $\mathbf{r}(u,v) = \langle x(u,v), y(u,v), z(u,v) \rangle$, we can again use the ideas and tools of vectors.

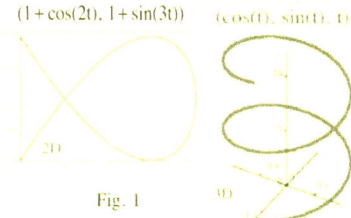

Fig. 1

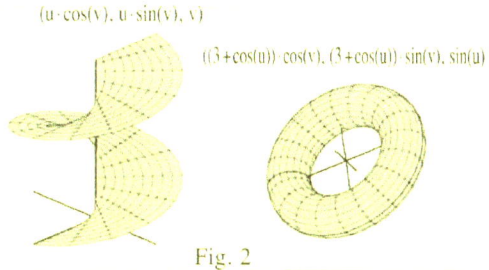

Fig. 2

Definitions: Parametric Function and Parametric Surface

Let x, y and z be functions of the parameters u and v for all (u, v) in a region D.
The vector-valued function $\mathbf{r}(u,v) = \langle x(u,v), y(u,v), z(u,v) \rangle$ is called a **parametric function** with domain D.
The set of points S = (x(u,v), y(u,v), z(u,v)) is called the **parametric surface** of the function **r** on domain D.

Note: As with the parametric functions of 1 variable, $\mathbf{r}(t) = \langle x(t), y(t), z(t) \rangle$, we will work with **r** as a vector $\langle x(u,v), y(u,v), z(u,v) \rangle$ but only plot the points S = (x(u,v), y(u,v), z(u,v)).

All of the rectangular, cylindrical and spherical coordinate functions we have used so far can be easily converted into parametric functions, and parametric functions give us even more freedom.

Example 1: (a) Convert $f(x,y) = x^2 + y^2$ with $-2 \le x \le 2$ and $-2 \le y \le 2$ into parametric form with u and v.

(b) Convert the spherical coordinate function (3, θ, φ), $0 \le \theta \le 2\pi$, $0 \le \varphi \le \pi/2$
(the top half of a sphere) into parametric form with u and v.

Solution: (a) Simply replace x with u and y with v and rewrite the z coordinate in terms of u and v: $(u, v, u^2 + v^2)$ with $-2 \leq u \leq 2$ and $-2 \leq v \leq 2$ (Fig. 3).

(b) $x = \rho \cdot \sin(\varphi) \cdot \cos(\theta)$, $y = \rho \cdot \sin(\varphi) \cdot \sin(\theta)$, $z = \rho \cdot \cos(\varphi)$ so we can replace

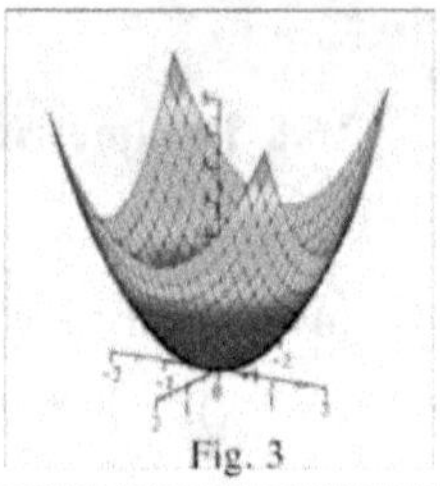

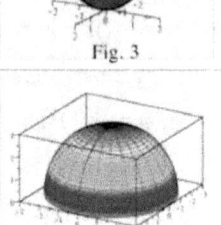

Fig. 3

θ and φ with u and v and rewrite x, y and z as

$x = 3 \cdot \sin(v) \cdot \cos(u)$, $y = 3 \cdot \sin(v) \cdot \sin(u)$, $z = 3 \cdot \cos(v)$

with $0 \leq u \leq 2\pi$, $0 \leq v \leq \pi/2$. (Fig. 4)

Fig. 4

Practice 1: Convert the cylindrical coordinate function

$(r, \theta, 1 + \sin(3\theta))$, $1 \leq r \leq 2$, $0 \leq \theta \leq 2\pi$ (Fig. 5) into parametric form with u and v.

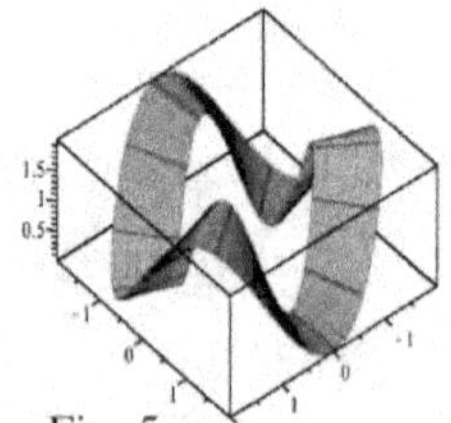

One common parametric surface is the torus which can be thought of as a small circular tube around a larger circle (Fig. 6). The parametric equation for a torus with large radius R and small radius r is

Fig. 5

$((R - r \cdot \cos(u)) \cdot \cos(v), (R - r \cdot \cos(u)) \cdot \sin(v), r \cdot \sin(u))$

The Appendix discusses the derivation of this surface as well as how to create a parametric representations of a small tubes around other curves in space.

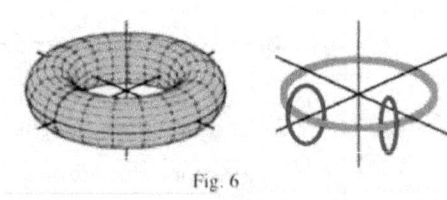

Fig. 6

Example 2: Write parametric equations for the surface generated by rotating the curve $z = \sqrt{y}$ around the y-axis for $0 \leq y \leq 4$. (Fig. 7)

Solution: Put y=u for $0 \leq u \leq 4$. Then the radius of the circle of revolution is $\sqrt{u}$ so $x = \sqrt{u} \cdot \cos(v)$ and $z = \sqrt{u} \cdot \sin(v)$ works.

Practice 2: Write parametric equations for the surface generated by rotating the curve $x = 2 + \sin(z)$ for $0 \leq z \leq 5$ around the z-axis.

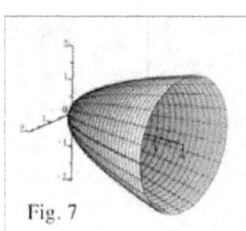

Fig. 7

Surface Area of a Parametric Surface

The derivation and result for the surface area of a parametric surface $\mathbf{r}(u,v)$ are similar to the method for a surface defined by $z = f(x,y)$ in Section 14.5. First partition the uv-domain D into small Δu by Δv rectangles (Fig. 8) with (u,v) at the lower left point of the rectangle. Call this rectangle R. Then $\mathbf{r}(u,v)$ maps this rectangle R onto a patch S on the parametric surface (Fig. 9). The bottom

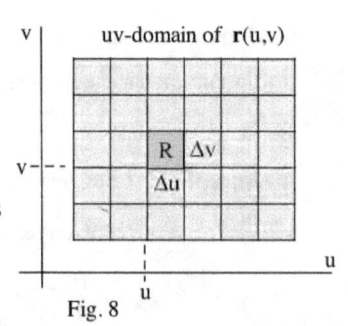

Fig. 8

corners of the R rectangle, (u,v) and (u+ Δu ,v), are mapped to **r**(u,v) and
r(u+ Δu,v). The left edge corners of R, (u,v) and (u,v+ Δv), are mapped
to **r**(u,v) and **r**(u, v+ Δv). The area of the patch S on the parametric surface
is approximated by the area of the rectangle whose corners are the images
under **r** of the corners of the R rectangle.

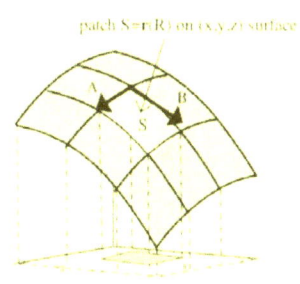

Fig. 9

Let A be the vector from **r**(u,v) and **r**(u+ Δu,v), and B be the vector from **r**
(u,v) and **r**(u, v+ Δv). Then (Fig. 10)

A= **r**(u+ Δu,v)− **r**(u,v) and **B**= **r**(u, v+ Δv)− **r** (u,v) so $\mathbf{A} = \dfrac{\mathbf{r}(u+\Delta u, v) - \mathbf{r}(u,v)}{\Delta u} \cdot \Delta u$ and

$\mathbf{B} = \dfrac{\mathbf{r}(u, v+\Delta v) - \mathbf{r}(u,v)}{\Delta v} \cdot \Delta v$. If Δu and Δv are small, then $\dfrac{\mathbf{r}(u+\Delta u, v) - \mathbf{r}(u,v)}{\Delta u} \approx \dfrac{\partial \mathbf{r}(u,v)}{\partial u} = \mathbf{r}_u(u,v)$

and $\dfrac{\mathbf{r}(u, v+\Delta v) - \mathbf{r}(u,v)}{\Delta v} \approx \dfrac{\partial \mathbf{r}(u,v)}{\partial v} = \mathbf{r}_v(u,v)$ so $\mathbf{A} \approx \mathbf{r}_u(u,v) \cdot \Delta u$ and $\mathbf{B} \approx \mathbf{r}_v(u,v) \cdot \Delta v$.

Finally, the area of the patch S is approximately $|\mathbf{r}_u \times \mathbf{r}_v| \cdot \Delta u \cdot \Delta v$. Summing over all of the
rectangles R in the uv-domain and taking limits as Δu and Δv both approach 0,

$$\sum_v \sum_u |\mathbf{r}_u \times \mathbf{r}_v| \cdot \Delta u \cdot \Delta v \to \iint_D |\mathbf{r}_u \times \mathbf{r}_v| \cdot dA .$$

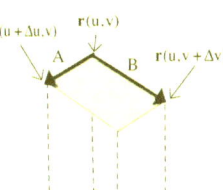

A = r(u+Δu,v) − r(u,v)
B = r(u,v+Δv) − r(u,v)
Fig. 10

Surface Area of a Parametric Surface

If S is a smooth surface given by **r** (u,v) = $\langle x(u,v), y(u,v), z(u,v) \rangle$ with uv-domain D,

then {surface area of S} = $\iint_D |\mathbf{r}_u \times \mathbf{r}_v| \cdot dA$.

(Note: The surface S may fold over on itself, but that will not happen if the R rectangle is very small.)

If z=f(x,y), then we could parameterize the surface by x=u, y=v and z=f(u,v), and it is straightforward to derive

that {Surface Area} = $\iint_R \sqrt{1 + \left(\dfrac{\partial z}{\partial x}\right)^2 + \left(\dfrac{\partial z}{\partial y}\right)^2} \cdot dA$ as we did in Section 14.5.

Proof: Parameterize this surface by x=u, y=v and z=f(u,v). Then **r**(u,v) = $\langle u, v, f \rangle$ so $\mathbf{r}_u = \dfrac{\partial \mathbf{r}}{\partial u} = \langle 1, 0, f_u \rangle$ and

$\mathbf{r}_v = \dfrac{\partial \mathbf{r}}{\partial v} = \langle 0, 1, f_v \rangle$. $\mathbf{r}_u \times \mathbf{r}_v = \langle -f_u, -f_v, 1 \rangle$ so $|\mathbf{r}_u \times \mathbf{r}_v| = \sqrt{(f_u)^2 + (f_v)^2 + 1}$ and the surface area

is $\iint_R \sqrt{1 + \left(\dfrac{\partial z}{\partial x}\right)^2 + \left(\dfrac{\partial z}{\partial y}\right)^2} \cdot dA$.

Example 3: Use the parametric surface form to find the surface area of the curve $z = \sqrt{y}$ around the y-axis for $0 \leq y \leq 4$.

Solution: This surface was parameterized in Example 2 by $y = u$ for $0 \leq u \leq 4$, $x = \sqrt{u} \cdot \cos(v)$ and $z = \sqrt{u} \cdot \sin(v)$ for $0 \leq v \leq 2\pi$. $\mathbf{r}_u = \dfrac{\partial \mathbf{r}}{\partial u} = \left\langle \dfrac{1}{2\sqrt{u}} \cdot \cos(v),\, 1,\, \dfrac{1}{2\sqrt{u}} \cdot \sin(v) \right\rangle$.

$\mathbf{r}_v = \dfrac{\partial \mathbf{r}}{\partial v} = \left\langle -\sqrt{u} \cdot \sin(v),\, 0,\, \sqrt{u} \cdot \cos(v) \right\rangle$, $\mathbf{r}_u \times \mathbf{r}_v = \left\langle \sqrt{u} \cdot \cos(v),\, \dfrac{1}{2},\, \sqrt{u} \cdot \sin(v) \right\rangle$ so

$|\mathbf{r}_u \times \mathbf{r}_v| = \sqrt{u \cdot \sin^2(v) + \dfrac{1}{4} + u \cdot \cos^2(v)} = \sqrt{\dfrac{1}{4} + u}$. Finally,

Surface area $= \iint\limits_D |\mathbf{r}_u \times \mathbf{r}_v| \cdot dA = \int\limits_{v=0}^{2\pi} \int\limits_{u=0}^{4} \sqrt{\dfrac{1}{4} + u}\; du\, dv$

$= \int\limits_{v=0}^{2\pi} \dfrac{2}{3}\left(\dfrac{1}{4} + u\right)^{3/2} \Big|_{u=0}^{4} dv = \left(\dfrac{2}{3}\left(\dfrac{17}{4}\right)^{3/2} - \dfrac{2}{3}\left(\dfrac{1}{4}\right)^{3/2} \right) \cdot 2\pi = \left(\dfrac{1}{12}(17)^{3/2} - \dfrac{1}{12} \right) \cdot 2\pi$

the same result as Practice 3 in Section 14.5.

And now we can extend surface area calculations to other systems such as cylindrical coordinates (r, θ, z).

Surface Area in Cylindrical Coordinates

If S is a smooth surface in cylindrical coordinates $(r, \theta, f(r, \theta))$ on domain D,

then $\{\text{surface area of S}\} = \iint\limits_D \sqrt{r^2(f_r)^2 + (f_\theta)^2 + r^2} \cdot dr \cdot d\theta$.

Proof: Parameterize this surface by $x = u \cdot \cos(v)$, $y = u \cdot \sin(v)$ (so $u = r$ and $v = \theta$) and $z = f(u \cdot \cos(v), u \cdot \sin(v))$. Then

$\mathbf{r}_u = \dfrac{\partial \mathbf{r}}{\partial u} = \langle \cos(v), \sin(v), f_u \rangle$ and $\mathbf{r}_v = \dfrac{\partial \mathbf{r}}{\partial v} = \langle -u \cdot \sin(v),\, u \cdot \cos(v),\, f_v \rangle$ so

$\mathbf{r}_u \times \mathbf{r}_v = \begin{vmatrix} \mathbf{i} & \mathbf{j} & \mathbf{k} \\ \cos(v) & \sin(v) & f_u \\ -u \cdot \sin(v) & u \cdot \cos(v) & f_v \end{vmatrix} = \langle f_v \cdot \sin(v) - f_u \cdot u \cdot \cos(v),\, -(f_v \cdot \cos(v) + f_u \cdot u \cdot \sin(v)),\, u \cdot \cos^2(v) + u \cdot \sin^2(v) \rangle$.

Finally, after some simplifying, $|\mathbf{r}_u \times \mathbf{r}_v| = \iint\limits_D \sqrt{u^2(f_u)^2 + (f_v)^2 + u^2} \cdot du \cdot dv$.

Example 4: Use the cylindrical coordinate integral form to calculate the surface area of the hemisphere $f(r, \theta) = \sqrt{R^2 - r^2}$ for $0 \leq r \leq R$ and $0 \leq \theta \leq 2\pi$.

15.8 Parametric Surfaces and Surface Area

Solution: $f_r = \dfrac{-r}{\sqrt{R^2 - r^2}}$ and $f_\theta = 0$ so

$$\iint_D \sqrt{r^2(f_r)^2 + (f_\theta)^2 + r^2} \cdot dr \cdot d\theta = \int_0^{2\pi} \int_0^R \dfrac{Rr}{\sqrt{R^2 - r^2}} \, dr \, d\theta = \int_0^{2\pi} -R \cdot \sqrt{R^2 - r^2} \Big|_0^R \, d\theta = \int_0^{2\pi} R^2 \, d\theta = 2\pi R^2.$$

The surface area of the entire sphere is $4\pi R^2$.

Practice 3: Use the parametric surface form to find the surface area of $z = x^2 - y^2$ for $0 \le x^2 + y^2 \le 4$.

Problems

For Problems 1 to 10, sketch the parametric surface.

1. $x = u$, $y = v$, $z = \sqrt{v}$, $0 \le u \le 2$, $0 \le v \le 4$.
2. $x = u$, $y = v$, $z = u^2$, $0 \le u \le 2$, $0 \le v \le 1$.
3. $x = 2 \cdot \cos(u)$, $y = 2 \cdot \sin(u)$, $z = v$, $0 \le u \le 2\pi$, $0 \le v \le 3$.
4. $x = \cos(u)$, $y = v$, $z = \sin(u)$, $0 \le u \le 2\pi$, $0 \le v \le 4$.
5. $x = u \cdot \cos(v)$, $y = u \cdot \sin(v)$, $z = u$, $0 \le u \le 2$, $0 \le v \le 2\pi$.
6. $x = u \cdot \cos(v)$, $y = u^2$, $z = u \cdot \sin(v)$, $0 \le u \le 2$, $0 \le v \le 2\pi$.
7. $x = u$, $y = \sqrt{u} \cdot \cos(v)$, $z = \sqrt{u} \cdot \sin(v)$, $0 \le u \le 4$, $0 \le v \le 2\pi$.
8. $x = \sqrt{u} \cdot \cos(v)$, $y = u$, $z = 2 + \sqrt{u} \cdot \sin(v)$, $0 \le u \le 4$, $0 \le v \le 2\pi$.
9. $x = u$, $y = v$, $z = \sin(v)$, $0 \le u \le 2$, $0 \le v \le 2\pi$.
10. $x = \sin(u)$, $y = v$, $z = u$, $0 \le u \le \pi$, $0 \le v \le 4$.

For problems 11 to 20, write the integral that represents the surface area of each surface in problems 1 to 10.

11. Write the surface area integral for the surface in problem 1.

12. Write the surface area integral for the surface in problem 2.

and so on for problems 13 to 20.

Practice Answers

Practice 1: $x = r \cdot \cos(\theta)$, $y = r \cdot \sin(\theta)$, $z = z$. Replace r and θ with u and v so we have ($u \cdot \cos(v)$, $u \cdot \sin(v)$, $1 + \sin(3v)$) with $1 \leq u \leq 2$, $0 \leq v \leq 2\pi$.

Practice 2: $x = (2 + \sin(u)) \cdot \cos(v)$, $y = (2 + \sin(u)) \cdot \sin(v)$, $z = u$ works.

Practice 3: Parameterize this surface by $x = r \cdot \cos(\theta)$, $y = r \cdot \sin(\theta)$ and
$f = r^2 \cdot \cos^2(\theta) - r^2 \cdot \sin^2(\theta) = r^2 \cdot \cos(2\theta)$ for $0 \leq r \leq 2$, $0 \leq \theta \leq 2\pi$. Then
$f_r = 2r \cdot \cos(2\theta)$, and $f_\theta = -2r^2 \cdot \sin(2\theta)$ so $r^2 (2r \cdot \cos(2\theta))^2 + (-4r^2 \cdot \sin(2\theta))^2 + r^2 = 4r^4 + r^2$

$$\text{Surface area} = \int_{\theta=0}^{2\pi} \int_{r=0}^{2} r \cdot \sqrt{4r^2 + 1} \, dr \, d\theta = \left[\frac{1}{12}(17)^{3/2} - \frac{1}{12} \right] \cdot 2\pi.$$

Strangely, the paraboloid of Example 3 and the hyperboloid of Practice 3 have the same surface area.

Appendix: Building a Torus and Other Tubes

A torus is the collection of little circles centered on a large circle where the plane of each small circle is perpendicular to the large circle (Fig. A1). We can extend this idea to the collection of small circles centered on any curve in 2D or 3D so that the plane of each small circle is perpendicular to the curve (Fig. A2).

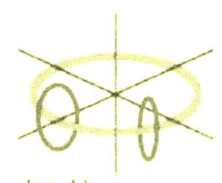

Fig. A1

In order to build these tubular surfaces (collections of small circles), we first need to know how to describe a circle in 3D at a given point, with a given radius and whose plane has a given normal vector (Fig. A3).

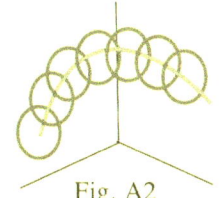

Fig. A2

Circle Algorithm: Center point at = (cx,cy,cz), radius R, and
normal vector V= <vx, vy, vz>.
Then the plane of this circle is $vx(x-cz)+vy(y-cy)+vz(z-cz)=0$.
Pick two non-colinear unit vectors A and B perpendicular to V.
Then the equation of the circle we want is

$$\mathbf{C}(t) = \langle \text{center point} \rangle + \mathbf{A} \cdot R \cdot \cos(t) + \mathbf{B} \cdot R \cdot \sin(t).$$

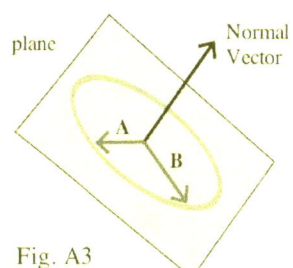

Fig. A3

Sometimes it is convenient for A and B to be perpendicular, but the only firm requirement on these unit vectors is that they are not co-linear.

Example A1: Find an equation for a circle with radius 2, center at P=(3,1,2) and normal vector $\mathbf{N} = \langle 4,2,1 \rangle$.

Solution: First we can find the equation of the plane that contains the point C and has normal vector N:
$4(x-3)+2(y-1)+1(z-2)=0$ so $z= 2-4(x-3)-2(y-1)$. (Fig A4)
Next we need two unit vectors perpendicular to N:
$\mathbf{A} = \langle 1,-1,-2 \rangle/\sqrt{6}$ works as a first one.

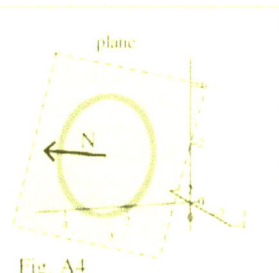

Fig. A4

For the second vector calculate
$$\mathbf{A} \times \mathbf{N} = \begin{vmatrix} \mathbf{i} & \mathbf{j} & \mathbf{k} \\ \frac{1}{\sqrt{6}} & \frac{-1}{\sqrt{6}} & \frac{-2}{\sqrt{6}} \\ 4 & 2 & 1 \end{vmatrix} = \frac{3}{\sqrt{6}} \langle 1,-3,2 \rangle$$

and divide by its magnitude to create the second unit vector $\mathbf{B} = \langle 1,-3,2 \rangle/\sqrt{14}$ that is perpendicular to both N and A. Then the parametric equation of the circle:

$$\mathbf{C}(u) = P + \mathbf{A} \cdot 2 \cdot \cos(u) + \mathbf{B} \cdot 2 \cdot \sin(u) = \langle 3,1,2 \rangle + \langle 1,-1,-2 \rangle \cdot \frac{2}{\sqrt{6}} \cdot \cos(u) + \langle 1,-3,2 \rangle \cdot \frac{2}{\sqrt{14}} \cdot \sin(u)$$

with $0 \le u \le 2\pi$. The component functions are: $x(t) = 3 + \frac{2}{\sqrt{6}} \cdot \cos(u) + \frac{2}{\sqrt{14}} \cdot \sin(u)$,

$y(t) = 1 - \dfrac{2}{\sqrt{6}} \cdot \cos(u) - \dfrac{6}{\sqrt{14}} \cdot \sin(u)$, and $z(t) = 2 - \dfrac{4}{\sqrt{6}} \cdot \cos(u) + \dfrac{4}{\sqrt{14}} \cdot \sin(u)$.

Fig. A4 shows the vector **N**, the plane normal to **N** at P and the parametric circle.

Practice A1: Find an equation for a circle with radius 2, center at P=(2,4,3) and normal vector **N** = $\langle -1, 3, -2 \rangle$.

Once we can create a circle at a point P with those properties, then we can create a tube around a curve simply by moving the point P along the curve using the tangent vector to the curve as the vector **N**.

Tube Algorithm: To create a tube of radius r along the curve **F**(v) = $\langle x(v), y(v), z(v) \rangle$ for a≤v≤b, apply the Circle Algorithm at each point **F**(v) using **N** = **F** '(v).

Example A2: Create parametric equations for a torus centered at the origin with large radius R and tube radius r . (Fig. A5)

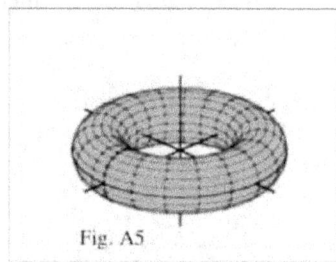

Fig. A5

Solution: For the large circle in the xy-plane, take

F(v) = $\langle R \cdot \cos(v), R \cdot \sin(v), 0 \rangle$ with $0 \leq v \leq 2\pi$. Then the unit tangent vector is **N** = **T** = $\langle -\sin(v), \cos(v), 0 \rangle$ and the unit vectors

A = **T** ' = $\langle -\cos(v), -\sin(v), 0 \rangle$ and **B** = **N**x**A** = $\langle 0, 0, 1 \rangle$ are each perpendicular to **N** and are not co-linear. Putting this together, the parametric equation of the torus is

$C(u,v) = F(v) + A \cdot r \cdot \cos(u) + B \cdot r \cdot \sin(u)$

$= \langle R \cdot \cos(v), R \cdot \sin(v), 0 \rangle + \langle -\cos(v), -\sin(v), 0 \rangle \cdot r \cdot \cos(u) + \langle 0, 0, 1 \rangle \cdot r \cdot \sin(u)$ for $0 \leq u, v \leq 2\pi$.

$x(u,v) = R \cdot \cos(v) - \cos(v) \cdot r \cdot \cos(u) + 0 = (R - r \cdot \cos(u)) \cdot \cos(v)$

$y(u,v) = R \cdot \sin(v) - \sin(v) \cdot r \cdot \cos(u) + 0 = (R - r \cdot \cos(u)) \cdot \sin(v)$

$z(u,v) = 0 + 0 + r \cdot \sin(u) = r \cdot \sin(u)$

Practice A2: Create parametric equations for a tube of radius r=1/2 centered on the curve **F**(v) = $\langle 0, v, v^2 \rangle$ for $-1 \leq v \leq 2$.

Appendix Practice Answers

Solution A1: Thee are many correct answers depending on the unit vectors **A** and **B**, both perpendicular to **N** and not co-linear, that are chosen: **A** = $\langle 1,1,1 \rangle / \sqrt{3}$ and **B** = $\langle 5,-1,-4 \rangle / \sqrt{35}$ work. Then the parametric equation of the circle is

$C(u) = P + A \cdot 2 \cdot \cos(u) + B \cdot 2 \cdot \sin(u)$

$= \langle 2,4,3 \rangle + \langle 1,1,3 \rangle \cdot \dfrac{2}{\sqrt{3}} \cdot \cos(u) + \langle 5,-1,-3 \rangle \cdot \dfrac{2}{\sqrt{35}} \cdot \sin(u)$ with $0 \leq u \leq 2\pi$.

Solution A2: For this curve $\mathbf{N} = \mathbf{T}(v)/|\mathbf{T}(v)| = \langle 0,1,2v \rangle / \sqrt{1+4v^2}$. Then $\mathbf{A} = \langle 1,0,0 \rangle$ and $\mathbf{B} = \langle 0, 2v, -1 \rangle$ are both perpendicular to $\mathbf{N}$ and to each other. Putting all of this together (Fig. A6), $x(u,v) = \frac{1}{2} \cdot \cos(u)$, $y(u,v) = v + v \cdot \sin(u)$, and $z(u,v) = v^2 - \frac{1}{2} \cdot \sin(u)$ for $-1 \leq v \leq 2$, $0 \leq u \leq 2\pi$.

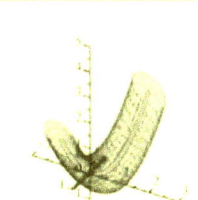

Fig. A6

15.9 Surface Integrals

Chapter 4 introduced integrals on intervals, Section 15.3 extended these ideas to integrals on paths in 2D and 3D, and Chapter 14 extended the ideas to integrals on 2D regions in the plane. This section goes one step further and considers integrals whose domains are parametric surfaces in 3D. In each previous situation the development was similar: partition, approximate on small pieces, sum, and take limits to achieve an integral. The approach here is the same. Surface integrals will be important in the coming sections on Stoke's Theorem and the Divergence Theorem and their applications.

Surface Integral for a Scalar Function

If S is a smooth surface in xyz-space parameterized by
$\mathbf{r}(u,v) = \langle x(u,v), y(u,v), z(u,v) \rangle$ with uv-domain R, then a partition of the uv-domain into small Δu by Δv rectangles ΔR (Fig. 1) is mapped by $\mathbf{r}$ to a partition of S into small patches in space with areas ΔS (Fig. 2), and the previous section showed that the area of each ΔS patch was
$\Delta S \approx |\mathbf{r}_u \times \mathbf{r}_v| \Delta u \cdot \Delta v$.

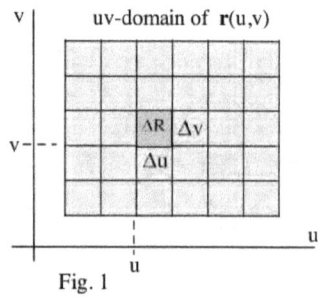

Fig. 1

Let (u^*, v^*) be a point in the uv-domain R. Then
$\mathbf{r}(u^*, v^*) = \langle x(u^*, v^*), y(u^*, v^*), z(u^*, v^*) \rangle = \langle x^*, y^*, z^* \rangle$ is a point on a patch S^*. If f(x,y,z) is a scalar-valued function on S, then the value of $f(\mathbf{r}(u^*, v^*)) \cdot \Delta S^*$ is approximately
$f(\mathbf{r}(u^*, v^*)) \cdot \Delta S^* \approx f(\mathbf{r}(u^*, v^*)) \cdot |\mathbf{r}_{u*} \times \mathbf{r}_{v*}| \Delta u \cdot \Delta v$. Adding these values together for each of the uv-rectangles we have the Riemann sum
$\sum_{u,v} f(x^*, y^*, z^*) \cdot \Delta S^* = \sum_{u,v} f(\mathbf{r}(u^*, v^*)) \cdot |\mathbf{r}_{u*} \times \mathbf{r}_{v*}| \Delta u \cdot \Delta v$. Taking limits as $\Delta u, \Delta v \to 0$, we get
$\iint_S f(x,y,z) \, dS = \iint_R f(\mathbf{r}(u,v)) \cdot |\mathbf{r}_u \times \mathbf{r}_v| \, dA$.

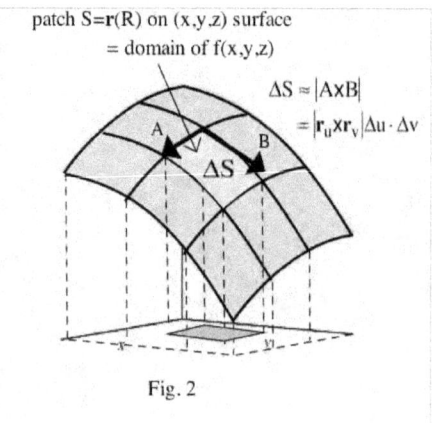

Fig. 2

If S is a smooth surface parameterized by $\mathbf{r}(u,v)$ on domain R in the uv-domain,
and f(x,y,z) is a scalar-valued function defined on S,
then $\iint_S f(x,y,z) \, dS = \iint_R f(\mathbf{r}(u,v)) \cdot |\mathbf{r}_u \times \mathbf{r}_v| \, dA$.

This result enables us to evaluate many surface integrals in 3D as iterated integrals in u and v.

15.9 Surface Integrals

Note: You should notice the similarity of this result with the result for a line integral of a scalar-valued function along a curve C: $\int_C \mathbf{f}\, ds = \int_{t=a}^{b} \mathbf{f}(\mathbf{r}(t)) \cdot |\mathbf{r}'(t)|\, dt$. In this new situation the curve C is replaced with the surface S, and $|\mathbf{r}'(t)|dt$ is replaced with $|\mathbf{r}_u \times \mathbf{r}_v|dA$.

Note: If $f(x,y,z)=1$ for all (x,y,z) on S, then $\iint_S 1\, dS = \iint_R |\mathbf{r}_u \times \mathbf{r}_v|\, dA$ is simply the surface area of S.

Example 1: Let $f(x,y,z)=1+z$ on the surface S parameterized by
$\mathbf{r}(u,v) = \langle u \cdot \cos(v),\ u \cdot \sin(v),\ 3-u \rangle$ with
$0 \le u \le 2$ and $0 \le v \le 2\pi$. (Fig. 3) (a) Evaluate $\iint_S f(x,y,z)\, dS$.

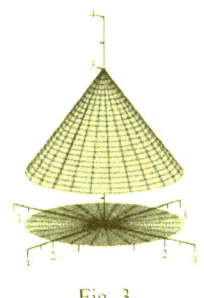

Fig. 3

(b) If the units of x, y and z are meters (m) and f is the surface density (g/m^2) at location (a,y,z), what are the units of $\iint_S f(x,y,z)\, dS$?

Solution: (a) $\iint_S f(x,y,z)\, dS = \iint_R f(\mathbf{r}(u,v)) \cdot |\mathbf{r}_u \times \mathbf{r}_v|\, dA$ so we need $f(\mathbf{r}(u,v))$, $\mathbf{r}_u$ and $\mathbf{r}_v$.

$\mathbf{r}_u = \langle \cos(v), \sin(v), -1 \rangle$, $\mathbf{r}_v = \langle -u \cdot \sin(v), u \cdot \cos(v), 0 \rangle$ and

$\mathbf{r}_u \times \mathbf{r}_v = \begin{vmatrix} \mathbf{i} & \mathbf{j} & \mathbf{k} \\ \cos(v) & \sin(v) & -1 \\ -u \cdot \sin(v) & u \cdot \cos(v) & 0 \end{vmatrix} = \langle -u \cdot \cos(v), u \cdot \sin(v), u \rangle$ so $|\mathbf{r}_u \times \mathbf{r}_v| = u\sqrt{2}$.

$\iint_S f(x,y,z)\, dS = \iint_R f(\mathbf{r}(u,v)) \cdot |\mathbf{r}_u \times \mathbf{r}_v|\, dA = \int_{v=0}^{2\pi} \int_{u=0}^{2} [1+(3-u)](u\sqrt{2})\, du\, dv$

$= \int_{v=0}^{2\pi} \sqrt{2}\left(-\frac{1}{3}u^3 + 2u^2\right)\Big|_{u=0}^{2} dv = \int_{v=0}^{2\pi} \frac{16}{3}\sqrt{2}\, dv = \frac{32}{3}\sqrt{2}\pi$.

(b) $\iint_S f(x,y,z)\, dS$ is the mass of the surface S, and the units are $(g/m^2)(m^2) = g$.

Practice 1: Evaluate $\iint_S (2+x)\, dS$ on the surface S in Example 1.

Example 2: Let $f(x,y,z)=xy$ on the surface S that is the part of the plane $z = 3-x-y$ that is in the first octant. Evaluate $\iint_S f(x,y,z)\, dS$.

Solution: S can be parameterized by $\mathbf{r}(u,v) = \langle u, v, 3-u-v \rangle$ for $0 \le u \le 3$ and $0 \le v \le 3-u$. Then

$\mathbf{r}_u = \langle 1, 0, -1 \rangle$, $\mathbf{r}_v = \langle 0, 1, -1 \rangle$, $\mathbf{r}_u \times \mathbf{r}_v = \langle 1, 1, 1 \rangle$ and $|\mathbf{r}_u \times \mathbf{r}_v| = \sqrt{3}$. $f(\mathbf{r}(u,v)) = uv$ so

$\iint_S f(x,y,z)\, dS = \int_0^3 \int_0^{u-3} uv\sqrt{3}\, dv\, du = \int_0^{u-3} \frac{\sqrt{3}}{2} u \cdot (3-u)^2\, dv = 3\sqrt{3}$.

The units of the answer are (units of f)(units of S).

Practice 2: Let $f(x,y,z)=x$ on the surface $S = \{(x,y,z): x^2 + y^2 = 3, 0 \leq z \leq 2\}$. Evaluate $\iint\limits_S f(x,y,z) \, dS$.

If the graph of $f(x,y,z)$ with $z=g(x,y)$ is a smooth surface S in xyz-space parameterized by $\mathbf{r}(u,v)$ on domain R in uv-space, ,

then $\iint\limits_S f(x,y,z) \, dS = \iint\limits_R f(\mathbf{r}(u,v)) \cdot \sqrt{1+(g_x)^2 + (g_y)^2} \, dA$.

Proof: We can parameterize this surface by setting $u=x$ and $v=y$ so $\mathbf{r}(u,v) = \langle u, v, g(u,v) \rangle$.

$\mathbf{r}_u = \langle 1, 0, g_x \rangle$ and $\mathbf{r}_v = \langle 0, 1, g_y \rangle$ so $\mathbf{r}_u \times \mathbf{r}_v = \begin{vmatrix} \mathbf{i} & \mathbf{j} & \mathbf{k} \\ 1 & 0 & g_x \\ 0 & 1 & g_y \end{vmatrix} = \langle -g_x, -g_y, 1 \rangle$ and $|\mathbf{r}_u \times \mathbf{r}_v| = \sqrt{1+(g_x)^2 + (g_x)^2}$

Then $\iint\limits_S f(x,y,g(x,y)) \, dS = \iint\limits_R f(\mathbf{r}(u,v)) \cdot \sqrt{1+(g_x)^2 + (g_x)^2} \, dA$.

Example 3: Let $f(x,y,z) = z\sqrt{x^2+y^2}$ on the heliocoid surface S parameterized by

$\mathbf{r}(u,v) = \langle u \cdot \cos(v), u \cdot \sin(v), v \rangle$ with $0 \leq u \leq 1$ and $0 \leq v \leq 2\pi$.

Evaluate $\iint\limits_S f(x,y,z) \, dS$. (Fig. 4 shows a heliocoid with $0 \leq v \leq 4\pi$.)

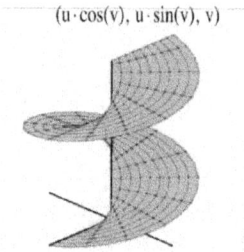

$(u \cdot \cos(v), u \cdot \sin(v), v)$

Fig. 4

Solution: $\iint\limits_S f(x,y,z) \, dS = \iint\limits_R f(\mathbf{r}(u,v)) \cdot \sqrt{1+(g_x)^2 + (g_y)^2} \, dA$ so we need

$f(\mathbf{r}(u,v))$, $\mathbf{r}_u$ and $\mathbf{r}_v$. $\mathbf{r}_u = \langle \cos(v), \sin(v), 1 \rangle$, $\mathbf{r}_v = \langle -u \cdot \sin(v), u \cdot \cos(v), 0 \rangle$ and

$\mathbf{r}_u \times \mathbf{r}_v = \begin{vmatrix} \mathbf{i} & \mathbf{j} & \mathbf{k} \\ \cos(v) & \sin(v) & 1 \\ -u \cdot \sin(v) & u \cdot \cos(v) & 0 \end{vmatrix} = \langle -u \cdot \cos(v), -u \cdot \sin(v), u \rangle$ so $|\mathbf{r}_u \times \mathbf{r}_v| = u\sqrt{2}$.

$f(x,y,z) = z\sqrt{x^2+y^2} = v \cdot u$ so

$\iint\limits_S f(x,y,z) \, dS = \iint\limits_R f(\mathbf{r}(u,v)) \cdot |\mathbf{r}_u \times \mathbf{r}_v| \, dA = \int_{v=0}^{2\pi} \int_{u=0}^{1} [v \cdot u](u\sqrt{2}) \, du \, dv = (2\pi^2)\left(\frac{1}{3}\sqrt{2}\right)$.

Practice 3: Evaluate $\iint\limits_S (1+y) \, dS$ on the surface S in Example 3.

Note: If $z=g(x,y)$ and $f(x,y,z)=1$ for all (x,y), then $\iint\limits_S f(x,y,z) \, dS$ is the surface area of S, and this area equals

$\iint\limits_R \sqrt{1+(g_x)^2 + (g_x)^2} \, dA$, then same result we saw in Section 14.5.

Oriented Surfaces and the Unit Normal Vector n

Before investigating surface integrals for vector-valued functions, some vocabulary and technical issues need to be considered: oriented surfaces and an orientation for vector **n** that is normal (perpendicular) to the surface.

A flat piece of paper in the xy-plane has a normal vector $\mathbf{n} = \langle 0,0,1 \rangle$ pointing upward at each point on the paper (Fig. 5). If we gently fold (but not crease) the paper, then the normal vector **n** will change continuously depending on its location on the paper (Fig. 6). If we follow a closed path on the paper that does not cross the paper's edge then the direction of the normal vector will change continuously and will return to the starting location pointing in the its original direction (Fig. 7). Such a surface is called oriented.

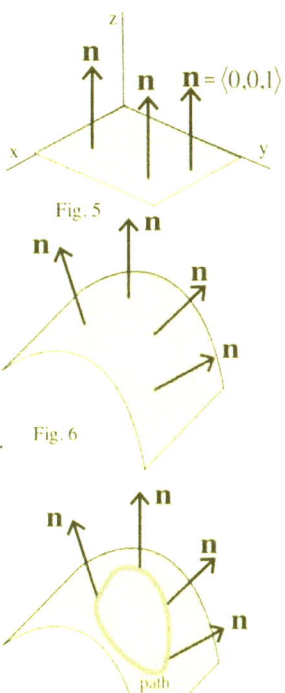

A smooth surface S is **oriented** if

* S has a non–zero normal vector at each point,
* the direction of the normal vector varies continuously as we move along S (not crossing an edge),
* and, a normal vector returns to its original orientation when it returns to its original position after moving along any closed path on S (not crossing an edge).

Fortunately, most surfaces are oriented. The most famous example of a non-oriented surface is a Mobius strip (Fig. 8). If we start with a normal vector **n** at any point and travel along the middle of the strip (not crossing an edge), then we end up at the starting point again but with the normal vector now pointing in the direction **–n** (Fig. 9).

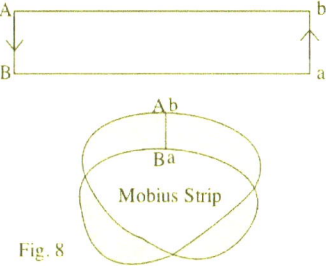

The results that follow require that our surfaces be oriented.

However, at a point A on an oriented surface there are two normal vectors, and we need to select one of them for the orientation. If the surface encloses a region of space, the convention is to pick the normal vector which points outward from the enclosed region (Fig. 10).

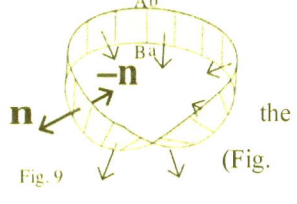

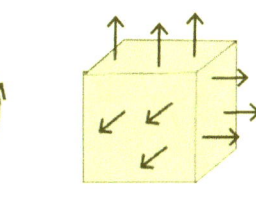

Fig. 10

Surface Integral for a Vector–Valued Function

Let ΔS be a small patch on the smooth, oriented surface S with oriented normal vector **n** (at some point of S). Then the magnitude of the vector F crossing the patch is the projection of F onto **n**. If we think of the vector field as water moving **F** at each point, then the amount of water passing through the patch ΔS in the direction of **n** is $(\mathbf{F} \bullet \mathbf{n})(\text{area of } \Delta S)$. Visually, that volume of that water (per unit of time) is the volume of the prism in Fig. 11. As before, adding the values of $(\mathbf{F} \bullet \mathbf{n})(\text{area of } \Delta S)$ for all of the patches we have the Riemann sum $\sum_{\Delta u, \Delta v} (\mathbf{F} \bullet \mathbf{n})(\text{area of } \Delta S)$.

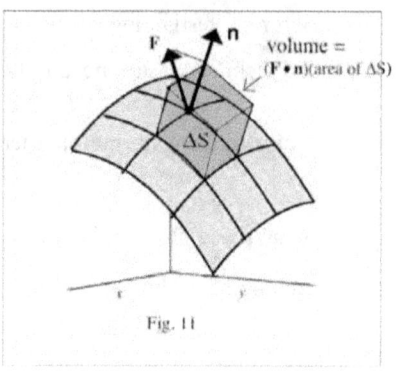

Fig. 11

Taking the limit as all of the Δu and Δv approach zero, we have the surface integral $\iint_S \mathbf{F} \cdot \mathbf{n} \, dS$. If the surface S is parameterized by the $\mathbf{r}(u,v) = (x(u,v), y(u,v), z(u,v))$ for (u,v) in a region R, then $\Delta S = |\mathbf{r}_u \times \mathbf{r}_v| \cdot \Delta u \cdot \Delta v$. The vector $\mathbf{r}_u \times \mathbf{r}_v$ is normal to the surface S and the unit normal vector is $\mathbf{n} = \dfrac{\mathbf{r}_u \times \mathbf{r}_v}{|\mathbf{r}_u \times \mathbf{r}_v|}$ so $\mathbf{F} \bullet \mathbf{n} \, dS = \mathbf{F} \bullet \dfrac{\mathbf{r}_u \times \mathbf{r}_v}{|\mathbf{r}_u \times \mathbf{r}_v|} \cdot |\mathbf{r}_u \times \mathbf{r}_v| \cdot \Delta u \cdot \Delta v$ and $\mathbf{F} \bullet \mathbf{n} \, dS = \mathbf{F} \bullet (\mathbf{r}_u \times \mathbf{r}_v) \, dA$.

Definition: Surface Integral of F over S

If **F** is a continuous vector field over the oriented surface S parameterized by $\mathbf{r}(u,v)$ and having unit normal vector **n**,

then the surface integral of **F** over S is $\iint_S \mathbf{F} \cdot \mathbf{n} \, dS = \iint_R \mathbf{F} \bullet (\mathbf{r}_u \times \mathbf{r}_v) \, dA$.

This integral is also called the **flux** of **F** across S.

Example 4: Suppose S is the part of the plane $3x+2y+6z=30$ with domain $0 \le x \le 4$ and $0 \le y \le 6$ (Fig. 12). This surface can be parameterized by $\mathbf{r}(u,v) = \left\langle u, v, 5 - \dfrac{u}{2} - \dfrac{v}{3} \right\rangle$ so $\mathbf{r}_u = \left\langle 1, 0, -\dfrac{1}{2} \right\rangle$, $\mathbf{r}_v = \left\langle 0, 1, -\dfrac{1}{3} \right\rangle$ and. If $\mathbf{F}(x,y,z) = \langle 0, 0, -2 \rangle$ then

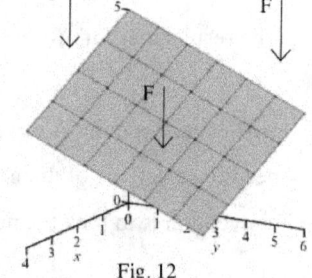

Fig. 12

$$\iint_S \mathbf{F} \cdot \mathbf{n} \, dS = \iint_R \mathbf{F} \bullet (\mathbf{r}_u \times \mathbf{r}_v) \, dA = \iint_R \langle 0, 0, -2 \rangle \bullet \left\langle \dfrac{1}{2}, \dfrac{1}{3}, 1 \right\rangle dA = \int_{u=0}^{4} \int_{v=0}^{6} -2 \, dv \, du$$

$$= (-2)(\text{area of R}) = (-2)(24) .$$

If $\mathbf{F} = \langle 0, 0, -2 \rangle$ is the velocity of water in m/s, and x and y are given in meters (m), then the units of $\iint_S \mathbf{F} \cdot \mathbf{n} \, dS$ are m^3/s : 48 m^3/s pass through the surface S. The negative sign in the answer results because the angle between $\mathbf{F}$ and $\mathbf{n}$ is greater than $90°$. If we had picked the opposite normal vector, then the answer would have been +48.

Practice 4: Use the surface S from Example 3 and calculate $\iint_S \mathbf{F} \cdot \mathbf{n} \, dS$ for $\mathbf{F} = \langle 0, -3, 0 \rangle$ and for $\mathbf{F} = \langle 1, 2, 3 \rangle$.

If F is not a constant vector field as in the previous Example and Practice problems, then we need to rewrite $\mathbf{F}(x,y,z)$ as $\mathbf{F}(x(u,v), y(u,v), z(u,v))$.

Practice 5: Suppose S is the part of the surface $z=f(x,y)$ above the region R in the xy-plane and that $\mathbf{F} = \langle M, N, P \rangle$.

Show that $\iint_S \mathbf{F} \cdot \mathbf{n} \, dS = \iint_R \{-P \cdot f_x - Q \cdot f_y + P\} \, dA$.

When S is a sphere

Spheres occur often in applications so it is worthwhile to see the calculations for a sphere. A sphere of radius R centered at the origin is easily described in spherical coordinates as (R, θ, φ) with $0 \leq \theta \leq 2\pi$ and $0 \leq \varphi \leq \pi$. Setting $u = \theta$ and $v = \varphi$ and then converting to rectangular coordinates we have (as in 15.7 Example 1)
$x(u,v) = R \cdot \sin(v) \cdot \cos(u)$, $y(u,v) = R \cdot \sin(v) \cdot \sin(u)$ and $z(u,v) = R \cdot \cos(v)$. Since $\mathbf{r}(u,v) = \langle x, y, z \rangle$,
$\mathbf{r}_u = \langle -R \cdot \sin(v) \cdot \sin(u), R \cdot \sin(v) \cdot \cos(u), 0 \rangle$ and $\mathbf{r}_v = \langle R \cdot \cos(v) \cdot \cos(u), R \cdot \cos(v) \cdot \sin(u), -R \cdot \sin(v) \rangle$.
Finally, $\mathbf{r}_u \times \mathbf{r}_v = -R^2 \langle \sin^2(v) \cdot \cos(u), \sin^2(v) \cdot \sin(u), \sin(v) \cdot \cos(v) \rangle$. For an outward facing normal vector $\mathbf{n}$, take $\mathbf{r}_u \times \mathbf{r}_v = R^2 \langle \sin^2(v) \cdot \cos(u), \sin^2(v) \cdot \sin(u), \sin(v) \cdot \cos(v) \rangle$.

Example 5: Suppose S is a sphere of radius 1 centered at the origin and $\mathbf{F}(x,y,z) = \langle x,y,z \rangle$ is a radial vector field.
(a) Determine the flux of $\mathbf{F}$ across S. (b) Determine the flux of $\mathbf{F}$ across S when S has radius R.

Solution: (a) $\mathbf{F}(x,y,z) = \langle x,y,z \rangle = \langle \sin(v) \cdot \cos(u), \sin(v) \cdot \sin(u), \cos(v) \rangle$. Then

$$\iint_S \mathbf{F} \cdot \mathbf{n} \, dS = \iint_R \mathbf{F} \bullet (\mathbf{r}_u \times \mathbf{r}_v) \, dA$$

$$= \iint_R \langle \sin(v) \cdot \cos(u), \sin(v) \cdot \sin(u), \cos(v) \rangle \bullet \langle \sin^2(v) \cdot \cos(u), \sin^2(v) \cdot \sin(u), \sin(v) \cdot \cos(v) \rangle \, dA$$

$$= \iint_R \sin^3(u) \cdot \cos^2(v) + \sin^3(v) \cdot \sin^2(u) + \sin(v) \cdot \cos^2(v) \, dA = \int_{u=0}^{2\pi} \int_{v=0}^{\pi} \sin(v) \, dv \, du = 4\pi.$$

(b) The only change in the calculation from part (a) is that now $\mathbf{r}_u \times \mathbf{r}_v$ has the factor R^2 so the result from part (a) needs to be multiplied by R^2: flux $= 4\pi R^2$.

Practice 6: Suppose S is the hemisphere $S = \{(x,y,z): x^2 + y^2 + z^2 = 1 \text{ and } 0 \leq z\}$ and $\mathbf{F}(x,y,z) = \langle z, x, y \rangle$. Determine the flux of F across S.

Connections with line integrals

There are nice parallels between the integrals of scalar and vector-valued functions along a curves C in 2D (section 15.3) and those on surfaces S in 3D.

scalar f on a curve C parameterized by $\mathbf{r}(t)$: $\displaystyle\int_C f \, ds = \int_{t=a}^{b} f(\mathbf{r}(t)) \cdot |\mathbf{r}'(t)| \, dt$

scalar f on a surface S parameterized by $\mathbf{r}(u,v)$: $\displaystyle\iint_S f(x,y,z) \, dS = \iint_R f(\mathbf{r}(u,v)) \cdot |\mathbf{r}_u \times \mathbf{r}_v| \, dA$

vector-valued F on a curve C parameterized by $\mathbf{r}(t)$: $\displaystyle\int_C \mathbf{F} \bullet \mathbf{T} \, ds = \int_{t=a}^{b} \mathbf{F}(\mathbf{r}(t)) \bullet \mathbf{r}'(t) \, dt$

vector-valued F on a surface S parameterized by $\mathbf{r}(u,v)$: $\displaystyle\iint_S \mathbf{F} \cdot \mathbf{n} \, dS = \iint_R \mathbf{F} \bullet (\mathbf{r}_u \times \mathbf{r}_v) \, dA$

Problems

1. $f(x,y,z) = x+y+z$ and $\mathbf{r}(u,v) = \langle u + 3v, 2u - v, 3u + v \rangle$. What is the value of $f(\mathbf{r}(u,v)) \cdot |\mathbf{r}_u \times \mathbf{r}_v| \Delta A$ when $u=1$, $v=2$, $\Delta u = 0.3$ and $\Delta v = 0.1$?

2. $f(x,y,z) = x+y+z$ and $\mathbf{r}(u,v) = \langle 2u + v, 3u - v, u + 2v \rangle$. What is the value of $f(\mathbf{r}(u,v)) \cdot |\mathbf{r}_u \times \mathbf{r}_v| \Delta A$ when $u=2$, $v=3$, $\Delta u = 0.1$ and $\Delta v = 0.2$?

3. $f(x,y,z) = 2x + y^2 - z$ and $\mathbf{r}(u,v) = \langle u^2, 3u + v, v^2 \rangle$. What is the value of $f(\mathbf{r}(u,v)) \cdot |\mathbf{r}_u \times \mathbf{r}_v| \Delta A$ when $u=2$, $v=1$, $\Delta u = 0.3$ and $\Delta v = 0.2$?

4. $f(x,y,z) = 2x + y^2 - z$ and $\mathbf{r}(u,v) = \langle u^2, 3u + v, v^2 \rangle$. What is the value of $f(\mathbf{r}(u,v)) \cdot |\mathbf{r}_u \times \mathbf{r}_v| \Delta A$ when $u=3$, $v=2$, $\Delta u = 0.3$ and $\Delta v = 0.2$?

5. $f(x,y,z) = x^2 + 4y + z$ on the surface $S = \{(x,y,z): 0 \leq x \leq 3, 0 \leq y \leq 2, z = 4\}$. Evaluate $\displaystyle\iint_S f(x,y,z) \, dS$.

6. $f(x,y,z) = x^2 + 4y + z$ on the surface $S = \{(x,y,z): 0 \leq x \leq 2, y = 3, 1 \leq z \leq 4\}$. Evaluate $\displaystyle\iint_S f(x,y,z) \, dS$.

7. $f(x,y,z) = y^2$ on the surface $S = \{(x,y,z): x+y+z = 4 \text{ in first octant}\}$ Evaluate $\displaystyle\iint_S f(x,y,z) \, dS$.

8. $f(x,y,z) = xy$ on the surface $S = \{(x,y,z): z = 1 + x^2 + y^2, 0 \leq x \leq 2, 0 \leq y \leq 2\}$. Evaluate $\iint_S f(x,y,z)\, dS$.

9. $\mathbf{F}(x,y,z) = \langle x, 2y, 3z \rangle$ on the surface $S = \{(x,y,z): 0 \leq x \leq 3, 0 \leq y \leq 2, z = 4\}$. Determine the flux of $\mathbf{F}$ across S. If the units of $\mathbf{F}$ are liters/second and the units of x, y and z are meters, what are the units of the flux?

10. $\mathbf{F}(x,y,z) = \langle x, -y, z \rangle$ on the surface $S = \{(x,y,z): z = x^2 + y^2, 0 \leq x \leq 2, 0 \leq y \leq 2\}$ Determine the flux of $\mathbf{F}$ across S. If the units of $\mathbf{F}$ are grams/meter^2 and the units of x, y and z are meters, what are the units of the flux?

11. $\mathbf{F}(x,y,z) = \langle x, y, z \rangle$ on the elliptical cylinder $S = \{(x,y,z): x^2 + 4y^2 = 4, 0 \leq z \leq 3\}$ Determine the flux of $\mathbf{F}$ across S.

12. $\mathbf{F}(x,y,z) = \langle 0, 0, K \rangle$ on the paraboloid $S = \{(x,y,z): x^2 + y^2 \leq A^2, z = A^2 - x^2 - y^2\}$ Determine the flux of $\mathbf{F}$ across S.

13. Suppose $\mathbf{F}$ is the same as in Problem 12 but now S is the "stretched" paraboloid $S = \{(x,y,z): x^2 + y^2 \leq A^2, z = C(A^2 - x^2 - y^2)\}$. Determine the flux of $\mathbf{F}$ across S.

Practice Answers

Practice 1: From Example 1, $|\mathbf{r}_u \times \mathbf{r}_v| = u\sqrt{2}$ and $x = u \cdot \cos(v)$ so

$$\iint_S (2+x)\, dS = \int_{v=0}^{2\pi} \int_{u=0}^{2} [2 + u \cdot \cos(v)] \cdot (u\sqrt{2})\, du\, dv = \int_{v=0}^{2\pi} \left(u^2\sqrt{2} + \frac{1}{3}u^3\sqrt{2} \cdot \cos(v) \right) \Big|_{u=0}^{2} dv = 8\sqrt{2}\pi.$$

Practice 2: S can be parameterized by $\mathbf{r}(u,v) = \langle 3\cdot\cos(u), 3\cdot\sin(u), v \rangle$ with $0 \leq u \leq 2\pi$ and $0 \leq v \leq 2$. Then

$\mathbf{r}_u = \langle -3\cdot\sin(u), 3\cdot\cos(u), 0 \rangle$, $\mathbf{r}_v = \langle 0, 0, 1 \rangle$, $\mathbf{r}_u \times \mathbf{r}_v = \langle 3\cos(u), 3\sin(u), 0 \rangle$ and $|\mathbf{r}_u \times \mathbf{r}_v| = 3$.

$f(\mathbf{r}(u,v)) = 3\cdot\cos(u)$ so $\iint_S f(x,y,z)\, dS = \int_0^{2\pi} \int_0^2 3\cdot\cos(u) \cdot 3\, dv\, du = \int_0^{2\pi} 18\cdot\cos(u)\, dv = 0$.

Practice 3: From Example 2, $|\mathbf{r}_u \times \mathbf{r}_v| = u\sqrt{2}$ and $1 + y = 1 + u\cdot\sin(v)$ so

$$\iint_S (1+y)\, dS = \int_{v=0}^{2\pi} \int_{u=0}^{1} [1 + u\cdot\sin(v)] \cdot (u\sqrt{2})\, du\, dv = \int_{v=0}^{2\pi} \left(\frac{1}{2}u^2\sqrt{2} - \frac{1}{3}u^3\sqrt{2} \cdot \cos(v) \right) \Big|_{u=0}^{1} dv = \sqrt{2}\pi.$$

Practice 4: For $\mathbf{F} = \langle 0, -3, 0 \rangle$, $\iint_S \mathbf{F} \cdot \mathbf{n}\, dS = \iint_R \langle 0, -3, 0 \rangle \cdot \langle \frac{1}{2}, \frac{1}{3}, 1 \rangle\, dA = \iint_R 1\, dA = 24$.

For $\mathbf{F} = \langle 1, 2, 3 \rangle$, $\iint_S \mathbf{F} \cdot \mathbf{n}\, dS = \iint_R \langle 1, 2, 3 \rangle \cdot \langle \frac{1}{2}, \frac{1}{3}, 1 \rangle\, dA = \iint_R \left(\frac{25}{6} \right) dA = 100$.

Practice 5: The surface S can be parameterized by $r(u,v) = \langle u, v, f(u,v) \rangle$.

Then $r_u = \langle 1, 0, f_u \rangle$, $r_v = \langle 0, 1, f_v \rangle$ and $r_u \times r_v = \langle -f_u, -f_v, 1 \rangle$ so

$$\iint_S F \cdot n \, dS = \iint_R F \bullet (r_u \times r_v) \, dA = \iint_R \langle M, N, P \rangle \bullet \langle -f_u, -f_v, 1 \rangle \, dA = \iint_R \{-P \cdot f_x - Q \cdot f_y + P\} \, dA.$$

Practice 6: As in the example, S can be parameterized by $x(u,v) = \sin(v) \cdot \cos(u)$, $y(u,v) = \sin(v) \cdot \sin(u)$ and $z(u,v) = \cos(v)$ with $0 \leq u \leq 2\pi$ and $0 \leq v \leq \pi/2$. Then $F(x,y,z) = \langle \cos(v), \sin(v) \cdot \cos(u), \sin(v) \cdot \sin(u) \rangle$ and

$$\iint_S F \cdot n \, dS = \iint_R F \bullet (r_u \times r_v) \, dA$$

$$= \iint_R \langle \cos(v), \sin(v) \cdot \cos(u), \sin(v) \cdot \sin(u) \rangle \bullet \langle \sin^2(v) \cdot \cos(u), \sin^2(v) \cdot \sin(u), \sin(v) \cdot \cos(v) \rangle \, dA$$

$$= \int_{v=0}^{\pi/2} \int_{u=0}^{2\pi} \cos(v) \cdot \sin^2(v) \cdot \cos(u) + \sin^3(v) \cdot \cos(u) \cdot \sin(u) + \sin^2(v) \cdot \sin(u) \cdot \cos(v) \, du \, dv.$$

But $\int_{u=0}^{2\pi}$ (each term) $du = 0$ so the flux $= 0$.

15.10 Stokes' Theorem

The circulation–curl form of Green's Theorem (section 15.5) says if $\mathbf{F} = \langle M, N \rangle$ is a 2D vector field and C is a simple, closed, piecewise smooth curve enclosing a region R then the integral of the curl of $\mathbf{F}$ on R is equal to the circulation of $\mathbf{F}$ around C (with a positive orientation): (Fig. 1)

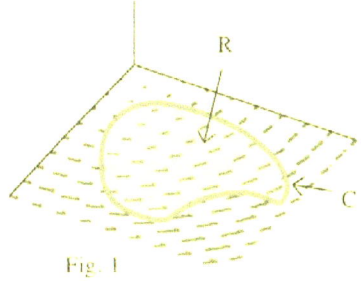

Fig. 1

$$\iint_R \text{curl } \mathbf{F} \ dA = \iint_R \left(\frac{\partial N}{\partial x} - \frac{\partial M}{\partial y} \right) dA = \oint_C \mathbf{F} \bullet \mathbf{T} \ ds = \text{circulation of } \mathbf{F} \text{ around C}.$$

Stokes' Theorem moves this result into 3D, and it has some very important consequences. If we think of Green's Theorem as applying to a flat soap film then we can think of Stokes' Theorem as giving the same result if we blow gently to create a soap bubble, a surface S in 3D. (Fig. 2)

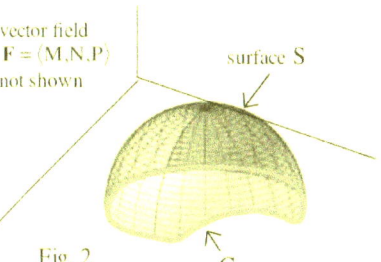

Fig. 2

Stokes' Theorem

If S is a connected, simply–connected, piecewise-smooth surface in 3D with piecewise–smooth boundary curve C, and $\mathbf{F}$ has continuous partial derivatives,

then $\iint_S \text{curl } \mathbf{F} \bullet d\mathbf{S} = \iint_S (\nabla \times \mathbf{F}) \bullet \mathbf{n} \ dS = \int_C \mathbf{F} \bullet d\mathbf{r} = \int_C \mathbf{F} \bullet \mathbf{T} \ ds = \text{circulation around C}$

Among the consequences of Stokes' Theorem:
* It allows us to trade 2D and 3D integrals – sometimes one of those is much easier than the other.
* It says that the integral of the curl of a vector field only depends on its values on the boundary.
* It says that the integral of the curl over a closed surface (like a sphere) is 0 since a closed surface has no boundary curve.
* It allows us to prove some of Maxwell's equations from physics (section 15.11).

The general proof of Stokes' Theorem is complicated. A proof of an easy special case is given next, and a proof for the common special case when the surface S has the form in Fig. 3 with
$S = \{(x,y,z): z = z(x,y) \text{ is a function of x and y}\}$ is given in the Appendix.

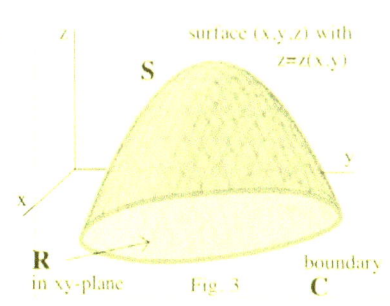

Fig. 3

Proof for an easy special case: S consists of a finite number of flat panels (Fig. 4) not necessarily in the same plane.

Then we can apply Green's Theorem to each panel and add the results together. This is very similar to the "finite Green's Theorem" we saw in section 15.5. The circulations along all of the interior edges cancel since adjacent panels have equal circulations going in opposite directions, and the only circulations remaining are those along the boundary of the region S. If S is the union of sub-regions $S_1, S_2, S_3 \ldots S_n$, then

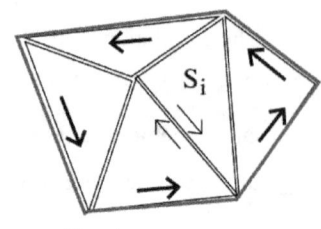

Fig. 4

$$\iint_S \text{curl } \mathbf{F} \bullet d\mathbf{S} = \sum_{i=1}^{n} \iint_{S_i} \text{curl } \mathbf{F} \bullet d\mathbf{S} = \sum_{i=1}^{n} \int_{C_i} \mathbf{F} \bullet \mathbf{T} \, dt = \int_C \mathbf{F} \bullet \mathbf{T} \, dt .$$

Example 1: Use Stokes' theorem to evaluate $\iint_S \text{curl } \mathbf{F} \bullet d\mathbf{S}$ where S is the hemisphere bounded by $x^2 + y^2 + z^2 = 9$ with $z \geq 0$ (Fig. 5) for the vector field $\mathbf{F} = \langle y, -x, 0 \rangle$.

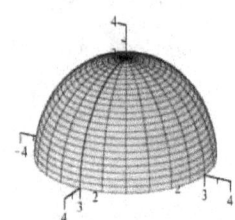

Fig. 5

Solution: By Stokes' theorem $\iint_S \text{curl } \mathbf{F} \bullet d\mathbf{S} = \int_C \mathbf{F} \bullet d\mathbf{r}$ where C is the bounding circle parameterized by $\mathbf{r}(t) = \langle 3 \cdot \cos(t), 3 \cdot \sin(t), 0 \rangle$. Then

$$\int_C \mathbf{F} \bullet d\mathbf{r} = \int_C \mathbf{F} \bullet \mathbf{T} \, dt = \int_{t=0}^{2\pi} \langle 3 \cdot \sin(t), -3 \cdot \cos(t), 0 \rangle \bullet \langle -3 \cdot \sin(t), 3 \cdot \cos(t), 0 \rangle \, dt$$

$$= \int_{t=0}^{2\pi} -9 \cdot \sin^2(t) - 9 \cdot \cos^2(t) \, dt = \int_{t=0}^{2\pi} -9 \, dt = (-9)(2\pi) = -18\pi .$$

It is also possible to evaluate $\iint_S \text{curl } \mathbf{F} \bullet d\mathbf{S}$ directly, but more difficult.

In this case the line integral around C was easier to evaluate than the surface integral on S.

Practice 1: Use Stokes' theorem to evaluate $\iint_S \text{curl } \mathbf{F} \bullet d\mathbf{S} = \int_C \mathbf{F} \bullet d\mathbf{r}$ where S is the paraboloid bounded by $2x^2 + 2y^2 + z = 18$ with $z \geq 0$ (Fig. 6) for the vector field $\mathbf{F} = \langle y, -x \rangle$.

Example 2: (a) Evaluate the line integral $\int_C \mathbf{F} \bullet d\mathbf{r}$ where $\mathbf{F} = \langle z, -z, x^2 - y^2 \rangle$ and C consists of the 3 line segments that bound the plane z=8–4x–2y in the first octant oriented as in Fig. 7.

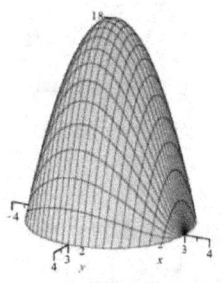

Fig. 6

15.10 Stokes' Theorem   Contemporary Calculus   389

(b) Evaluate $\int_C \mathbf{F} \cdot d\mathbf{r}$ for the line segment from (2, 0, 0) to (0, 4, 0).

Solution: (a) Rather than parameterizing the 3 line segments and evaluating the line integral along each of them, we can use Stoke's theorem and instead evaluate $\iint_S \text{curl } \mathbf{F} \cdot d\mathbf{S} = \iint_S (\nabla \times \mathbf{F}) \cdot \mathbf{n} \, dS$.

$$\nabla \times \mathbf{F} = \begin{vmatrix} \mathbf{i} & \mathbf{j} & \mathbf{k} \\ \dfrac{\partial}{\partial x} & \dfrac{\partial}{\partial y} & \dfrac{\partial}{\partial z} \\ z & -z & x^2 - y^2 \end{vmatrix} = \langle 1 - 2y,\ 1 - 2x,\ 0 \rangle \quad \text{and} \quad \mathbf{n} = \langle 4,\ 2,\ 1 \rangle$$

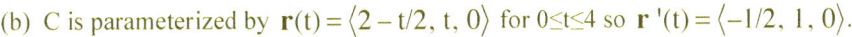

is normal to the plane. Then

$$\iint_S (\nabla \times \mathbf{F}) \cdot \mathbf{n} \, dS = \iint_S \langle 1 - 2y,\ 1 - 2x,\ 0 \rangle \cdot \langle 4,\ 2,\ 1 \rangle \, dS$$

Fig. 7

$$= \int_{x=0}^{2} \int_{y=0}^{4-2x} 6 - 8y + 2x \, dy \, dx = \text{(just a standard double integral for Maple)} = -\frac{88}{3}.$$

(b) C is parameterized by $\mathbf{r}(t) = \langle 2 - t/2,\ t,\ 0 \rangle$ for $0 \le t \le 4$ so $\mathbf{r}'(t) = \langle -1/2,\ 1,\ 0 \rangle$.

$$\int_C \mathbf{F} \cdot d\mathbf{r} = \int_{t=0}^{4} \mathbf{F} \cdot \mathbf{r}' \, dt = \int_{t=0}^{4} \langle 0,\ 0,\ (2-t/2)^2 - t^2 \rangle \cdot \langle -1/2,\ 1,\ 0 \rangle \, dt = \int_{t=0}^{4} 0 \, dt = 0.$$

Practice 2: Calculate the circulation of $\mathbf{F} = \langle xy,\ xz,\ -2yz \rangle$ around the curve C that consists of the 3 line segments that bound the plane $x + 2y + 2z = 4$ in the first octant oriented as in Fig. 8.

Example 3: Use Stokes' theorem to evaluate $\iint_S \text{curl } \mathbf{F} \cdot d\mathbf{S}$ where S is the "cap" on the hemisphere bounded by $x^2 + y^2 + z^2 = 25$ with $z \ge 3$ (Fig. 9) for the vector field $\mathbf{F} = \langle z - y,\ x,\ -x \rangle$.

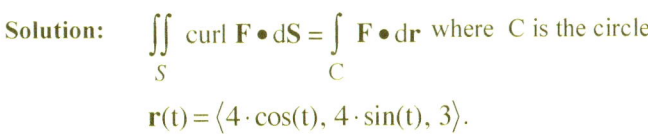

Fig. 8

Solution: $\iint_S \text{curl } \mathbf{F} \cdot d\mathbf{S} = \int_C \mathbf{F} \cdot d\mathbf{r}$ where C is the circle $\mathbf{r}(t) = \langle 4 \cdot \cos(t),\ 4 \cdot \sin(t),\ 3 \rangle$.

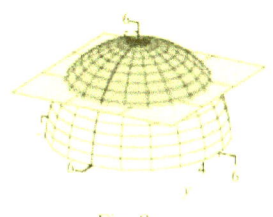

Fig. 9

Then $\int_t \mathbf{F} \cdot \mathbf{r}' \, dr = \int_{t=0}^{2\pi} \langle 3 - 4 \cdot \sin(t),\ 4 \cdot \cos(t),\ -4 \cdot \cos(t) \rangle \cdot \langle -4 \cdot \sin(t),\ 4 \cdot \cos(t),\ 0 \rangle \, dr$

$$= \int_{t=0}^{2\pi} -12 \cdot \sin(t) + 16 \cdot \sin^2(t) + 16 \cos^2(t) \, dt = 12\cos(t) + 16t \Big|_0^{2\pi} = 32\pi.$$

Practice 3: Use Stokes' theorem to evaluate $\iint_S \text{curl } \mathbf{F} \bullet d\mathbf{S}$ where F is the same as in Example 3, but now S is the cap with z≥4 on the same hemisphere.

Example 4: Evaluate $\iint_S \text{curl } \mathbf{F} \bullet d\mathbf{S}$ when S is the surface of the unit cube $0 \le x \le 1$, $0 \le y \le 1$, $0 \le z \le 1$.

Solution: Since S is a piecewise **closed** smooth surface, then S has no boundary curve C and
$$\iint_S \text{curl } \mathbf{F} \bullet d\mathbf{S} = 0 \ .$$

Practice 4: Evaluate $\iint_S \text{curl } \mathbf{F} \bullet d\mathbf{S}$ when S is the ellipsoid $x^2 + 2y^2 + 4z^2 = 16$.

If S has holes

If the surface S has a hole (Fig. 10) then the boundary of S has an additional boundary curve, and we can treat that new boundary in the same way we treated the boundary of a hole using Green's Theorem. We can create a single boundary for S by adding a path along S to the hole and then back from the hole (Fig. 11). Then the integral along this total new path will be

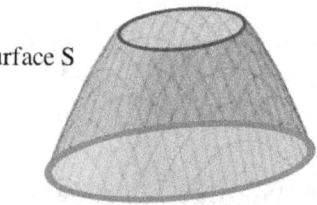

Fig. 10

the sum of the counterclockwise integrals around the outer boundary of S minus the sum of the counterclockwise integral around the hole. The integrals along the added paths sum to 0 since they are traveled once in each direction. Remember, for a counterclockwise orientation the region is always on our left hand side.

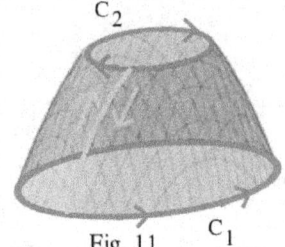

Fig. 11

Example 5: Evaluate $\iint_S \text{curl } \mathbf{F} \bullet d\mathbf{S}$ for $\mathbf{F} = \langle xy, x+z^2, y^3 \rangle$ with
$$S = \{(x,y,z): x^2 + y^2 + z^2 = 25, 0 \le z \le 4\}$$

Solution: This is the situation in Fig. 11. C_1 is parameterized by $\mathbf{r}_1(t) = \langle 5 \cdot \cos(t), 5 \cdot \sin(t), 0 \rangle$ and C_2 by $\mathbf{r}_2(t) = \langle 3 \cdot \cos(t), 3 \cdot \sin(t), 4 \rangle$ for $0 \le t \le 2\pi$ (both are counterclockwise).

$$\int_{C_1} \mathbf{F} \bullet d\mathbf{r} = \int_0^{2\pi} \langle 25 \cdot \sin(t) \cdot \cos(t), 5 \cdot \cos(t) + 0, 125 \cdot \sin^3(t) \rangle \bullet \langle -5\sin(t), 5\cos(t), 0 \rangle \, dt$$

$$= \int_0^{2\pi} -125 \cdot \sin^2(t) \cdot \cos(t) + 25 \cdot \cos^2(t) + 0 \, dt = 25\pi.$$

$$\int_{C_2} \mathbf{F} \bullet d\mathbf{r} = \int_0^{2\pi} \langle 9 \cdot \sin(t) \cdot \cos(t), \, 3 \cdot \cos(t) + 4, \, 27 \cdot \sin^3(t) \rangle \bullet \langle -3 \cdot \sin(t), \, 3 \cdot \cos(t), \, 0 \rangle \, dt$$

$$= \int_0^{2\pi} -27 \cdot \sin^2(t) \cdot \cos(t) + 9 \cdot \cos^2(t) + 12 \cdot \cos(t) + 0 \, dt = 9\pi.$$

So $\iint_S \text{curl } \mathbf{F} \bullet d\mathbf{S} = \int_{C_1} \mathbf{F} \bullet d\mathbf{r} - \int_{C_2} \mathbf{F} \bullet d\mathbf{r} = 16\pi.$

Practice 5: Evaluate $\iint_S \text{curl } \mathbf{F} \bullet d\mathbf{S}$ for $\mathbf{F} = \langle xy, \, x+z^2, \, y^3 \rangle$ with $S = \{(x,y,z): x^2 + y^2 + z^2 = 25, \, 0 \leq z \leq 3\}$.

Meaning of the curl

In section 15.6 we claimed that the curl vector had two important properties:
* the magnitude of the curl gives the rate of the fluid's rotation, and
* the direction of the curl is normal to the plane of greatest circulation and points in the direction so that the circulation at the point has a right hand orientation (Fig. 12).

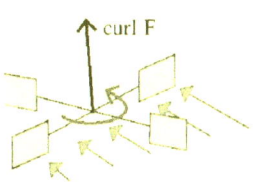

If we have a small paddle wheel at point P and tilt it in different directions, then the claims say that the wheel will spin fastest with a right-hand orientation when the axis points in the direction of the curl vector.

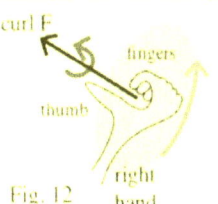

Fig. 12

Now we can use Stokes' Theorem to justify those claims.

Let P be a point in the vector field **F**, and let **u** be any unit vector. Suppose S is a small disk that has center at point P and radius r and that lies in the plane determined by P and u. S has a boundary circle C oriented positively so that S is always on the left side as we move along C.

Since S is small, the value of $\nabla \times \mathbf{F}$ is almost constant on S and so

$$\iint_S (\nabla \times \mathbf{F}) \bullet \mathbf{u} \, dS \approx (\nabla \times \mathbf{F}) \bullet \mathbf{u} \iint_S 1 \, dS = (\nabla \times \mathbf{F}) \bullet \mathbf{u} \, (\text{area of } S) = |\nabla \times \mathbf{F}| \cdot \cos(\theta) \cdot (\pi \cdot r^2)$$

But by Stokes' Theorem, $\iint_S (\nabla \times \mathbf{F}) \bullet \mathbf{u} \, dS = \oint_C \mathbf{F} \bullet \mathbf{T} \, dt = \text{circulation of F around C}$ so

$|\nabla \times \mathbf{F}| \cdot \cos(\theta) = \dfrac{\text{circulation of } \mathbf{F} \text{ around } C}{\pi \cdot r^2}$ which is maximum when $\theta = 0$ and **u** has the same direction as $\nabla \times \mathbf{F}$.

> Together these statements say that an axis in the direction of $\nabla \times \mathbf{F}$ gives the maximum circulation, and the magnitude of $\nabla \times \mathbf{F}$ is the maximum rate of circulation per unit of area.

In section 15.6 we also stated the following theorem and said that a proof needed to wait until we had Stokes' Theorem.

> **Theorem:** If $\mathbf{F}$ is defined and has continuous partial derivatives at every point in 3D and curl $\mathbf{F} = 0$,
> then $\mathbf{F}$ is a conservative field.

Proof: With Stokes' Theorem this is easy. If curl $\mathbf{F} = 0$ then $0 = \iint_S \text{curl } \mathbf{F} \, d\mathbf{S}$ for every simply-connected region S so by Stokes' Theorem $\int_C \mathbf{F} \bullet \mathbf{T} \, ds = \iint_S \text{curl } \mathbf{F} \, d\mathbf{S} = 0$ for every simple, closed, piece-wise smooth curve C. That means that $\mathbf{F}$ is path independent and conservative.

Problems

For problems 1 to 12 find the circulation of vector field $\mathbf{F}$ around the positively oriented curve C by using Stokes' Theorem and evaluating the surface integral.

1. $\mathbf{F} = \langle y, 2x, -z^2 \rangle$ and C is the ellipse $x^2 + 4y^2 = 4$.

2. $\mathbf{F} = \langle y, 2x, -z^2 \rangle$ and C is the circle $x^2 + y^2 = 9$.

3. $\mathbf{F} = \langle z, y^2, xy \rangle$ and C is the boundary of the triangle 2x+2y+2z=6 in the first octant.

4. $\mathbf{F} = \langle yz, xz, xy \rangle$ and C is the boundary of a simple closed curve in the yz=plane.

5. $\mathbf{F} = \langle z-y, x-z, x-y \rangle$ and C is the boundary of a simple closed curve in the x+y+z=5 plane.

6. $\mathbf{F} = \langle x+y^2, 3x, 2z \rangle$ and C is the boundary of the rectangle $R = \{(x,y,z): 0 \le x \le 3, 1 \le y \le 3, z = 0\}$ oriented in the counterclockwise direction.

7. $\mathbf{F} = \langle y^2, 3x+z, 2z+y \rangle$ and C is the boundary of the circle $R = \{(x,y,z): x^2+y^2 \le 4, z = 0\}$ oriented in the counterclockwise direction.

8. $\mathbf{F} = \langle -y^2, x+z, z^2+y \rangle$ and C is the boundary of the circle $R = \{(x,y,z): x^2+y^2 \le 9, z = 2\}$ oriented in the counterclockwise direction.

9. $\mathbf{F} = \langle y, -x, z \rangle$ and $S = \{(x,y,z): x^2+y^2+z^2 = 16, 0 \le z\}$ is a hemisphere.

10. $\mathbf{F} = \langle y, -x, z \rangle$ and $S = \{(x,y,z): x^2+y^2+z^2 = 16, 0 \le y\}$ is a hemisphere.

11. $\mathbf{F} = \langle \sin(x), \cos(y), \sin(z) \rangle$ and S is the solid tour with large radius 3 and small radius 1.

12. $\mathbf{F} = \langle xy, x^2+z^2, y^3 \rangle$ and S is the solid cube with vertices $x = \pm 1, y = \pm 1, z = \pm 1$.

In problems 13 to 18 use Stokes' Theorem to determine which of these fields are conservative.

13. $\mathbf{F} = \langle yz, xz, xy \rangle$

14. $\mathbf{F} = \langle yz+a, xz+b, xy+c \rangle$ a, b and c are constants.

15. $\mathbf{F} = \langle yza, xzb, xyc \rangle$ a, b and c are constants.

16. $\mathbf{F} = \langle y+z, x+z, x+y \rangle$

17. $\mathbf{F} = \langle x+z, y+z, x+y \rangle$

18. $\mathbf{F} = \langle yz, x-y, -x \rangle$

For problems 19 to 26 evaluate $\iint\limits_{S} \text{curl } \mathbf{F} \bullet d\mathbf{S}$ for the given field $\mathbf{F}$ and surface S.

19. $\mathbf{F} = \langle z-y, x-z, y-x \rangle$ and $S = \{(x,y,z): x^2+y^2 = 16, 0 \le z \le 3\}$ is a cylinder.

20. $\mathbf{F} = \langle y, -x, z \rangle$ and $S = \{(x,y,z): x^2+z^2 = 9, 1 \le y \le 5\}$ is a cylinder.

21. $\mathbf{F} = \langle x, y^2, z^3 \rangle$ and $S = \{(x,y,z): x^2+y^2+z^2 = 25, -3 \le y\}$.

22. $\mathbf{F} = \langle \sin(x), \cos(y), \sin(z) \rangle$ and $S = \{(x,y,z): x^2++y^2+z^2 = 25, -3 \le z \le 4\}$.

23. $\mathbf{F} = \langle -y, z, x \rangle$ and $S = \{(x,y,z): \dfrac{x^2}{4}+\dfrac{y^2}{9}+\dfrac{z^2}{25} = 1, 0 \le y\}$ a truncated ellipsoid.

24. $\mathbf{F} = \langle -y, z, x \rangle$ and $S = \{(x,y,z) : \frac{x^2}{4} + \frac{y^2}{9} + \frac{z^2}{25} = 1, 0 \le z\}$ a truncated ellipsoid.

25. $\mathbf{F} = \langle y, z, x \rangle$ and $S = \{(x,y,z) : x^2 + y^2 + z = 25, 0 \le z \le 16\}$ a truncated paraboloid.

26. $\mathbf{F} = \langle y, z, x \rangle$ and $S = \{(x,y,z) : x^2 + y^2 + z = 25, 0 \le z \le 9\}$.

Practice Answers

Practice 1: The field F and the boundary C (a circle in the xy-plane with radius 3) are the same as in Example 1 so $\iint_S \text{curl } \mathbf{F} \cdot d\mathbf{S} = \int_C \mathbf{F} \cdot d\mathbf{r} = -18\pi$ just as in Example 1.

Practice 2: By Stokes' theorem $\int_C \mathbf{F} \cdot d\mathbf{r} = \iint_S (\nabla \times \mathbf{F}) \cdot \mathbf{n} \, dS$ and the second integral is easier to evaluate then doing 3 line integrals.

$$\nabla \times \mathbf{F} = \begin{vmatrix} \mathbf{i} & \mathbf{j} & \mathbf{k} \\ \frac{\partial}{\partial x} & \frac{\partial}{\partial y} & \frac{\partial}{\partial z} \\ xy & xz & -2yz \end{vmatrix} = \langle -2z - x, 0, z - x \rangle \text{ and } \mathbf{n} = \langle 1, 2, 2 \rangle \text{ is normal to the plane.}$$

Then $\iint_S (\nabla \times \mathbf{F}) \cdot \mathbf{n} \, dS = \iint_S \langle 1-2y, 1-2x, 0 \rangle \cdot \langle 4, 2, 1 \rangle \, dS = \int_{x=0}^{4} \int_{y=0}^{2-x/2} -3x \, dy \, dx$

$= \int_{x=0}^{4} -6x + \frac{3}{2}x^2 \, dx = -12$.

Practice 3: Now $\mathbf{r}(t) = \langle 3 \cdot \cos(t), 3 \cdot \sin(t), 4 \rangle$ so $\mathbf{r}'(t) = \langle -3 \cdot \sin(t), 3 \cdot \cos(t), 0 \rangle$ and

$\int_t \mathbf{F} \cdot \mathbf{r}' \, dr = \int_{t=0}^{2\pi} \langle 4 - 3 \cdot \sin(t), 4 \cdot \cos(t), -3 \cdot \cos(t) \rangle \cdot \langle -3 \cdot \sin(t), 3 \cdot \cos(t), 0 \rangle \, dr$

$= \int_{t=0}^{2\pi} -12 \cdot \sin(t) + 9 \cdot \sin^2(t) + 9\cos^2(t) \, dt = \int_{t=0}^{2\pi} -12 \cdot \sin(t) + 9 \, dt = 18\pi$.

Practice 4: S is a smooth **closed** surface so $\iint_S \text{curl } \mathbf{F} \cdot d\mathbf{S} = 0$.

Practice 5: C_1 and $\mathbf{r}_1$ are the same as in Example 5 but now C_2 is parameterized by
$\mathbf{r}_2(t) = \langle 4 \cdot \cos(t), 4 \cdot \sin(t), 3 \rangle$ for $0 \le t \le 2\pi$ (counterclockwise).

$$\int_{C_2} \mathbf{F} \bullet d\mathbf{r} = \int_0^{2\pi} \langle 16 \cdot \sin(t) \cdot \cos(t), 4 \cdot \cos(t) + 3, 27 \cdot \sin^3(t) \rangle \bullet \langle -4 \cdot \sin(t), 4 \cdot \cos(t), 0 \rangle \, dt$$

$$= \int_0^{2\pi} -64 \cdot \sin^2(t) \cdot \cos(t) + 16 \cdot \cos^2(t) + 12 \cdot \cos(t) + 0 \, dt = 16\pi.$$

Then $\iint_S \text{curl } \mathbf{F} \bullet d\mathbf{S} = \int_{C_1} \mathbf{F} \bullet d\mathbf{r} - \int_{C_2} \mathbf{F} \bullet d\mathbf{r} = 25\pi - 16\pi = 9\pi$.

Appendix: Proof of Stokes' Theorem for a common special case

Stoke's Theorem: $\iint_S \text{curl } \mathbf{F} \cdot d\mathbf{S} = \iint_S (\nabla \times \mathbf{F}) \cdot \mathbf{n} \; dS = \int_C \mathbf{F} \cdot d\mathbf{r} = \int_C \mathbf{F} \cdot \mathbf{T} \; dt = $ circulation around C

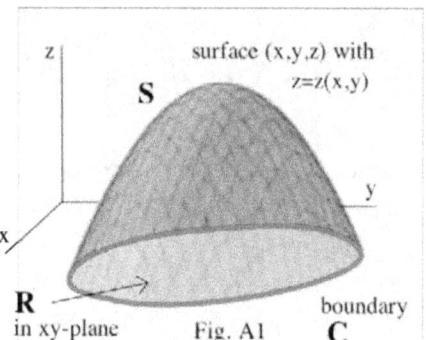

Common special case: S is a surface of the form $(x, y, z(x,y))$ (Fig. A1) This requires a lot of calculations and very careful attention to details. Let $\mathbf{F} = \langle P, Q, R \rangle$ and $S = \{(x,y,z): z = z(x,y) \text{ is a function of } x \text{ and } y\}$. We will evaluate $\int_C \mathbf{F} \cdot d\mathbf{r}$ and $\iint_S (\nabla \times \mathbf{F}) \cdot \mathbf{n} \; dS$ separately and show that they are equal.

$\iint_S (\nabla \times \mathbf{F}) \cdot \mathbf{n} \; dS$: Thinking of S as a parameterized surface (but using x and y instead of u and v) for x and y in the xy-region R (Fig. A2), then $\iint_S (\nabla \times \mathbf{F}) \cdot \mathbf{n} \; dS = \iint_R (\nabla \times \mathbf{F}) \cdot (\mathbf{r}_x \times \mathbf{r}_y) \; dA$. $\mathbf{r}_x = \langle 1, 0, z_x \rangle$, $\mathbf{r}_y = \langle 0, 1, z_y \rangle$ and $\mathbf{r}_x \times \mathbf{r}_y = \langle -z_y, -z_y, 1 \rangle$. Then

$$\iint_S (\nabla \times \mathbf{F}) \cdot \mathbf{n} \; dS = \iint_R \langle R_y - Q_z, P_z - R_x, Q_x - P_y \rangle \cdot \langle -z_y, -z_y, 1 \rangle \; dA .$$

$\int_C \mathbf{F} \cdot d\mathbf{r}$: This one is more complicated and requires the Chain Rule for functions of several variables.

$$\int_C \mathbf{F} \cdot d\mathbf{r} = \int_C \mathbf{F} \cdot \mathbf{r}' \; dt = \int_C \langle P, Q, R \rangle \cdot \langle dx, dy, dz \rangle \; dt = \int_C P \cdot dx + Q \cdot dy + R \cdot dz .$$

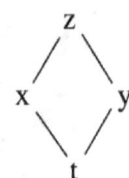

Fig. A2

But $z=z(x,y)$ so (Fig. A2) $dz = z_x \cdot dx + z_y \cdot dy$, and

$$\int_C \mathbf{F} \cdot d\mathbf{r} = \int_C (P + R \cdot z_x) dx + (Q + R \cdot z_y) \cdot dy = \int_C M \; dx + N \; dy .$$

By Green's Theorem $\int_C M \; dx + N \; dy = \iint_R N_x - M_y \; dA$ with $M = P + R \cdot z_x$ and $N = Q + R \cdot z_y$.

$$M_y = \frac{\partial}{\partial y}(P + R \cdot z_x) = P_y + P_z \cdot z_y + z_x \cdot R_y + R \cdot \frac{\partial}{\partial y} z_x$$

$$= P_y + P_z \cdot z_y + R \cdot z_{xy} + z_x \cdot (R_y + R_z \cdot z_y) .$$

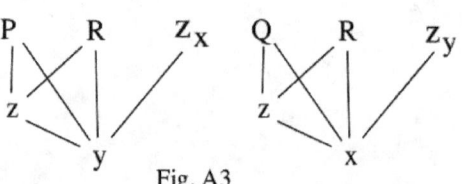

Fig. A3

Similarly, $N_x = \dfrac{\partial}{\partial x}(Q + R \cdot z_y) = Q_x + Q_z \cdot z_x + R \cdot z_{xy} + z_y \cdot (R_x + R_z \cdot z_x)$.

Then, after substituting and simplifying,

$$\int_C \mathbf{F} \bullet d\mathbf{r} = \iint_R N_x - M_y \; dA = \iint_R z_x(Q_z - R_y) + z_y(R_x - P_z) + (Q_x - P_y) \; dA$$

$$= \iint_R \langle R_y - Q_z, P_z - R_x, Q_x - P_y \rangle \bullet \langle -z_y, -z_y, 1 \rangle \; dA = \iint_S (\nabla \times \mathbf{F}) \bullet \mathbf{n} \; dS$$

so Stokes' Theorem is also true for surfaces of the form $S = \{(x,y,z) : z = z(x,y)\}$.

Suppose we cut a hole in the surface S and attach a smooth "bump" to cover the hole (Fig. A4). Is Stokes' Theorem still true for this new surface consisting of the old S minus the hole plus the bump? You should be able to justify your answer.

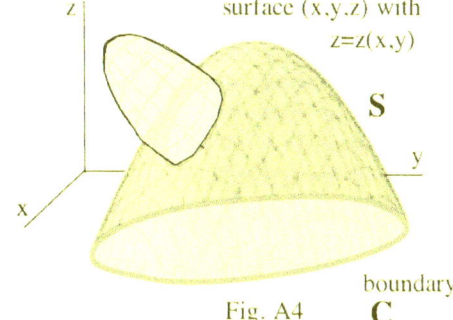

Fig. A4

15.11 Gauss/Divergence Theorem

The Gauss/Divergence Theorem is the final fundamental theorem of calculus and the final mathematical piece needed to create Maxwell's equations. Like each of the previous fundamental theorems, it relates an accumulation (integral) in some dimension to the values of a related function in a lower dimension.

The Fundamental Theorem of Calculus (section 4.5) and the Fundamental Theorem of Line Integrals (section 15.4) said that the accumulation (integral) of a function over an interval or a line is equal to a related function (the antidarivative or potential function) evaluated at the boundary (endpoints) of the interval or line.

Stoke's Theorem (15.9), like the curl-circulation form of Green's Theorem (section 15.5), said that the accumulation of a function (curl) on a 2D surface is equal to a related function (the circulation) evaluated on the boundary curve of the surface.

Finally, the Gauss/Divergence Theorem, like the divergence-flux form of Green's Theorem, says that the accumulation of the divergence in a solid 3D region is equal to a related function (flux) evaluated on the boundary surface of the region.

Gauss/Divergence Theorem

If   E is a solid, closed, simple 3D region with a piecewise-smooth boundary surface S, and **n** is the outward unit normal vector to S

then   for the vector field $\mathbf{F} = \langle M, N, P \rangle$ whose components have continuous partial derivatives in a open region containing E

$$\iiint_E \text{div } \mathbf{F} \bullet dV = \iint_S \mathbf{F} \bullet d\mathbf{S} = \iint_S \mathbf{F} \bullet \mathbf{n} \, dS = \text{flux across S}$$

A proof for a common special case is given in the Appendix.

The theorem is rather obvious in the case where the 3D region E contains a finite number if sources and sinks. If the region E inside of the boundary surface S contains a number of springs (pipes adding water) and sinks (pipes removing water), then the flux of water across the boundary surface S is the (springs input)–(sinks outputs). If several pipes inside E are adding water at a total rate of 5 m^3/s and several pipes inside E are removing water at a rate of 2 m^3/s then the (outward) flux across the boundary S is 3 m^3/s. The sum of the pipes adding water (positive divergence) plus the sum of the pipes removing water (negative divergence) equals the outward flux across the boundary of the region. The Gauss/Divergence Theorem extends this finite case idea to the case where potentially every point in the region E is a source or sink.

Example 1: Suppose pipes P1 at (1,0,0) and P2 at (2,2,1) are adding water at the rates of 5 m^3/s and 3 m^3/s, respectively, and P3 at (0,2,0) and P4 at (0,0,4) are removing water at the rates of 2 m^3/s and 1 m^3/s, respectively. What is the outward flux of water across (a) the sphere S with center at (0,0,0) and radius 3.5 m, and (b) a tiny cube with center at (2,2,1)?

Solution: (a) Only P1, P2 and P3 are inside S so the flux is (5)+(3)–(2)= 6 m^3/s.

(b) Only P2 is inside this tiny cube so flux = 3 m^3/s.

Practice 1: If the same pipes as in Example 1 are adding and removing grams of water per second, what is the flux across (a) the sphere S with center at (1,1,1) and radius 5 m and (b) a tiny cube with center at (1,2,3)?

Example 2: Calculate the flux across the sphere $x^2 + y^2 + z^2 = R^2$ for the radial vector field $\mathbf{F} = \langle x, y, z \rangle$.

Solution: E = the 3D sphere with radius R so the volume of E is $\frac{4}{3}\pi R^3$.

$$\text{div } \mathbf{F} = \nabla \bullet \mathbf{F} = \frac{\partial}{\partial x}(x) + \frac{\partial}{\partial y}(y) + \frac{\partial}{\partial z}(z) = 3 \text{ so}$$

$$\text{flux across S} = \iiint_E \text{div } \mathbf{F} \bullet dV = 3 \cdot (\text{volume of the sphere}) = 4\pi R^3.$$

Practice 2: Calculate the flux across the sphere $x^2 + y^2 + z^2 = R^2$ for the more general radial vector field $\mathbf{F} = \langle ax, by, cz \rangle$.

Example 3: Calculate the outward flux across the boundary D of the solid unit cube
$E = \{(x,y,z): 0 \leq x \leq 1, 0 \leq y \leq 1, 0 \leq z \leq 1\}$ for the field $\mathbf{F} = \langle xy, yz, xz \rangle$.

Solution: E = the solid cube, and $\text{div } \mathbf{F} = \nabla \bullet \mathbf{F} = \frac{\partial}{\partial x}(xy) + \frac{\partial}{\partial y}(yz) + \frac{\partial}{\partial z}(xz) = y + z + x$ so

$$\text{flux across D} = \iiint_E \text{div } \mathbf{F} \bullet dV = \int_0^1 \int_0^1 \int_0^1 x + y + z \, dz \, dy \, dz$$

$$= \int_0^1 \int_0^1 \left(xz + yz + \frac{1}{2}z^2 \right) \bigg|_{z=0}^1 dy \, dz = \int_0^1 \int_0^1 \left(x + y + \frac{1}{2} \right) dy \, dz = \ldots = \frac{3}{2}.$$

Practice 3: Calculate the outward flux across the boundary D of the solid unit cube
$E = \{(x,y,z): 0 \leq x \leq 1, 0 \leq y \leq 1, 0 \leq z \leq 1\}$ for the field $\mathbf{F} = \langle xyz, xyz, xyz \rangle$.

Flux for the inverse–square vector field $\mathbf{F} = \dfrac{\langle x, y, z \rangle}{\sqrt{x^2 + y^2 + z^2}} = \dfrac{\mathbf{r}}{|\mathbf{r}|^3}$

We cannot apply the Divergence Theorem to this field on a solid that includes the origin since the field is not defined at the origin, but we can apply it to a region between two spheres (or other smooth surfaces) so that the region doe not include the origin. Let $D = \{(x,y,z): 0 < A^2 < x^2 + y^2 + z^2 < B^2\}$ be the region outside of a sphere centered at the origin with radius A but inside a sphere centered at the origin with radius B. The boundary of D consists of the two spheres $S_1 = \{(x,y,z): A^2 = x^2 + y^2 + z^2\}$ and $S_2 = \{(x,y,z): x^2 + y^2 + z^2 = B^2\}$ but the unit normal vectors, pointing outward from the region D, have opposite directions (Fig. 1). Then

$$\iiint_D \text{div } \mathbf{F} \, dV = \iint_S \mathbf{F} \bullet \mathbf{n} \, dS = \iint_{S_1} \mathbf{F} \bullet \mathbf{n} \, dS - \iint_{S_2} \mathbf{F} \bullet \mathbf{n} \, dS \ .$$ But in Section 15.6 we determined that $\text{div } \mathbf{F} = 0$ for this field so

$$\iint_{S_1} \mathbf{F} \bullet \mathbf{n} \, dS = \iint_{S_2} \mathbf{F} \bullet \mathbf{n} \, dS$$ meaning that the outward flux (away from the origin) across S_2 equals the inward flux (toward the origin) across S_1.

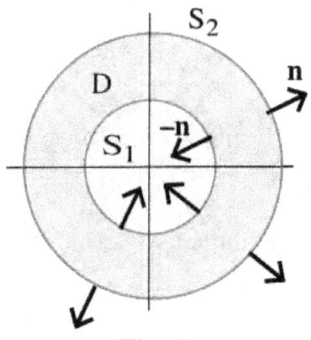

Fig. 1

We can evaluate $\iint_{S_1} \mathbf{F} \bullet \mathbf{n} \, dS$ by noting that on S_1 we have $|\mathbf{r}| = A$ and

$$\text{flux} = \iint_{S_1} \mathbf{F} \bullet \mathbf{n} \, dS = \iint_{S_1} \left(\frac{\mathbf{r}}{|\mathbf{r}|^3}\right) \bullet \left(\frac{\mathbf{r}}{|\mathbf{r}|}\right) dS = \iint_{S_1} \left(\frac{\mathbf{r} \bullet \mathbf{r}}{|\mathbf{r}|^4}\right) dS = \iint_{S_1} \frac{1}{|\mathbf{r}|^2} dS = \iint_{S_1} \frac{1}{A^2} dS$$

$$= \frac{1}{A^2}(\text{surface area of } S_1) = \frac{1}{A^2}(4\pi A^2) = 4\pi \ .$$

The resulting flux is the same 4π for a any region between two smooth surfaces that surround the origin.

Example 4: Determine the flux for the solid region D outside the sphere $S = \{(x,y,z): x^2 + y^2 + z^2 = 1\}$ and inside the ellipsoid $E = \left\{(x,y,z): \dfrac{x^2}{4} + \dfrac{y^2}{9} + \dfrac{z^2}{16} = 1\right\}$ for $\mathbf{F} = \dfrac{\langle x, y, z \rangle}{\sqrt{x^2 + y^2 + z^2}} = \dfrac{\mathbf{r}}{|\mathbf{r}|^3}$.

Solution: Based on the previous discussion we can immediately conclude that flux across D is 4π .

Practice 4: Determine the flux for the solid region D between a sphere centered at (2,3,4) with radius 4 and another sphere centered at (2,3,4) with radius 1 for $\mathbf{F} = \dfrac{\langle x, y, z \rangle}{\sqrt{x^2 + y^2 + z^2}} = \dfrac{\mathbf{r}}{|\mathbf{r}|^3}$.

Units of flux

Suppose **v** is a velocity vector field in meters per second (m/s) of a material that has constant density δ given in grams per cubic meter (g/m^3). Then the field $\mathbf{F} = \delta \cdot \mathbf{v}$ has units $\left(\dfrac{g}{m^3}\right)\left(\dfrac{m}{s}\right) = \left(\dfrac{g}{m^2 \cdot s}\right)$ which measures the amount of material per a square meter flowing past a point each second. This field **F** is sometimes called the **flux density**. But in mathematics flux is defined as a surface integral, and then the units of flux become

$$\text{flux} = \iint_S \mathbf{F} \bullet \mathbf{n} \, dS = \{\text{amount per square meter per second}\}\{\text{area of S in square meters}\}$$

$$= \left\{\dfrac{g}{m^2 \cdot s}\right\}(m^2) = \dfrac{g}{s} = \text{the amount (mass) per second flowing across the surface S.}$$

> "In the case of fluxes, we have to take the integral, over a surface, of the flux through every element of the surface. The result of this operation is called the surface integral of the flux. It represents the quantity which passes through the surface."
> —James Clerk Maxwell

In other fields the flux density **F** has other units as does the surface integral:

Heat flux density has units $\dfrac{J}{m^2 \cdot s}$ so the units of the surface integral (our flux) are $\dfrac{J}{s}$ (Joules per second).

Magnetic flux density has units $\mathbf{B} = \dfrac{Wb}{m^2}$ (Weber per square meter =Tesla) so the units of magnetic flux are

$$\Phi_B = \iint_S \mathbf{B} \bullet d\mathbf{S} = \text{Weber.}$$

> **Gauss's Law for magnetism** states that the total magnetic flux through a closed surface is 0:
> $$\Phi_B = \iint_S \mathbf{B} \bullet d\mathbf{S} = 0$$ for any closed surface S (since every magnetic north pole is attached to a magnetic south pole).

Electric flux density has units $\mathbf{E} = \dfrac{\mathbf{F}}{q}$ (force/charge= Newtons/coulomb=volts/meter) the magnetic flux is

$$\Phi_E = \iint_S \mathbf{E} \bullet d\mathbf{S} = \dfrac{Q}{4\pi\varepsilon_0}(4\pi) = \dfrac{Q}{\varepsilon_0} \, .$$

> **Gauss' Law for electric fields** states that the total electric flux through a closed surface S is
> $$\Phi_E = \iint_S \mathbf{E} \bullet d\mathbf{S} = \dfrac{Q}{4\pi\varepsilon_0}(4\pi) = \dfrac{Q}{\varepsilon_0}$$ where Q is the total electric charge inside S and ε_0 is a (very small) constant called the electric constant or permittivity of free space. The total electric flux need not equal 0 for a closed surface S since the divergence at a point charge inside S may not equal 0 and the flux is a constant times the total of the charges inside S.

Problems

For problems 1 to 6 pipe P1=(1,2,3) is inputting 7 m^3/s of water, P2=(−2,1,1) is removing 3 m^3/s of water, P3=(2,2,0) is inputting 2 m^3/s of water, P4=(1,1,0) is inputting 6 m^3/s of water, P5=(2,2,2) is removing 4 m^3/s of water, and P6=(3,1,3) is inputting 6 m^3/s of water,

1. What is the net flux across the sphere with center at the origin and (a) radius 2 and (b) radius 4?
2. What is the net flux across the sphere with center at the origin and (a) radius 3 and (b) radius 5?
3. What is the net flux across the sphere with center at (2,2,0) and (a) radius 1 and (b) radius 3?
4. What is the net flux across the sphere with center at (2,2,0) and (a) radius 4 and (b) radius 5?
5. What is the net flux across the boundary of the region outside the sphere with center at the origin and radius 3 and inside the sphere with center at the origin and radius 5?
6. What is the net flux across the boundary of the region outside the sphere with center at the origin and radius 1 and inside the sphere with center at the origin and radius 3?

In problems 7 to 14 find the outward flux of the field F across the boundary surface of the given solid.

7. $\mathbf{F} = \langle x, y, z \rangle$ across the solid sphere $E = \{(x,y,z): x^2 + y^2 + z^2 \leq 4\}$.
8. $\mathbf{F} = \langle -y, x, z \rangle$ across the solid sphere $E = \{(x,y,z): x^2 + y^2 + z^2 \leq 4\}$.
9. $\mathbf{F} = \langle x, 2y, -z \rangle$ across the solid sphere $E = \{(x,y,z): x^2 + y^2 + z^2 \leq 9\}$.
10. $\mathbf{F} = \langle -x, 2y, -z \rangle$ across the solid sphere $E = \{(x,y,z): x^2 + y^2 + z^2 \leq 9\}$.
11. $\mathbf{F} = \langle x, 2y, z \rangle$ across the solid box $E = \{(x,y,z): 0 \leq x \leq 2, 0 \leq y \leq 3, 0 \leq z \leq 4\}$.
12. $\mathbf{F} = \langle 2x, 3y, 4z \rangle$ across the solid box $E = \{(x,y,z): 1 \leq x \leq 2, 1 \leq y \leq 3, 1 \leq z \leq 4\}$.
13. $\mathbf{F} = \langle x^2, y^2, z^2 \rangle$ across the solid box $E = \{(x,y,z): 0 \leq x \leq 2, 0 \leq y \leq 3, 0 \leq z \leq 4\}$.
14. $\mathbf{F} = \langle x^2, -y^2, z^2 \rangle$ across the solid box $E = \{(x,y,z): 1 \leq x \leq 2, 1 \leq y \leq 3, 1 \leq z \leq 4\}$.

15. If div $\mathbf{F} = 0$ for the region between two concentric spheres, what is the relationship between the fluxes across the surfaces of the spheres?
16. E is a convex solid with volume V. Find the flux of F across the boundary of E (a) if div $\mathbf{F} = 0$ at every point, and (b) if div $\mathbf{F} = 3$ at every point.
17. $\mathbf{F} = \langle 1, -2, 3 \rangle$. Without doing any calculations determine whether the flux across each sphere is positive, negative or zero. (a) Sphere with radius 2 and center at the origin. (b) Sphere with radius 1 and center at (1, 2, 3) (c) Sphere with radius 1 and center at (3, −2, 1).

18. $\mathbf{F} = \langle a, b, c \rangle$. Without doing any calculations determine whether the flux across each sphere is positive, negative or zero. (a) Sphere with radius 2 and center at the origin. (b) Sphere with radius 1 and center at (1, 2, 3) (c) Sphere with radius 1 and center at (3, –2, 1).

19. $\mathbf{F} = \langle 0, y, 0 \rangle$. Without doing any calculations determine whether the flux across each sphere is positive, negative or zero. (a) Sphere with radius 2 and center at the origin. (b) Sphere with radius 1 and center at (1, 2, 3) (c) Sphere with radius 1 and center at (3, –2, 1).

20. $\mathbf{F} = \langle 0, 1, z \rangle$. Without doing any calculations determine whether the flux across each sphere is positive, negative or zero. (a) Sphere with radius 2 and center at the origin. (b) Sphere with radius 1 and center at (1, 2, 3) (c) Sphere with radius 1 and center at (3, –2, 1).

21. $\mathbf{F} = \langle x^2, y^2, z^2 \rangle$. Without doing any calculations determine whether the flux across each sphere is positive, negative or zero. (a) Sphere with radius 2 and center at the origin. (b) Sphere with radius 1 and center at (1, 2, 3) (c) Sphere with radius 1 and center at (3, –2, 1).

22. Suppose **v** is a velocity vector field in meters per second (m/s) and δ is a density given in cows per cubic meter (c/m^3), and F is the vector field $\mathbf{F} = \delta \cdot \mathbf{v}$. What are the units of flux for this field across the boundary of a solid region?

Practice Answers

Practice 1: (a) All of the pipes are inside this sphere so flux = (5)+(3)–(2)–(1) = 5 g/s.

(b) None of the pipes are inside this tiny cube so flux = 0 g/s.

Practice 2: $\text{div } \mathbf{F} = \nabla \bullet \mathbf{F} = \frac{\partial}{\partial x}(ax) + \frac{\partial}{\partial y}(by) + \frac{\partial}{\partial z}(cz) = a + b + c$ so

$$\text{flux across S} = \iiint_E \text{div } \mathbf{F} \bullet dV = (a+b+c) \cdot (\text{volume of the sphere}) = (a+b+c) \cdot \frac{4}{3}\pi R^3 \ .$$

Practice 3: E = the solid cube, and $\text{div } \mathbf{F} = \nabla \bullet \mathbf{F} = yz + xz + xy$ so

$$\text{flux across D} = \iiint_E \text{div } \mathbf{F} \bullet dV = \int_0^1 \int_0^1 \int_0^1 (yz + xz + xy) \, dz \, dy \, dz = \ldots = \frac{3}{4} \ .$$

Practice 4: The smaller sphere is inside the larger one and the larger one does not contain the origin so, as was shown in section 15.5, div $\mathbf{F} = 0$ inside the larger sphere. Then by the Divergence Theorem, the flux across each sphere is 0 so the flux across the boundary of D is 0.

Appendix: Proof of the Gauss/Divergence Theorem

There are two intuitive ways to think of the Gauss/Divergence Theorem that make the result seem "obvious." These are not proofs (see later in this Appendix), but they contain the essence of the theorem.

First intuitive approach: Imagine that the interior of the solid region E is partitioned into lots of cells and that the divergence of each cell is the total outward flow from that cell. If two cells share a boundary surface then the outward flow from one (a positive divergence) is the inward flow into the other (a negative divergence). So adding the divergences for all of the cells (a triple sum in x, y and z), all of the inside divergences sum to 0, and the final result is just the sum of those divergences on the surface S of the solid region E: $\sum\sum\sum_{x,y,z} \text{div } \mathbf{F} = \sum\sum_{\text{surface}} \mathbf{F} \bullet \mathbf{n}$

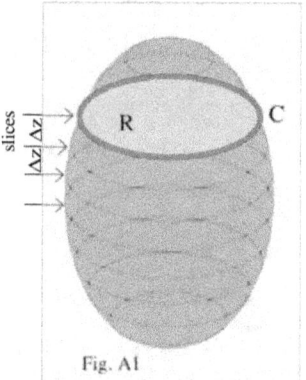

Second intuitive approach: Partition the solid region E into thin slices parallel to the xy-plane (Fig. A1). Then each slice will contain a 2D region R with boundary S. Applying Green's Theorem to this slice we have

$$\iint_R \text{div } \mathbf{F} \, dA = \oint_C \mathbf{F} \bullet \mathbf{n} \, ds.$$ And by adding these results together

$$\sum_{\Delta z} \left(\iint_R \text{div } \mathbf{F} \, dA \right) dz = \sum_{\Delta z} \left(\oint_C \mathbf{F} \bullet \mathbf{n} \, ds \right) dz \quad \text{we expect the Divergence Theorem}$$

$$\iiint_E \text{div } \mathbf{F} \, dV = \iint_S \mathbf{F} \bullet \mathbf{n} \, dS.$$

Fig. A1

Neither of these is a proof, but they might give you a better understanding of the "why" of the Divergence Theorem.

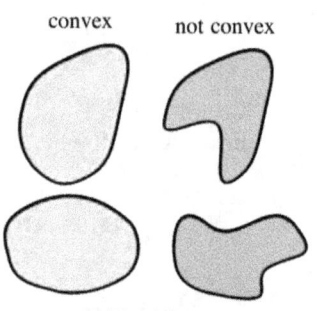

Proof for a common special case: E is a convex region

A region E is called **convex** if a straight line connecting any two points in E lies in E. Fig. A2 shows some convex and non-convex 2D regions.

Assume that E is a convex region with a piecewise-smooth boundary S, that $\mathbf{F} = \langle M, N, P \rangle$ has continuous partial derivatives, and that $\mathbf{n} = \langle n_1, n_2, n_3 \rangle$ is the unit, outward pointing normal vector to S.

Fig. A2

If E is convex, then the projection of E onto the xy, xz and yz-planes is a 2D convex region. Let D be the projection of E onto the xy-plane (Fig. A3). Then for each (x,y)

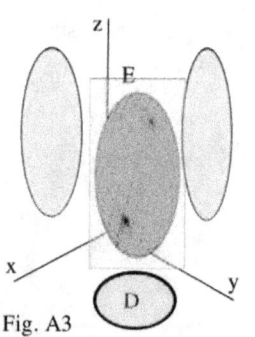

Fig. A3

in D the solid E is bounded by a top surface
ST = $\{(x,y,z) : (x,y)$ is in D and $z = f(x,y)\}$ and a bottom surface
SB = $\{(x,y,z) : (x,y)$ is in D and $z = g(x,y)\}$ (Fig. A4). If we can show that

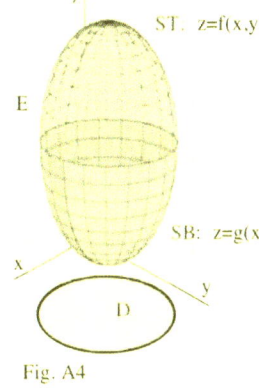

Fig. A4

$$\iiint_E \frac{\partial M}{\partial x} \, dV = \iint_S M \, dS = \iint_S \mathbf{F} \bullet \mathbf{n}_1 \, dS,$$

$$\iiint_E \frac{\partial N}{\partial y} \, dV = \iint_S N \, dS = \iint_S \mathbf{F} \bullet \mathbf{n}_2 \, dS, \text{ and}$$

$$\iiint_E \frac{\partial P}{\partial z} \, dV = \iint_S P \, dS = \iint_S \mathbf{F} \bullet \mathbf{n}_3 \, dS \text{ then}$$

$$\iiint_E \text{div } \mathbf{F} \, dV = \iiint_E \frac{\partial M}{\partial x} + \frac{\partial N}{\partial y} + \frac{\partial P}{\partial z} \, dV = \iint_S (M+N+P) \, dS = \iint_S \mathbf{F} \bullet \mathbf{n}_1 + \mathbf{F} \bullet \mathbf{n}_2 + \mathbf{F} \bullet \mathbf{n}_3 \, dS$$

$$= \iint_S \mathbf{F} \bullet \mathbf{n} \, dS$$

which is the Divergence Theorem.

Working to show $\iiint_E \frac{\partial P}{\partial z} \, dV = \iint_S P \, dS$:

$\iiint_E \frac{\partial P}{\partial z} \, dV$: By the Fundamental Theorem of Calculus,

$$\iiint_E \frac{\partial P}{\partial z} \, dV = \iint_D \left(\int_{g(x,y)}^{f(x,y)} \frac{\partial P}{\partial z} \, dz \right) dx \, dy = \iint_D P(x,y,f(x,y)) - P(x,y,g(x,y)) \, dx \, dy.$$

$\iint_S P \, dS$: The surface S consists of the two surfaces ST where $z=f(x,y)$ and SB where $z=g(x,y)$. On the top surface ST, the normal $\mathbf{n}$ points up and $\mathbf{n} = \langle -f_x, -f_y, 1 \rangle$. On the bottom surface SB, the normal vector $\mathbf{n}$ points down and $\mathbf{n} = \langle g_x, g_y, -1 \rangle$.

$$\iint_{ST} P \, dS = \iint_{ST} P(x,y,z) \, dS = \iint_D \langle 0,0,P(x,y,f(x,y)) \rangle \bullet \langle -f_x, -f_y, 1 \rangle \, dx \, dy = \iint_D P(x,y,f(x,y)) \, dx \, dy.$$

$$\iint_{SB} P \, dS = \iint_{SB} P(x,y,z) \, dS = \iint_D \langle 0,0,P(x,y,g(x,y)) \rangle \bullet \langle g_x, g_y, -1 \rangle \, dx \, dy$$

$$= \iint_D -P(x,y,g(x,y)) \, dx \, dy.$$

Putting these last two results together,

$$\iint_S P \, dS = \iint_{ST} P \, dS + \iint_{SB} P \, dS = \iint_D P(x,y,f(x,y)) \, dx \, dy - \iint_D P(x,y,g(x,y)) \, dx \, dy$$

$$= \iint_D P(x,y,f(x,y)) - P(x,y,g(x,y)) \, dx \, dy \text{ which is the same result we got for } \iiint_E \frac{\partial P}{\partial z} \, dV.$$

If the ST and SB surfaces are connected by vertical walls SV, the result is still true since **n** has the form $\mathbf{n} = \langle a, b, 0 \rangle$ so $\iint_{SV} P \, dS = \iint_{D} \langle 0, 0, P(x,y,g(x,y)) \rangle \bullet \langle a, b, 0 \rangle \, dx \, dy = 0$.

Similarly $\iiint_{E} \frac{\partial M}{\partial x} \, dV = \iint_{S} M \, dS$ and $\iiint_{E} \frac{\partial N}{\partial y} \, dV = \iint_{S} N \, dS$, and then the Divergence Theorem is proven for convex solid regions.

Te proof for more general solid regions is more complicated and is not given here.

15.0 Odd Answers

1. A small eddy or whirlpool.
3. Easier from D to San Juan Island.
5. Easier from A to B.
7. Easier from E to F.

15.1 Odd Answers

9. a-C, b-A, c-B
11. a-B, b-A, c-C
13. $\mathbf{F}(x,y) = \langle 2x, -2y \rangle$
15. $\mathbf{F}(x,y) = \langle y^2 - 2x \cdot y, 2x \cdot y - x^2 \rangle$
17. $\mathbf{F}(x,y,z) = \langle 3, 2, -1 \rangle$
19. $\mathbf{F}(x,y,z) = \langle x,y,z \rangle / \sqrt{x^2 + y^2 + z^2}$
21. $\mathbf{F}(x,y,z) = \langle y \cdot z \cdot \cos(x \cdot y \cdot z), x \cdot z \cdot \cos(x \cdot y \cdot z), x \cdot y \cdot \cos(x \cdot y \cdot z) \rangle$

23.
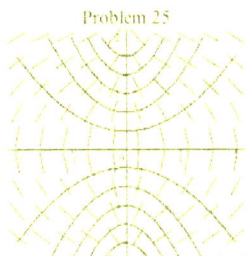

25.
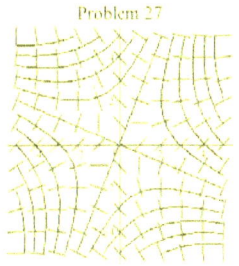

27.

15.2 Odd Answers

1. At A div $\mathbf{F} > 0$, at B div $\mathbf{F} < 0$, at C div $\mathbf{F} = 0$.
3. div $\mathbf{F} = 2x + 2$. At A div $\mathbf{F} = 4$, at B div $\mathbf{F} = 6$ and at C div $\mathbf{F} = 0$.
5. div $\mathbf{F} = 5 + 2$. At A div $\mathbf{F} = 7$, at B div $\mathbf{F} = 7$ and at C div $\mathbf{F} = 7$.
7. See Fig. 12. More going out than coming in.
9. At A curl $\mathbf{F} > 0$, at B curl $\mathbf{F} < 0$, at C curl $\mathbf{F} = 0$.
11. curl $\mathbf{F} = 1 - 3$. At A curl $\mathbf{F} = -2$, at B curl $\mathbf{F} = -2$, at C curl $\mathbf{F} = -2$.
13. curl $\mathbf{F} = 1 - (-3)$. At A curl $\mathbf{F} = 4$, at B curl $\mathbf{F} = 4$, at C curl $\mathbf{F} = 4$.
15. See Fig. 13. Enough to rotate counterclockwise.
17. All of the partial derivatives are 0 so both the div and the curl are 0.
19. No. The curl could be anything. See Fig. 14.

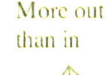
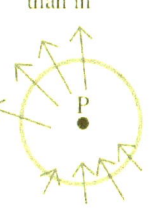

Fig. 12

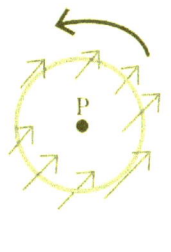

Fig. 13

In 14a the curl is positive. In 14b the curl is negative.

21. curl $\mathbf{F} = -\dfrac{\partial M}{\partial y}$ might be any value.

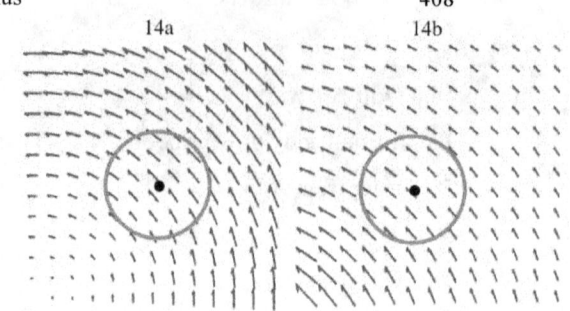

14a 14b

15.3 Odd Answers

1. $\displaystyle\int_0^3 \left((2t+1)^2 - 4(3+t^2) + 11\right)\sqrt{(2)^2 + (2t)^2}\, dt = \int_0^3 (4t)\sqrt{4+4t^2}\, dt = \dfrac{8}{3}(t^2+1)^{3/2}\Big|_0^3 = \dfrac{8}{3}(10\sqrt{10}-1)$

3. $\displaystyle\int_{t=a}^{t=b} y(t)\cdot\sqrt{\left(\dfrac{dx}{dt}\right)^2 + \left(\dfrac{dz}{dt}\right)^2}\cdot dt$

4. $\displaystyle\int_{t=a}^{t=b} x(t)\cdot\sqrt{\left(\dfrac{dy}{dt}\right)^2 + \left(\dfrac{dz}{dt}\right)^2}\cdot dt$

5. mass = $\displaystyle\int_0^{\pi/2} (1+3\cos(t)+6\sin(t))(3)\, dt = 27 + \dfrac{3}{2}\pi$

7. mass = $\displaystyle\int_0^2 (2+4t)(\sqrt{11})\, dt = 12\sqrt{11}$

9. $\displaystyle\int_0^2 (2t+9)(5)\, dt = 110$

11. $\displaystyle\int_0^3 (10t^2+10)(2t)\, dt = 495$

13. $\displaystyle\int_0^\pi (3-\sin(t)+\cos(t))(1)\, dt = -2 + 3\pi$

15. Along A work > 0. Along B work < 0.

17. Along A flow > 0. Along B flow > 0.

19. work = $\displaystyle\int_0^3 (14+37t)\, dt = 417/2$

21. work = $\displaystyle\int_0^\pi (-\sin(t)+2+3\cdot\cos(t))\, dt = 2\pi - 2$

23. flow = $\displaystyle\int_1^2 (3t^2+32t)\, dt = 55$

25. flux = $\displaystyle\int_0^{2\pi} (2\cdot\cos(t))(2\cdot\cos(t)) - (1+4\cdot\sin(t))(-\sin(t))\, dt = 8\pi$

27. All of them are possible for the flux.

For 29 to 34 consider the integral formula used for each calculation.

29. Area is unchanged.

31. Sign of work is changed.

33. Sign of circulation is changed.

15.4 Odd Answers

1. $\mathbf{F} = \langle 2x, 6y \rangle$

3. $\mathbf{F} = \langle 3\cos(3x+2y), 2\cos(3x+2y) \rangle$

5. $\mathbf{F} = \left\langle \dfrac{2}{2x+5y}, \dfrac{5}{2x+5y} + e^y \right\rangle$

7. $f(x,y) = x^2 + y^2$ $f(5,1) - f(1,2) = 21$

9. $M_y = 0$, $N_x = 1$ so F is not conservative. $\mathbf{r}(t) = \langle 3t, 2+4t \rangle$ (t from 0 to 1).

$\mathbf{F}(\mathbf{r}(t))\bullet \mathbf{r}'(t) = \langle 3t, 3t \rangle \bullet \langle 3, 4 \rangle = 21t$ so work = $\displaystyle\int_{t=0}^1 \mathbf{F}(\mathbf{r}(t))\bullet \mathbf{r}'(t)\, dt = \int_{t=0}^1 21t\, dt = \dfrac{21}{2}$

11. $f(x,y,z) = xyz$ $f(4,2,1) - f(1,0,0) = 8$

13. A to B: 5 C to D: 2 E to F: −3

15. A to B: 6 C to D: 0 E to F: -2

17. $f(x,y) = x^3 + 4x + 6y$

19. $f(x,y) = \sin(xy) + x^3 y$ 21. $f(x,y) = x \cdot \sin(y)$

23. $N_z \neq P_y$

15.6 Odd Answers

1. $M = x^2 y$, $N = 3y$, $\dfrac{\partial N}{\partial x} - \dfrac{\partial M}{\partial y} = -x^2$ so $\iint\limits_R -x^2 \, dA = \int_0^1 \int_0^2 -x^2 \, dx \, dy = -\dfrac{8}{3}$

3. $M = 3xy$, $N = 2x^2$, $\dfrac{\partial N}{\partial x} - \dfrac{\partial M}{\partial y} = 4x - 3x$ so $\iint\limits_R x \, dA = \int_0^1 \int_0^{2x} x \, dx \, dy = \dfrac{2}{3}$

5. $M = ax$, $N = by$, $\dfrac{\partial N}{\partial x} - \dfrac{\partial M}{\partial y} = 0$ so $\iint\limits_R 0 \, dA = 0$

7. circulation $= \iint\limits_R \left(\dfrac{\partial N}{\partial x} - \dfrac{\partial M}{\partial y} \right) dA = \iint\limits_R (-1-2) \, dA = \int_0^2 \int_0^2 -3 \, dx \, dy = -12$

   flux $= \iint\limits_R \left(\dfrac{\partial M}{\partial x} + \dfrac{\partial N}{\partial y} \right) dA = \iint\limits_R (1+1) \, dA = \int_0^2 \int_0^2 2 \, dx \, dy = 8$

9. circulation $= \iint\limits_R \left(\dfrac{\partial N}{\partial x} - \dfrac{\partial M}{\partial y} \right) dA = \iint\limits_R (2x - 2y) \, dA = \int_0^2 \int_0^x 2x - 2y \, dy \, dx = \dfrac{8}{3}$

   flux $= \iint\limits_R \left(\dfrac{\partial M}{\partial x} + \dfrac{\partial N}{\partial y} \right) dA = \iint\limits_R (2x + 2y) \, dA = \int_0^2 \int_0^x 2x + 2y \, dy \, dx = 8$

11. C_1 is from -2 to 2: $\mathbf{r}(t) = \langle t, t^2 \rangle$ so $\dfrac{1}{2} \int_{-2}^{2} (t)(2t) - (t^2)(1) \, dt = \dfrac{1}{2} \int_{-2}^{2} t^2 \, dt = \dfrac{8}{3}$

    C_2 is from 2 to -2: $\mathbf{r}(t) = \langle t, 8 - t^2 \rangle$ so $\dfrac{1}{2} \int_{2}^{-2} (t)(-2t) - (8 - t^2)(1) \, dt = \dfrac{1}{2} \int_{-2}^{2} (-t^2 - 8) \, dt = \dfrac{56}{3}$

    So the total area is $\dfrac{64}{3}$.

    Note: $\int_{-2}^{2} ((8 - x^2) - (x^2)) \, dx$ is much easier. Green's Theorem does not make everything easier.

13. Call D the boundary of R so $\int_D \mathbf{F} \cdot d\mathbf{r} = \int_{C_1} \mathbf{F} \cdot d\mathbf{r} - \int_{C_2} \mathbf{F} \cdot d\mathbf{r} = \int_{C_1} \mathbf{F} \cdot d\mathbf{r} - 20$.

    But $\int_D \mathbf{F} \cdot d\mathbf{r} = \iint\limits_R \left(\dfrac{\partial N}{\partial x} - \dfrac{\partial M}{\partial y} \right) dA = \iint\limits_R 5 \, dA = 500$ so $\int_{C_1} \mathbf{F} \cdot d\mathbf{r} = 520$.

15. Call D the boundary of R so $\int_D \mathbf{F} \cdot d\mathbf{r} = \int_{C_1} \mathbf{F} \cdot d\mathbf{r} - \int_{C_2} \mathbf{F} \cdot d\mathbf{r} - \int_{C_3} \mathbf{F} \cdot d\mathbf{r} = \int_{C_1} \mathbf{F} \cdot d\mathbf{r} - 3\pi - 4\pi$.

    Area of R $= = 42 - 2\pi$.

But $\int_D \mathbf{F} \cdot d\mathbf{r} = \iint_R \left(\frac{\partial N}{\partial x} - \frac{\partial M}{\partial y}\right) dA = \iint_R 9 \, dA = 9(42 - 2\pi) = 378 - 18\pi$ so $\int_{C_1} \mathbf{F} \cdot d\mathbf{r} = 378 - 11\pi$.

17. All of the partial derivatives are 0 so the double integrals for circulation and flux are 0.

19. circulation $= \oint_C M \, dx + N \, dy = \iint_R \left(\frac{\partial N}{\partial x} - \frac{\partial M}{\partial y}\right) dA = \iint_R (c - b) \, dA = (c - b) \cdot (\text{area of R})$

15.7 Odd Answers

1. div $\mathbf{F} = 3xz^2 + 2xy + xz$, curl $\mathbf{F} = \langle -xy, -z^3, -x^2 + yz \rangle$
3. div $\mathbf{F} = e^z + z \cdot e^y$, curl $\mathbf{F} = \langle e^x - e^y, xe^z - ye^x, 0 \rangle$
5. div $\mathbf{F} = 3$, curl $\mathbf{F} = \langle -1, -1, -1 \rangle$

In problems 7 to 10, X = not meaningful.

7. div f =X , div F =scalar, div (curl F)=scalar , gradient (curl F)=X, curl F=vector
9. gradient f =vector, curl (curl F)=vector , curl (div (gradient f))=X
11. Not conservative. 13. Not conservative.
15. Conservative. $f(x,y,z) = x \cdot \sin(yz) + 2z$
17. Your location = (5, 10, 5)= (1, 2, 3) + (2)(2, 4, 1) so you have moved in the direction of the curl. Looking back, you will see a Counterclockwise rotation.
19. Your location = (4, 4, 3)= (6, 3, 2) + (–1)(2, –1, 1) so you have moved in the opposite direction of the curl. Looking toward (6, 3, 2), you will see a Clockwise rotation.

15.8 Odd Answers

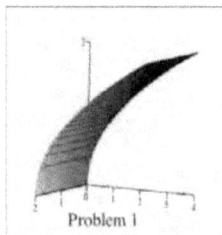
Problem 1
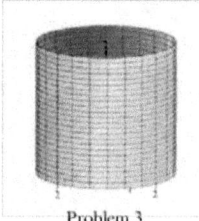
Problem 3
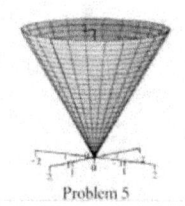
Problem 5
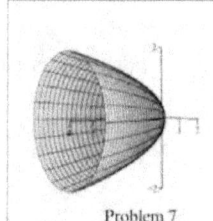
Problem 7
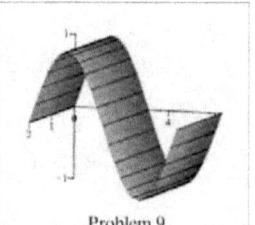
Problem 9

11. $\mathbf{r}(u,v) = \langle u, v, \sqrt{v} \rangle$, $\mathbf{r}_u = \langle 1, 0, 0 \rangle$, $\mathbf{r}_v = \langle 0, 1, \frac{1}{2\sqrt{v}} \rangle$, $\mathbf{r}_u \times \mathbf{r}_v = \langle 0, \frac{1}{2\sqrt{v}}, 1 \rangle$

Surface area $= \iint_R |\mathbf{r}_u \times \mathbf{r}_v| \, dA = \int_{u=0}^{2} \int_{v=0}^{4} \sqrt{\frac{1}{4v} + 1} \, dv \, du$.

13. $\mathbf{r}(u,v) = \langle 2 \cdot \cos(u), 2 \cdot \sin(u), v \rangle$, $\mathbf{r}_u = \langle -2 \cdot \sin(u), 2 \cdot \cos(u), 0 \rangle$, $\mathbf{r}_v = \langle 0, 0, 1 \rangle$,

$\mathbf{r}_u \times \mathbf{r}_v = \langle 2 \cdot \cos(u), 2 \cdot \sin(u), 0 \rangle$. Surface area $= \iint_R |\mathbf{r}_u \times \mathbf{r}_v| \, dA = \int_{u=0}^{2\pi} \int_{v=0}^{3} 2 \, dv \, du$.

15. $r(u,v) = \langle u \cdot \cos(v), u \cdot \sin(v), u \rangle$, $\mathbf{r}_u = \langle \cos(v), \sin(v), 1 \rangle$, $\mathbf{r}_v = \langle -u \cdot \sin(v), u \cdot \cos(v), 0 \rangle$

$\mathbf{r}_u \mathbf{x} \mathbf{r}_v = \langle -u\cos(v), u \cdot \sin(v), u \rangle$. Surface area $= \iint\limits_R |\mathbf{r}_u \mathbf{x} \mathbf{r}_v| \, dA = \int_{u=0}^{2} \int_{v=0}^{2\pi} u\sqrt{2} \, dv \, du$.

17. Surface area $= \int_{u=0}^{4} \int_{v=0}^{2\pi} \sqrt{u + \frac{1}{4}} \, dv \, du$. 19. Surface area $= \int_{u=0}^{2} \int_{v=0}^{2\pi} \sqrt{1 + \cos^2(v)} \, dv \, du$.

15.9 Odd Answers

1. $\mathbf{r}_u = \langle 1,2,3 \rangle$, $\mathbf{r}_v = \langle 3,-1,1 \rangle$, $\mathbf{r}_u \mathbf{x} \mathbf{r}_v = \langle 5,8,-7 \rangle$ and $|\mathbf{r}_u \mathbf{x} \mathbf{r}_v| = \sqrt{138}$. $f(\mathbf{r}(1,2)) = f(7,0,5) = 12$,

    $\Delta A = \Delta u \cdot \Delta v = 0.03$ so $f(\mathbf{r}(u,v)) \cdot |\mathbf{r}_u \mathbf{x} \mathbf{r}_v| \Delta A = (12)(\sqrt{138})(0.03) \approx 4.23$.

3. When u=2 and v=1, then $\mathbf{r}_u = \langle 2u, 3, 0 \rangle = \langle 4,3,0 \rangle$, $\mathbf{r}_v = \langle 0, 1, 2v \rangle = \langle 0,1,2 \rangle$,

    $\mathbf{r}_u \mathbf{x} \mathbf{r}_v = \langle 6v, -4uv, 2u \rangle = \langle 6,-8,4 \rangle$ and $|\mathbf{r}_u \mathbf{x} \mathbf{r}_v| = \sqrt{116}$. $f(\mathbf{r}(2,1)) = f(4,7,1) = 56$,

    $\Delta A = \Delta u \cdot \Delta v = 0.06$ so $f(\mathbf{r}(u,v)) \cdot |\mathbf{r}_u \mathbf{x} \mathbf{r}_v| \Delta A = (56)(\sqrt{116})(0.06) \approx 36.19$.

5. S can be parameterized by $r(u,v) = \langle u,v,4 \rangle$ with $0 \le u \le 3$ and $0 \le v \le 2$. Then $\mathbf{r}_u = \langle 1,0,0 \rangle$,

    $\mathbf{r}_v = \langle 0,1,0 \rangle$, $\mathbf{r}_u \mathbf{x} \mathbf{r}_v = \langle 0,0,1 \rangle$ and $|\mathbf{r}_u \mathbf{x} \mathbf{r}_v| = 1$. $f(r(u,v)) = u^2 + 4v + 4$.

    $\iint\limits_S f(x,y,z) \, dS = \iint\limits_R f(\mathbf{r}(u,v)) \cdot |\mathbf{r}_u \mathbf{x} \mathbf{r}_v| \, dA = \int_{v=0}^{2} \int_{u=0}^{3} u^2 + 4v + 4 \, du \, dv = \int_{v=0}^{2} 9 + 12v + 12 \, dv = 66$.

7. S can be parameterized by $r(u,v) = \langle u, v, 4-u-v \rangle$ with $0 \le u \le 2$ and $0 \le v \le 2$. Then $\mathbf{r}_u = \langle 1, 0, -1 \rangle$,

    $\mathbf{r}_v = \langle 0, 1, -1 \rangle$, $\mathbf{r}_u \mathbf{x} \mathbf{r}_v = \langle 1, 1, 1 \rangle$ and $|\mathbf{r}_u \mathbf{x} \mathbf{r}_v| = \sqrt{3}$. $f(r(u,v)) = v^2$.

    $\iint\limits_S f(x,y,z) \, dS = \iint\limits_R f(\mathbf{r}(u,v)) \cdot |\mathbf{r}_u \mathbf{x} \mathbf{r}_v| \, dA = \int_0^2 \int_0^2 v^2 \sqrt{3} \, dv \, du = \int_0^2 \frac{8}{3}\sqrt{3} \, du = \frac{16}{3}\sqrt{3}$.

9. S can be parameterized by $r(u,v) = \langle u,v,4 \rangle$ with $0 \le u \le 3$ and $0 \le v \le 2$. Then $\mathbf{r}_u = \langle 1,0,0 \rangle$,
    $\mathbf{r}_v = \langle 0,1,0 \rangle$, and $\mathbf{r}_u \mathbf{x} \mathbf{r}_v = \langle 0,0,1 \rangle$. $\mathbf{F}(r(u,v)) = \langle u, 2v, 3 \cdot 4 \rangle$.

    Flux $= \iint\limits_S \mathbf{F} \cdot \mathbf{n} \, dS = \iint\limits_R \mathbf{F} \bullet (\mathbf{r}_u \mathbf{x} \mathbf{r}_v) \, dA = \iint\limits_R \langle u, 2v, 3 \cdot 4 \rangle \bullet \langle 0, 0, 1 \rangle \, dA = \int_0^2 \int_0^3 12 \, du \, dv = 72$.
    The flux units are (liters/second)(meter^2).

11. S can be parameterized by $\mathbf{r}(u,v) = \langle 2 \cdot \cos(u), \sin(u), v \rangle$ with $0 \le u \le 2\pi$ and $0 \le v \le 3$. Then

    $\mathbf{r}_u = \langle -2 \cdot \sin(u), \cos(u), 0 \rangle$, $\mathbf{r}_v = \langle 0, 0, 1 \rangle$ and $\mathbf{r}_u \mathbf{x} \mathbf{r}_v = \langle \cos(u), 2 \cdot \sin(u), 0 \rangle$. $\mathbf{F}(\mathbf{r}(u,v)) = \langle 2 \cdot \cos(u), \sin(u), v \rangle$.

    Flux $= \iint\limits_S \mathbf{F} \cdot \mathbf{n} \, dS = \iint\limits_R \mathbf{F} \bullet (\mathbf{r}_u \mathbf{x} \mathbf{r}_v) \, dA = \iint\limits_R \langle 2 \cdot \cos(u), \sin(u), v \rangle \bullet \langle \cos(u), 2 \cdot \sin(u), 0 \rangle \, dA$

    $= \int_0^3 \int_0^{2\pi} 2 \, du \, dv = 12\pi$.

13. S can be parameterized by $r(u,v) = \langle v \cdot \cos(u), v \cdot \sin(u), A^2 - v^2 \rangle$ with $0 \le u \le 2\pi$ and $0 \le v \le A$. Then

$r_u = \langle -v \cdot \sin(u), v \cdot \cos(u), 0 \rangle$, $r_v = \langle \cos(u), \sin(u), -2Cv \rangle$ and $r_u \times r_v = \langle -2v^2 C \cdot \cos(u), -2v^2 C \cdot \sin(u), -v \rangle$.

$F(r(u,v)) = \langle 0, 0, K \rangle$.

$$\text{Flux} = \iint_S F \cdot n \, dS = \iint_R F \cdot (r_u \times r_v) \, dA = \iint_R \langle 0, 0, K \rangle \cdot \langle -2v^2 C \cdot \cos(u), -2v^2 C \cdot \sin(u), -v \rangle \, dA$$

$$= \int_0^A \int_0^{2\pi} -Kv \, du \, dv = -KA^2 \pi.$$

This is the same flux as in Problem 12. In fact, the flux is the same even if C=0 so that S is just the disk $x^2 + y^2 \le A^2$ in the xy-plane.

15.10 Odd Answers

1. curl $F = \nabla \times F = \langle 0, 0, 1 \rangle$, $R = \{(x,y): x^2 + 4y^2 \le 4\}$, area of $R = \pi(1)(2) = 2\pi$ and $n = \langle 0, 0, 1 \rangle$.

    circulation $= \oint_C F \cdot T \, ds = \iint_R (\nabla \times F) \cdot n \, dA = \iint_R \langle 0, 0, 1 \rangle \cdot \langle 0, 0, 1 \rangle \, dS$ = area of $R = 2\pi$.

3. curl $F = \nabla \times F = \langle x, 1 - y, 0 \rangle$, $n = \langle 2, 2, 2 \rangle$, $R = \{(x,y): 0 \le x \le 3, 0 \le y \le 3 - x\}$

    circulation $= \oint_C F \cdot T \, ds = \iint_R (\nabla \times F) \cdot n \, dA = \iint_R \langle x, 1 - y, 0 \rangle \cdot \langle 2, 2, 2 \rangle \, dS$

    $= \int_0^3 \int_0^{3-x} 2x + 2 - 2y \, dy \, dx = 9$.

5. curl $F = \nabla \times F = \langle 0, 0, 2 \rangle$, and $n = \langle 1, 1, 1 \rangle$.

    circulation $= \oint_C F \cdot T \, ds = \iint_R (\nabla \times F) \cdot n \, dA = \iint_R \langle 0, 0, 2 \rangle \cdot \langle 1, 1, 1 \rangle \, dA = 2 \cdot \iint_R 1 \, dA$

    = 2(area of R) where R is the area enclosed by C.

7. curl $F = \nabla \times F = \langle 0, 0, 3 - 2y \rangle$ and $n = \langle 0, 0, 1 \rangle$.

    circulation $= \oint_C F \cdot T \, ds = \iint_R (\nabla \times F) \cdot n \, dA = \iint_R \langle 0, 0, 3 - 2y \rangle \cdot \langle 0, 0, 1 \rangle \, dA$

    $= \int_0^2 \int_0^{2\pi} (3 - 2r \cdot \sin(\theta)) \cdot r \, d\theta \, dr = \int_0^2 6\pi r \, dr = 12\pi$.

9. The boundary of S is the circle C parameterized by $r(t) = \langle 4 \cdot \cos(t), 4 \cdot \sin(t), 0 \rangle$.

    $$\iint_S \text{curl } F \cdot dS = \int_C F \cdot dr = \int_{t=0}^{2\pi} \langle 4 \cdot \sin(t), -4 \cdot \cos(t), 0 \rangle \cdot \langle -4 \cdot \sin(t), 4 \cdot \cos(t), 0 \rangle \, dt$$

    $$= \int_{t=0}^{2\pi} -16\sin^2(t) - 16\cos^2(t) \, dt = -32\pi.$$

11. The solid torus is a closed surface so it has no boundary curve so $\iint_S \text{curl } F \cdot dS = 0$.

13. $\nabla \times \mathbf{F} = \langle 0,0,0 \rangle$ so F is a conservative field.

15. $\nabla \times \mathbf{F} = \langle cx - bx, ay - cy, bx - az \rangle$ so F is not a conservative field.

17. $\nabla \times \mathbf{F} = \langle 0,0,0 \rangle$ so F is a conservative field.

19. Quick check: $\nabla \times \mathbf{F} = \langle 2, 2, 2 \rangle$.

    The top and bottom boundaries of S can be parameterized by $\mathbf{r}_1 = \langle 4 \cdot \cos(t), 4 \cdot \sin(t), 0 \rangle$ and $\mathbf{r}_2 = \langle 4 \cdot \cos(t), 4 \cdot \sin(t), 3 \rangle$ for $0 \le t \le 2\pi$ (both going counterclockwise).

    $$\int_{C_1} \mathbf{F} \bullet \mathbf{r}_1' \, dt = \int_0^{2\pi} \langle 0 - 4 \cdot \sin(t), 4 \cdot \cos(t) - 0, 4 \cdot \sin(t) - 4 \cdot \cos(t) \rangle \bullet \langle -4 \cdot \sin(t), 4 \cdot \cos(t), 0 \rangle \, dt$$

    $$= \int_0^{2\pi} 16 \cdot \sin^2(t) + 16 \cdot \cos^2(t) \, dt = 32\pi.$$

    $$\int_{C_2} \mathbf{F} \bullet \mathbf{r}_2' \, dt = \int_0^{2\pi} \langle 3 - 4 \cdot \sin(t), 4 \cdot \cos(t) - 3, 4 \cdot \sin(t) - 4 \cdot \cos(t) \rangle \bullet \langle -4 \cdot \sin(t), 4 \cdot \cos(t), 0 \rangle \, dt$$

    $$= \int_0^{2\pi} -12 \cdot \sin(t) + 16 \cdot \sin^2(t) + 16 \cdot \cos^2(t) - 12 \cdot \cos(t) \, dt = 32\pi.$$

    Finally, $\iint_S \text{curl } \mathbf{F} \bullet d\mathbf{S} = \int_{C_1} F \bullet \mathbf{r}_1' \, dt - \int_{C_2} F \bullet \mathbf{r}_1' \, dt = 0$.

21. Quick check $\nabla \times \mathbf{F} = \langle 0, 0, 0 \rangle$ so $\iint_S \text{curl } \mathbf{F} \bullet d\mathbf{S} = \iint_S \langle 0, 0, 0 \rangle \bullet d\mathbf{S} = 0$.

23. Quick check $\nabla \times \mathbf{F} = \langle -1, -1, 1 \rangle$.

    The boundary curve of S can be parameterized by $\mathbf{r} = \langle 2 \cdot \cos(t), 0, -5 \cdot \sin(t) \rangle$.
    (goes counterclockwise looking back along the y-axis).

    $$\iint_S \text{curl } \mathbf{F} \bullet d\mathbf{S} = \int_C \mathbf{F} \bullet d\mathbf{r} = \int_{t=0}^{2\pi} \langle 0, -5 \cdot \sin(t), 2 \cdot \cos(t) \rangle \bullet \langle -2 \cdot \sin(t), 0, -5 \cdot \cos(t) \rangle \, dt$$

    $$= \int_{t=0}^{2\pi} -10 \cdot \cos^2(t) \, dt = -10\pi.$$

25. Quick check $\nabla \times \mathbf{F} = \langle -1, 0, 1 \rangle$.

    When z=0, $x^2 + y^2 = 25$ is a circle of radius 5: $\mathbf{r}_1 = \langle 5 \cdot \cos(t), 5 \cdot \sin(t), 0 \rangle$.

    $$\int_{C_1} \mathbf{F} \bullet \mathbf{r}_1' \, dt = \int_0^{2\pi} \langle 5 \cdot \sin(t), 0, 5 \cdot \cos(t) \rangle \bullet \langle -5 \cdot \sin(\theta), 5 \cdot \cos(t), 0 \rangle \, dt = \int_0^{2\pi} -25 \cdot \sin^2(t) \, dt = -25\pi.$$

    When z=16, $x^2 + y^2 = 9$ is a circle of radius : $\mathbf{r}_2 = \langle 3 \cdot \cos(t), 3 \cdot \sin(t), 16 \rangle$.

    $$\int_{C_2} \mathbf{F} \bullet \mathbf{r}_2' \, dt = \int_0^{2\pi} \langle 3 \cdot \sin(t), 16, 3 \cdot \cos(t) \rangle \bullet \langle -3 \cdot \sin(t), 3 \cdot \cos(t), 0 \rangle \, dt$$

    $$= \int_0^{2\pi} -9 \cdot \sin^2(t) + 48 \cdot \cos(t) \, dt = -9\pi. \text{ So } \iint_S \text{curl } \mathbf{F} \bullet d\mathbf{S} = \int_{C_1} F \bullet \mathbf{r}_1' \, dt - \int_{C_2} F \bullet \mathbf{r}_1' \, dt = -16\pi.$$

15.11 Odd Answers

1. Distance to the origin of P1 is $\sqrt{14} \approx 3.7$, of P2 is $\sqrt{6} \approx 2.4$, of P3 is $\sqrt{8} \approx 2.8$, of P4 is $\sqrt{2} \approx 1.4$, of P5 is $\sqrt{12} \approx 3.5$, and of P6 is $\sqrt{19} \approx 4.4$.

    (a) Only P4: net flux = 6 m^3/s. (b) P1, P2, P3, P4 and P5: net flux=(7)+(−3)+(2)+(6)+(−4)= 8 m^3/s.

3. Dist to center of P1 is $\sqrt{10} \approx 3.2$, of P2 is $\sqrt{18} \approx 4.2$, of P3 is 0, of P4 is $\sqrt{6} \approx 2.4$, of P5 is 2, and of P6 is $\sqrt{11} \approx 3.3$.

    (a) Only P3: net flux= 2 m^3/s. (b) P3, P4, P5: net flux=(2)+(−6)+(−4)= −8 m^3/s.

5. Outward flux sphere 1: 5 m^3/s. outward flux of sphere 2: 14 m^3/s. Net flux = (14)−(5)= 9 m^3/s.

7. div **F** = 3 so flux = $\iiint_E \text{div } \mathbf{F} \, dV = 3(\text{volume of E}) = 3\left(\frac{4}{3}\pi(2)^3\right) = 32\pi$.

9. div **F** = 2 so flux = $\iiint_E \text{div } \mathbf{F} \, dV = 2(\text{volume of E}) = 2\cdot\left(\frac{4}{3}\pi(3)^3\right) = 72\pi$.

11. div **F** = 4 so flux = $\iiint_E \text{div } \mathbf{F} \, dV = (4)\iiint_E 1 \, dV = 4\cdot(\text{volume of E}) = (4)\cdot(2)(3)(4) = 96$.

13. div **F** = $2x+2y+2z$ so flux = $\iiint_E \text{div } \mathbf{F} \, dV = \int_0^4 \int_0^3 \int_0^2 (2x+2y+2z)\,dx\,dy\,dz = 216$.

15. $0 = \iiint_D \text{div } \mathbf{F} \, dV = \iint_S \mathbf{F}\bullet\mathbf{n}\,dS = \iint_{S_1} \mathbf{F}\bullet\mathbf{n}\,dS - \iint_{S_2} \mathbf{F}\bullet\mathbf{n}\,dS$ so the outward flux from the outer sphere equals the outward flux from the inner sphere.

17. Flux in each case is zero.

19. (a) div **F** = 0 (b) div **F** is negative (c) div **F** is negative

21. (a) div **F** is positive (b) div **F** is positive (c) div **F** is positive

www.ingramcontent.com/pod-product-compliance
Lightning Source LLC
Chambersburg PA
CBHW080904170526
45158CB00008B/1989